AF410852

Soil Erosion and Sustainable Land Management (SLM)

Soil Erosion and Sustainable Land Management (SLM)

Editors

Atsushi Tsunekawa
Nigussie Haregeweyn

MDPI • Basel • Beijing • Wuhan • Barcelona • Belgrade • Manchester • Tokyo • Cluj • Tianjin

Editors
Atsushi Tsunekawa
Arid Land Research Center,
Tottori University
Japan

Nigussie Haregeweyn
International Platform for
Dryland Research and
Education, Tottori University
Japan

Editorial Office
MDPI
St. Alban-Anlage 66
4052 Basel, Switzerland

This is a reprint of articles from the Special Issue published online in the open access journal *Sustainability* (ISSN 2071-1050) (available at: https://www.mdpi.com/journal/sustainability/special_issues/Soil_Erosion_Sustainable_Land_Management).

For citation purposes, cite each article independently as indicated on the article page online and as indicated below:

LastName, A.A.; LastName, B.B.; LastName, C.C. Article Title. *Journal Name* **Year**, *Volume Number*, Page Range.

ISBN 978-3-0365-0786-6 (Hbk)
ISBN 978-3-0365-0787-3 (PDF)

Contents

About the Editors

Atsushi Tsunekawa graduated from the Graduate School of Agricultural Sciences, University of Tokyo, after which he joined the National Institute for Environmental Studies, Japan. Currently, he serves as a professor at the Arid Land Research Center, Tottori University, and also serves as a Head of the Strategic Management Office of the International Platform for Dryland Research and Education, Tottori University. His primary scholarly interests are developing sustainable land management (SLM) technologies and approaches to restore degraded land and improve farmers' livelihood and monitoring and modeling of terrestrial ecosystems under climate change using remote sensing and GIS. He is selected as a Science and Technology Correspondent from Japan to the Committee on Science and Technology of the United Nations Convention to Combat Desertification.

Nigussie Haregeweyn received his PhD in Physical Geography from K.U Leuven (Belgium). He started his professional career in Ethiopia as an irrigation engineer in 1995 before joining Mekelle University in 2001 as a lecturer. Since 2010, he has been serving in various research capacities at Tottori University (Japan), where he currently holds a professor position at the International Platform for Dryland Research and Education. His research interests include land degradation and development with emphasis on soil erosion, land use and climate change and sustainable land management through integrating a range of tools and techniques covering various spatiotemporal scales.

Preface to "Soil Erosion and Sustainable Land Management (SLM)"

Soil erosion-induced land degradation is one of the topsoil threats identified in the 2015 Status of the World's Soil Resources report. This concern is expected to be intensified as recent projections show global soil erosion rates to increase by up to 66% during the period 2015–2070 due to the unsustainable use of the limited land and water resources under scenarios of climate and land use changes and population growth. Several studies stress the significance of sustainable land management (SLM) for achieving the UNCCD's (United Nations Convention to Combat Desertification) target of land degradation neutrality by 2030. For this, proper identification and implementation of sustainable land management (SLM) practices are believed to offer a triple-win solution of conserving the resources, increasing agricultural productivity and improving human livelihood and well-being.

However, success in fighting land degradation through SLM requires improved understanding and documentation of land degradation processes and their risks as well as development and adoption of mitigation techniques. There is therefore a strong need for the development of systematic, robust and validated methods to better understand and support the development and adoption of SLM measures at various spatiotemporal scales.

This Special Issue presents 13 case studies conducted in Africa, Europe, North America and Asia. The studies specifically help to improve assessment of the causes and impacts of soil erosion at various spatiotemporal scales; clarify the human and natural drivers of soil erosion; and propose promising land management technologies to mitigate soil erosion while improving land productivity. These would ultimately help in the formulation of appropriate policies towards improving the adoption of SLM and solving associated socio-economic problems. Such comprehensive topical coverage places the book at a unique position that would be useful for researchers, planners and practitioners of SLM.

Atsushi Tsunekawa, Nigussie Haregeweyn
Editors

Article

The Recent Sediment Record as an Indicator of Past Soil-Erosion Dynamics. Study in Dehesa Areas in the Province of Cáceres (Spain)

María Teresa de Tena Rey *, Agustín Domínguez Álvarez and Lorenzo García-Moruno

Department of Graphical Expression, University of Extremadura, 06800 Mérida, Spain; adomguez@unex.es (A.D.Á.); lgmoruno@unex.es (L.G.-M.)
* Correspondence: mtdetena@unex.es; Tel.: +34-924-28-93-00

Received: 14 September 2019; Accepted: 28 October 2019; Published: 2 November 2019

Abstract: The work presented is a study of the recent sediment deposits in a pilot basin in dehesa areas in the province of Cáceres (Spain) through analysis of the sediment record, radiocarbon dating and correlation with historic data to assess the factors that conditioned the deposit in these areas over time. It is a qualitative study based on the important role of sediments as recorders of history, given that sediment facies and their architecture provide one of the best records of past processes and environmental factors. For the study, sediment profile surveys were used to determine the configuration and characteristics of the infill and its chronology. The sediment model of the facies studied is associated with a context of slope water erosion that led to the infill of the watercourse areas, mainly sand and fine gravel, where alterations in the normal rate were detected due to the insertion of a thicker level of materials (soil stoniness) that was able to be dated. The sediment and chronological results obtained can be used to determine the historical events in the area that could have affected the erosion and deposit processes in the basin for the estimated period, from the late 18th to the early 19th century. During this period, pastureland that maintained the ecological balance of the dehesa, with a balanced, stable displacement of soil particles, was converted to cropland, in most cases resulting in soil with a limited profile, overuse and the consequent loss of structure and texture, making it more vulnerable to erosion. Greater remobilisation would have carried thicker material to the watercourses than the material deposited as a result of limited ploughing. This study provides data for the dehesa areas studied with regard to their hydrogeomorphological dynamics, from which past environmental impacts due to tillage can be inferred.

Keywords: sediment; land use; erosion crises; environmental impact

---

1. Introduction

Sediment studies are a source of information for inferring, through sediment sequence analysis, the conditions that have governed sedimentation processes in an area over time. Sediment facies and their architecture provide one of the best archives of the evolution and transformation of terrestrial systems, because they preserve the record of past processes and environmental factors that have occurred naturally or, in recent times, through human intervention [1,2] including soil erosion. Historical reconstruction of the effects humans have on their surrounding environment shows that the period of greatest impact was the last few centuries, and in recent decades the role of sediments as recorders of history has become increasingly more valued [3–5]. The sediment record can indicate the nature and level of the environmental impact of past events, because although we can use landscapes to deduce the relationship between humans and the environment [6], the traces of these actions can be blurred or erased. The sediment record, however, does not suffer the same fate. By characterising the sediment record, we can reconstruct the erosion processes that occurred in the source area in the past.

Starting from this premise, and applying it to dehesas (agrosilvopastoral areas), the study focused on recent sediment infill in a pilot basin in the province of Cáceres established by the Physical Geography department of the University of Extremadura to study the current hydrogeomorphological processes operating in dehesas in this region, including soil erosion, hydrological balance and rain interception by the holm oak [7–11] and sediment volume measurement and characterisation [12].

The objective of the work was not limited to a contemporary perspective, in which it is difficult know at what stage of the dynamic the system is in. It is a historical study of accumulated sediment, using the interpretation of the sequence to determine the circumstances that have conditioned the deposition processes in the basin over time and the possible influence of humans through varied land-use management practices.

Dehesas make up more than half the useful agrarian area in the west–southwest provinces of Spain and occur in 53% of municipalities in Extremadura. These ecosystems combine forestry, agriculture and livestock raising, and are a paradigm of sustainable development. Rational harvesting of a dehesa's natural resources can optimise production yield and ensure the ecological stability of the system [13]. Any study of this regionally important, extensive ecosystem is relevant, in particular those that provide a greater understanding of its functioning and limitations, to avoid clearly unsustainable practices. In this context, it is particularly important to study sediment deposits as a historic record of the events that have shaped degradation processes in the source area, because soil erosion is one of the issues that the dehesas are now facing.

The small basins of the streams that drain the peneplain dehesas collect sediment from the slopes, and the deposits from past erosion processes are, therefore, stored in the watercourses. The watercourse areas in the study basin and the adjacent areas form paleovalleys that have been filled in by materials from the surrounding areas. These materials can be associated with historic erosion crises [7]. Sediment production on slopes is determined to a large extent by how the land is used. Degradation of dehesas in semi-arid areas is triggered by continued harvesting and overgrazing [14]. Knowledge of how these processes occur today is essential to interpret past events in the sediment record. Other experimental studies have addressed aspects such as management of agricultural land, land management and soil erosion and the impact of land abandonment [15–17]. Changes in erosion rates due to different land management practices in the past must be recorded in some way in the deposits that occurred under the conditions at the time.

Under this initial premise, the study comprises an analysis and interpretation of the sediment sequences. After the entire deposit is studied, any alterations must be detected and dated for correlation with the historical and documentary data available. This is, therefore, an interpretive work, based on the data obtained from the sediment and chronological study.

2. Study Area

The study area is in the southeast of the province of Cáceres, in the municipality of Trujillo, on the Trujillo peneplain created by erosion of slate and granite material from the Central Iberian Hercynican basement [18].

Upper Precambrian-Lower Cambrian successions occupy extensive outcrops of detrital sediment series made up primarily of slate and greywacke, ranging in altitude from 400–500 m (Figure 1).

The area presents a topography of gentle forms, with slight variations because the current hydrographic network is firmly established in the peneplain. The most significant feature is the absence of any high relief due to the predominance of Precambrian materials with medium to low resistance and a long history of erosion. The small pilot basin is at the head of the Magasca river, a tributary of the Almonte, part of the network of the Tajo. It has an area of approximately 35.4 ha and includes the upper basin of the 813 m-long Guadalperalón stream, whose downstream limit is at the measuring station used to study the erosion and hydrological processes operating in these areas.

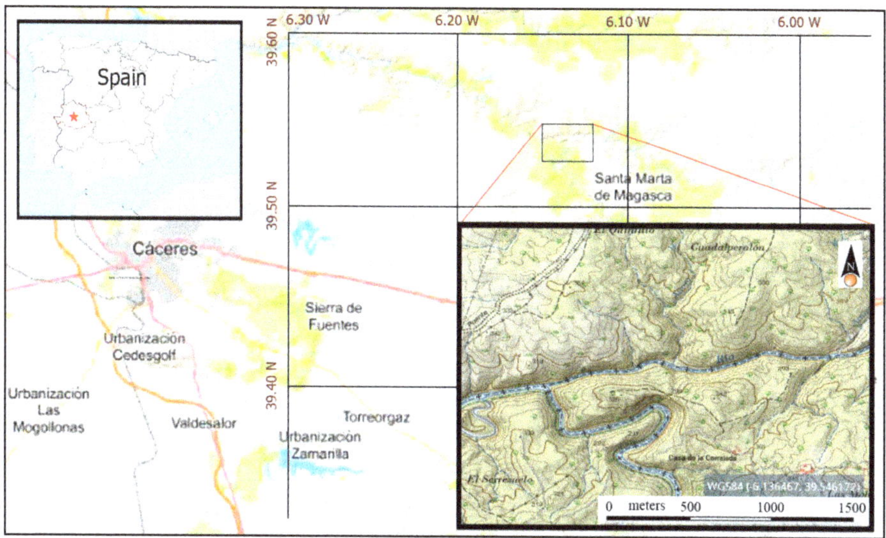

Figure 1. Study area location. Source: authors' own figure.

It can be considered a typical dehesa, with similar features to many others in the region in terms of topography, rock substratum, soil, vegetation and current and past uses, all of which can be extrapolated to other areas of Extremadura [19].

Following the Food and Agriculture Organization (FAO) nomenclature, the soil in the area is dystric cambisol, with significant slate outcrops. In slope areas it is degraded to leptosols. In most of the area, the soil is less than 30 cm thick, with deeper soils occurring at the bottom of the watercourses [20], on alluvial deposits.

In these conditions, the most rational land use from a conservation perspective is livestock grazing in combination with controlled cultivation. While this would maintain the balance of the structure and low potential of the area, tillage erosion increases with the number of tillage operations [21], and the lack of an alternative livelihood or opportunity to obtain maximum return in the short term may have led to intensive cropping or overgrazing of the land in the past.

3. Methodology

The procedure used to recreate the conditions of the deposits in the study area was based on sediment and geochronological studies and correlation with historic and documentary data. This is a qualitative study, based on the important role of sediments as recorders of history.

After the deposit areas in the basin had been identified and defined, the study was conducted by surveying small sediment columns throughout the available sections in the basin. The levels and structures appearing in the sequence were examined, measured and described. The initial premise was that the conditions occurring during the sedimentation process must be reflected in some way in the sediment sequence.

The laboratory work comprised textural analysis of the sediment samples from each level and radiocarbon dating of charcoal samples collected in these levels.

3.1. Sediment Study

The deposits accumulated throughout the watercourse areas occupy an area of around 18,800m^2, representing approximately 5% of the total area of the basin [22], and are strongly incised by gullies. In the middle and lower section of the bed, where the incisions reach the substratum, sediment thickness can be measured directly. These areas provide the sections where the sediment profiles can be surveyed, because the entire sediment infill can be easily observed to describe the vertical and lateral variations it has experienced. In the upper section of the bed, the material observed is almost homogeneous and the sediment infill thickness is limited. Using geophysical methods and applying vertical electric soundings (VES) when the conditions of the deposit outcrop prevented the use of other methods, we were able to study the sediment sequence indirectly and correlate the levels detected in the gully areas [22].

The representation of the sediment profiles shows the type of material, the thickness of the deposit, and the recognisable sections or levels.

Seven columns were surveyed across an area of approximately 140 m, starting from the first profile (P1) 180 m upstream from the final part of the study area (location of the measuring station) and continuing downstream (Figure 2).

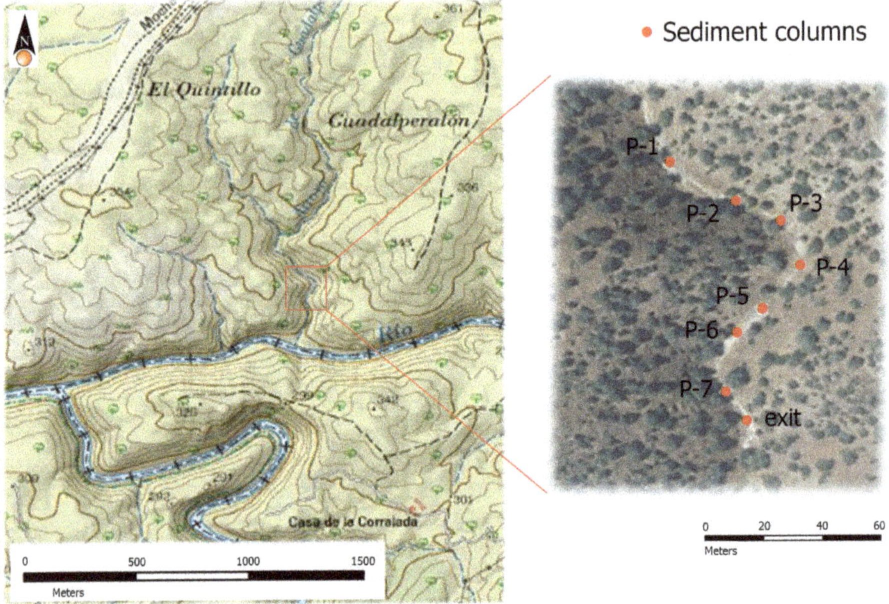

Figure 2. Location map of column survey areas. Source: authors' own work.

No distance was set between the profiles. The survey points were chosen in the field after identifying the sections where it would be easiest to measure and describe the levels, following the evolution of the sediment sequence. The final column (P7) was surveyed in the area next to the measuring station. The strong incision by gullies down to the rock bed permitted a complete survey of the sediment columns. Because of the continuity and quality of the sections in the study area, it was possible to establish correlations and identify variations observed in the deposit sequences. The levels identified in the deposit sequences were shown in a correlation diagram.

The sediment was texturally characterised by taking 12 samples of approximately 1 kg across the three levels differentiated in the surveyed columns. The textural analysis was performed at the Extremadura Regional Government Agricultural Laboratory. The content of coarse elements (expressed in %) was determined and the fine elements (sand, silt and clay) were used to define the texture class.

3.2. Sample Dating

Radiocarbon (or carbon-14) dating was used to date the sediment deposits. The material chosen was charcoal, because it was likely to occur in the deposits. Seven samples were collected, corresponding to the lower area, intermediate area (location of larger grain pebbles) and upper area in each profile to determine the age of each level. Dating by the conventional method was unsuccessful because of the small amount of carbon in the samples.

Accelerator mass spectrometry (AMS), which requires 10 mg carbon, was attempted. This method dated sample No. 1 (M-1), collected at the intermediate level of largest grain size (Figure 3) and identified as Ua-22602, but obtained no results for the other samples because they contained insufficient carbon.

Figure 3. Profile (**A**) and detail of level (**B**) where sample M-I was collected. Source: authors' own work.

It was possible to date only one sample, collected at a depth of 50 cm, in the level with the elements of largest grain size detected in the sediment sequence. No results were obtained for the samples taken in the base and the upper part of the deposit. However, because the dated sample was found at a level that can be followed throughout much of the basin (guide level), it was possible to correlate the age obtained throughout this level and relate its occurrence to the processes that shaped its formation.

4. Results

The overall deposit sequence occurring in the study basin was determined through a complete survey of the sediment infill, by direct observation throughout each profile. The sediment profiles were indicated by drawing all the levels distinguished and the thickness of each one, as shown in the example of profile 1 (Figure 4).

The data for the thickness of the levels differentiated in each profile are shown in Table 1. The lower thickness of the deposit closer to the measuring station can be seen. Level 2 is absent in the final profile surveyed.

With regard to texture, the predominant sediment in the overall sequence comprises around 62% coarse elements, mainly fine gravel. The texture of the fine fraction (sand, silt and clay) present in most of the samples analysed is sandy loam.

These results indicate that the infill materials in the watercourse areas are mainly sand and fine gravel with a silty matrix and scattered pebbles. A level of mainly medium-coarse gravel of 1.6 to 3.2 cm occurs in the mid-upper level of the deposit, and the sediment below and above this level has similar characteristics. The sequence was divided into three sections and their thickness was measured and described in each column surveyed. The upper level, named level 1, is located above the section with a predominance of medium gravel, named level 2. Level 3 is located in the lower part of the sequence, above the slate bed.

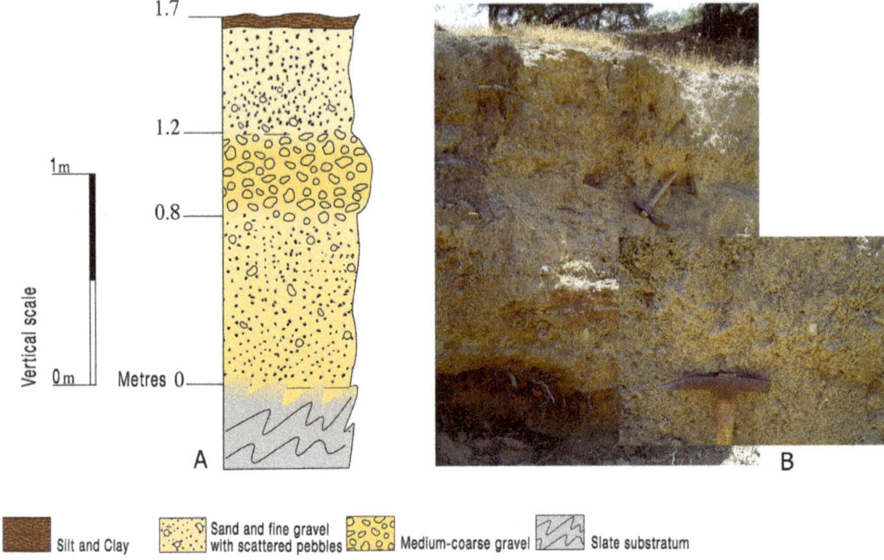

Figure 4. Sediment profile 1 (**A**). Detail of the medium-coarse gravel level (**B**). Source: authors' own work.

Table 1. Thickness of each level in the sediment profiles surveyed in the deposit areas.

	PROFILE 1	PROFILE 2	PROFILE 3	PROFILE 4	PROFILE 5	PROFILE 6	PROFILE 7
Level 1 (m)	0.5	0.47	0.30	0.35	0.35	0.30	
Level 2 (m)	0.4	0.28	0.25	0.20	0.15	0.15	-
Level 3 (m)	0.8	0.75	0.85	0.65	0.50	0.55	0.8
Total thickness	1.70	1.50	1.40	1.20	1.00	1.00	0.8

The upper part of level 1 culminates in a level of fine materials enriched with organic matter, the base of the herbaceous layer.

Contact between level 2 and the upper and lower levels is mainly irregular and at times poorly defined, but at some points it is well defined.

The correlation of the sediment columns is shown in the drawing in Figure 5, which shows the decreased thickness of the deposit further downstream. In the last profile surveyed it was not possible to distinguish the larger grain size level.

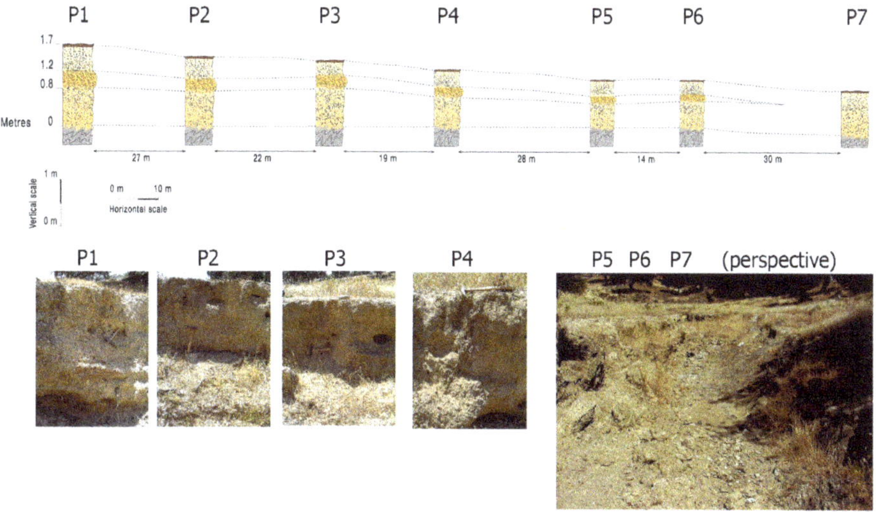

Figure 5. Correlation of the sediment profiles surveyed in the study area. Different images are shown corresponding to points of the deposit area where the profiles have been carried out (P1-P7). Source: authors' own work.

4.1. Overall Deposit Sequence

The sediments that make up the watercourse areas are terrigenous, with a predominance of sand and fine gravel in which facies of sand, clay and silt were identified. These materials tend to accumulate, creating a deposit with a thickness ranging from a few centimetres in some areas to 1.80 m in others. The silt and clay in the upper part of the profile are dark, with abundant edaphic features.

A level with a higher percentage of coarse elements with medium-coarse gravel particles is inserted in the mid-upper part of the profile. In situ measuring revealed sizes of 16 to 32 mm. These elements are mainly angular slate and quartz pebbles and scattered boulders. This level can be followed throughout most of the sediment infill of the bed, primarily in the lower section, where the greater thicknesses were recorded. The direct data from the survey of the columns indicated that it is located at a depth of 14 cm to 50 cm, occurring at lower depths in the thicker profiles. The minimum thickness measured in this level was 10 cm and the maximum was 50 cm, although the level was not identifiable when the deposit was less than 30–40 cm thick.

The sediment model that makes up these facies is associated with a context of slope erosion that resulted in the dense fraction and the sand. The soil corresponding to the fine fraction was displaced by rainwater.

Rainfall and runoff provide the initial energy necessary for the process of dislodging and transporting soil particles, and the topographical features determine the energy of the water current for transporting the particles. Similarly, soil composition and structure (particle size, cohesion, etc.) affect the capacity of the rainfall and runoff energy to carry soil particles. Therefore, human activity, through cropping, is determinant in altering soil erosion susceptibility. The flow erodibility coefficient depends on the type of soil, and changes in this parameter produce considerable changes in sediment production [23].

The sediment of the sequence studied represents the material corresponding to soil erosion in the area where the basin is located; i.e., soil of the Cáceres peneplain occurring on slate, with shallow depths of 25 to 50 cm. In the slope erosion process, a greater transport capacity was registered for rainfall, seen in the level of larger grain pebbles identified in the sequence. This could be due to the higher amount of stones in the soil profile caused by more intense tillage than in an earlier period. Soil redistribution by mechanical displacement during tillage has been recognised as a process of intense soil degradation (mechanical erosion, also known as tillage erosion). Empirical models describing the mechanisms of mechanical soil redistribution have shown that most agricultural tools used in very diverse farming conditions generate very high rates of soil remobilisation [24–29].

The deposit analysed can be interpreted as the effect of soil redistribution caused by past agricultural practices, entailing a modification or interaction in the water-erosion processes that altered the normal rhythm of the deposit of the sediment sequence in the study area, displacing larger grain size material.

4.2. Radiocarbon Dating Results

The results of dating sample M-1, identified as Ua-22606, provided a conventional radiocarbon age of 180 ± 30 BP, corresponding to a calibrated result (2 sigma, 95% probability) of Cal 1650 to 1950. The calibration graph is shown in Figure 6.

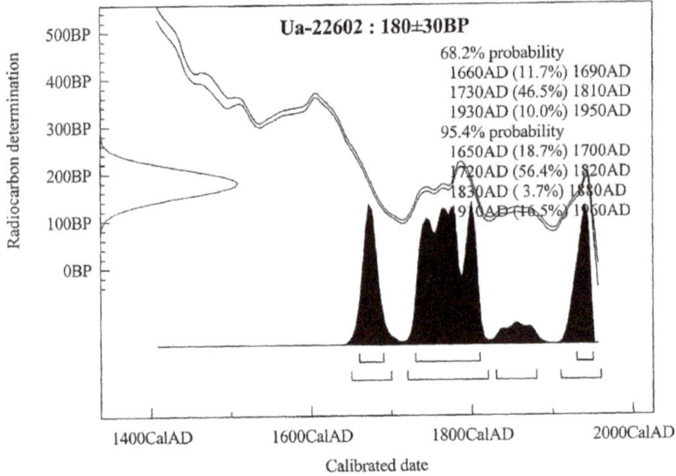

Figure 6. Calibration curve for sample Ua-22606.

This calibrated age range indicates that the materials deposited above the level where the sample is located are later than 1650 and have a 56% probability of being later than 1720–1820.

Because the results were obtained from the sample taken at the bottom of the level of coarse pebbles (indicative of a change in the conditions of the deposit), a connection can be made to events in the environment that are documented, occurred over time, and could have influenced the variation in the conditions of the deposit. Although carbon 14 does not allow a specific date to be assigned to an occurrence, it helps to identify a possible period for investigation.

If the dated sample had been taken at a level with continuity in the rhythm, it would be possible only to indicate the interval in which the deposit might have occurred. In a deposit where the sequence shows no alteration or variation, it is difficult to determine the time frame if there is no known external episode that it can be referenced to. Unfortunately, no other examples were found for dating to determine the age of the sediment at the lowest or the uppermost part of the deposit and extend the chronology of all the materials. However, dating provided an important reference to study the recent evolution of these spaces.

5. Discussion

The sediment study shows that the predominant sediments in the overall sequence are sand and fine gravel with a silty matrix, with the insertion of a level with a high percentage of coarse elements (close to 80%) of medium-sized gravel. After this level was measured in each column, it was shown to have a variable thickness (maximum 50 cm, minimum 14 cm). It is located at a depth of 50 to 20 cm, decreasing in thickness and depth as the deposit becomes less thick, and is not identifiable when the profile has a limited thickness. Therefore, it is not a lens without lateral continuity, because it is distinguishable and continues in each column surveyed. Interpretation of vertical electric soundings [22] also shows these variations in deposit grain size at other parts of the basin where the sequence was not observed directly.

At the level with the largest grain size identified, remains of carbonised organic matter collected at a depth of 0.5 m show a conventional radiocarbon age of 180 ± 30 BP. This age corresponds to a calibrated result (2 sigma, 95% probability) Cal of 1650–1950, with a higher probability in the interval 1720–1820. It is therefore likely that the higher grain size level is later than this time interval.

The sediment model that makes up these facies is associated with a context of slope water erosion that caused the infill in the watercourse areas. The infill is terrigenous, with a predominance of sand and fine gravel in which the level of coarser pebbles must be associated with a variation in the normal rhythm of sedimentation, resulting in displacement of larger grain elements. This greater displacement could be associated with agricultural tasks, because tillage has considerable impact on water erosion processes and greatly increases soil particle mobilisation [29].

Studies of gully erosion in an experimental basin (Parapuños) near the study area reported a similar level of material of fluvio-colluvial origin, with a percentage of coarse fragments inserted in the infill, indicating a variation in the soil erodibility coefficient comparable to that of the study area. After field observation followed by grain size analysis [30] reported a level of coarse material (>2 mm) in all profiles. This layer was located at different depths of each profile and at least 40% of the sample material was coarse. However, in most of the profiles, the coarse material it contained was more than 70% of the total sample. This level mainly comprised fragments with a diameter of 2 to 6 cm, but in some cases as much as 20 cm.

Obtaining similar results in areas close to the study basin was a very significant finding, because they have common source areas in which the same model of sediment production must have occurred.

After analysing all the data, it was necessary to return to the study objective of assessing the deposit conditions in the basin to determine whether the sedimentation process was natural or human-accelerated.

Using the sedimentation and chronological results obtained, it was possible to determine the historical events that might have influenced the erosion and deposit processes in the basin. The data for the land-use changes in the dehesa in the estimated period were studied, focusing on any modifications caused by human activity. The documentary sources for the estimated period do not refer to any exceptional rainfall events associated with crises or epidemics that could have caused anomalous transportation of materials or high erosion rates.

In the late 18th and early 19th centuries, historical events led to changes in land use in the region. This period was the culmination of a policy that began in 1770 to reallocate land for crops. A Royal Decree of 1793 declared that extensive tracts of land previously available for transhumance were to be used for grazing and cropping [31].

The land reform, initiated by Godoy, was preceded by strong complaints about practitioners of transhumance, who were blamed for the lack of cropland, the decline of agriculture and the debilitation of the region [32].

Pastureland that kept the dehesa in ecological balance was converted to cropland, in most cases resulting in soil with a limited profile, overuse and the consequent loss of structure and texture, making the land more vulnerable to erosion. The rock fragments characteristic of the soil were more exposed to displacement processes after the land was tilled [21,28]. Soil disintegration and redistribution due to agricultural tasks have been recognised as a process of intense soil degradation [29,33]. The accumulated results of this process are evidenced in recent deposits in the watercourse areas.

During this phase of extensive, widespread ploughing, a considerable amount of materials would have reached the watercourses and gradually filled in their beds. However, the infill would have had little stability, because the beds were exposed to active processes from the concentration of surface runoff [21] that created deep gullies along the watercourses.

To a certain extent, the introduction of the agricultural reform, followed by the decline of long-distance transhumance, marked a turning point in land use in the region. The reform changed agriculture by increasing the area under crops after the granting of wasteland and uncropped land, although some of the land was cultivated only for a short time or left untouched because of its poor quality or the high costs of farming it.

It is likely that the dehesa where the Guadalperalón and Parapuños basins are located was no exception to these land-use changes, which must have had a repercussion or impact in the short to medium term. The results of analysing the sediments corresponding to this historical period in a typical dehesa such as the pilot hydrographic basin studied should be considered in this light.

6. Conclusions

Using the data provided in this study, we can assess the possible contribution of tillage erosion in the past to the evolution of the landscape. Studies of this kind, which are lacking for Extremadura, can be used to reconstruct many aspects of past actions in these spaces with regard to their hydrogeomorphological dynamics, and have the added value of completing the findings of other studies, thus helping to increase knowledge about recent processes in these areas of high regional significance.

The sediment study of the watercourse areas indicates an alteration in the sediment sequence (level of coarse elements of larger grain size) attributable to changes in the deposit conditions. In this qualitative study, the materials in the level where the grain size changed were dated, providing an age range in which there were no data about the chronology. This is important for identifying the historical context in which the deposit occurred; i.e., the late 18th century.

The continuity of this level of coarser pebbles detected in the deposit sequence (continuing through a large part of the bed) and in other watercourses nearby indicates that the origin of the change was a process associated with the entire source area rather than an isolated process. It must have continued over a sufficient period to make it significant in the overall sequence, interpreted as slope erosion. After the change in grain size occurred, there was a further period of stabilisation.

Although it is difficult to assess the scope of the historical change in land use and determine whether it can be considered an environmental crisis, the almost constant appearance of this level of larger grain size materials in the infill sequence of a nearby experimental basin with similar environmental features [30] supports the idea of a repercussion on the environment caused by historic events that occurred during the period studied. However, more detailed studies of the agricultural history are needed to verify the agricultural transformations in the study areas.

With regard to the limitations of the study, the pilot basin was chosen as the experimental area despite its limited scope because its uses and physical and geological features make it a typical dehesa. It is also located in an extensively studied seminatural ecosystem that forms part of the drainage network degrading the Cáceres peneplain. Because of this, it can be considered representative of the evolution of the area it belongs to and optimum for possible data extrapolation.

It should be noted that although it was possible to date only one sample from the sediment level where the alteration in the rhythm of the sequence occurred, the dating can be considered valid because the sample is located at an extensive guide level that continues along much of the deposit. The carbon dating indicates the interval from which sedimentation of this level occurred and identifies the period, allowing the deposition to be related to documented historical events. Although the exact period is not defined, the range is important because it provides sedimentary evidence to investigate the cause of the larger grain-size level. If no alteration in the rhythm of deposition had been detected in the sequence, it would not have been possible to relate the deposit of the material to external processes that took place in the area during the period when the deposit occurred.

Despite these limitations, the sedimentary evidence and radiocarbon dating revealed increased erosion of coarse material in the study basin that could have undergone the same historical transformations occurring in similar areas. This concurs with the historical sources cited, probably in the late 18th century and early 19th century, due to human intervention in the region after changes in land use. The results shed light on the environmental implications of past human activity, such as tillage operations, and have future applications for conserving this dominant ecosystem in a region of high economic and environmental importance, in which humans play a significant role in its regeneration and conservation.

Author Contributions: Formal analysis, M.T.T.R., A.D.Á. and L.G.-M.; Funding acquisition, L.G.-M.; Investigation, M.T.T.R., A.D.Á. and L.G.-M.; Methodology, M.T.T.R.; Project Administration, M.T.T.R.; Supervision, M.T.T.R., A.D.Á. and L.G.-M.; Writing—original draft, M.T.T.R.; Writing—review and editing, M.T.T.R., A.D.Á. and L.G.-M.

Funding: This article has been funded by the Government of Extremadura (Ref. GR18176).

Acknowledgments: Publication of this article has been possible thanks to the funding of the Government of Extremadura and the European Regional Development Fund (ERDF), reference GR18176, granted to the research team INNOVA (Diseño, Sostenibilidad y Valor Añadido) of the University of Extremadura.

Conflicts of Interest: The authors declare no conflict of interest

References

1. Brown, A.; Toms, P.; Carey, C.; Rhodes, E. Geomorphology of the Anthropocene: Time-transgressive discontinuities of human-induced alluviation. *Anthropocene* **2013**, *1*, 3–13. [CrossRef]
2. Syvitski, J.P.M.; Kettner, A. Sediment flux and the Anthropocene. *Philos. Trans. R. Soc. Lond. Math. Phys. Eng. Sci.* **2011**, *369*, e975. [CrossRef] [PubMed]
3. Santisteban, J.I.; Mediavilla, R.; Gil García, M.J.; Domínguez Castro, F.; Ruiz Zapata, M.B. La historia a través de los sedimentos: Cambios climaticos y de uso del suelo en el registro reciente de un humedal mediterráneo (Las Tablas de Daimiel, Ciudad Real). *Bol. Geol. Min.* **2009**, *120*, 497–508.
4. Arnaud, F.; Poulenard, J.; Giguet-Covex, C.; Wilhelm, B.; Révillon, S.; Jenny, J.P.; Revel, M.; Enters, D.; Bajard, M.; Fouinat, L.; et al. Erosion under climate and human pressures: An alpine lake sediment perspective. *Quat. Sci. Rev.* **2016**, *152*, 1–18. [CrossRef]
5. Dearing, J.A.; Jones, R.T. Coupling temporal and spatial dimensions of global sediment flux through lake and marine sediment records. *Glob. Planet. Chang.* **2003**, *39*, 147–168. [CrossRef]
6. Plieninger, T.; Schair, H. *Elementos Estructurales del Paisaje Adehesado Tradicional en Monroy y Torrejón el Rubio (Cáceres) y su Importancia Para la Conservación de la Naturaleza y el Desarrollo Rural*; Revista de Estudios Extremeños I, Departamento de Publicaciones, Diputación de Badajoz: Badajoz, Spain, 2006.
7. Gómez Amelia, D.; Schnabel, S. *Procesos Sedimentológicos e Hidrogeológicos en Una Pequeña Cuenca Bajo Explotación de Dehesa en Extremadura*; López Bermúdez, F., Conesa García, C., Romero Díaz, M.A., Eds.; II Reunión Nacional de Geomorfología, Sociedad Española de Geomorfología: Murcia, Spain, 1992; pp. 55–63.
8. Schnabel, S. *Soil Erosion and Runoff Production in a Smallwatershed Under Silvo-Pastoral Landuse (Dehesas) in Extremadura, Spain*; Geoforma Ediciones: Logroño, Spain, 1997.
9. Ceballos Barbancho, A. *Procesos Hidrológicos en Una Pequeña Cuenca Hidrográfica Bajo Explotación de Dehesa en Extremadura*; Servicio de Publicaciones; Dpto de Geografía y O.T. Facultad de Filosofía y Letras Universidad de Extremadura: Badajoz, Spain, 1999.
10. Mateos Rodríguez, A.B. Interceptación de la Lluvia por la Encina en Espacios Adehesados. Ph.D. Thesis, Universidad de Extremadura, Cáceres, Spain, 2003.
11. Schnabel, S.; Ceballos Barbancho, A.; Gómez Gutiérrez, Á. *Erosión Hídrica en la Dehesa Extremeña. En: Aportaciones a la Geografía Física de Extremadura con Especial Referencia a las Dehesas*; Schnabel, S., Lavado Contador, J.F., Gómez Gutiérrez, Á., García Marín, R., Eds.; Fundicotex: Cáceres, Spain, 2010; pp. 153–185.
12. De Tena Rey, M.T. *Caracterización y Análisis de los Depósitos Sedimentarios de Áreas de Vaguada en Dehesas de Extremadura. Arroyo de Gadalperalón (Cáceres)*; Universidad de Extremadura: Badajoz, Spain, 2008.
13. Puerto, A. La dehesa. *Investig. Cienc.* **1997**, *253*, 66–73.
14. Pulido, M.; Schnabel, S.; Contador, J.F.L.; Lozano-Parra, J.; Gómez-Gutiérrez, Á.; Brevik, E.C.; Cerdà, A. Reduction of the Frequency of Herbaceous Roots as an Effect of Soil Compaction Induced by Heavy Grazing in Rangelands of SW Spain. *Catena* **2017**, *158*, 381–389. [CrossRef]

15. García-Ruiz, J.M.; Berguería, S.; Lana-Renault, N.; Nadal-Romero, E.; Cerdà, A. Ongoing and emergingquestions in water erosion studies. *Land Degrad. Develop.* **2017**, *28*, 5–21. [CrossRef]
16. Novara, A.; Stallone, G.; Cerdà, A.; Gristina, L. The Effect of Shallow Tillage on Soil Erosion in a Semi-Arid Vineyard. *Agronomy* **2019**, *9*, 257. [CrossRef]
17. Cerdà, A.; Rodrigo-Comino, J.; Novara, A.; Brevik, E.C.; Vaezi, A.R.; Pulido, M.; Keesstra, S.D. Longterm impact of rainfed agricultural land abandonment on soil erosion in the Western Mediterranean basin. *Prog. Phys. Geogr. Earth Environ.* **2018**, *42*, 202–219. [CrossRef]
18. De Tena Rey, M.T. *Penillanura Trujillanos-Cacereña*; Muñoz Barco, P., Martínez Flores, E., Eds.; Patrimonio Geológico de Extremadura; Junta de Extremadura; Conserjería de Industria, Energía y Medio Ambiente: Mérida, España, 2010; p. 478.
19. De Tena Rey, M.T.; Gutiérrez Gallego, J.A.; Martín Nogales, E. Aplicación de un SIG a la Cartografía Regional de Áreas con los Parámetros Físicos-Geológicos y Usos de Una Cuenca Experimental (Provincia de Cáceres). *Mapping* **2007**, *115*, 64–68.
20. Gómez Amelia, D.; Schnabel, S. *Hidrología y Erosión en Ambientes de Pastoreo Extensivo*; Lasanta, T., García Ruiz, J.M., Eds.; Erosión y Recuperación de Tierras en Áreas Marginales, Instituto de Estudios Riojanos y Sociedad Española de Geomorfología: Logroño, España, 1996; pp. 137–154.
21. De Alba, S.; Lindstrom, M.; Schumacher, T.E.; Malo, D.D. Soil landscape evolution due to soil redistribution by tillage: A new conceptual model of soil catena evolution in agricultural landscape. *Catena* **2004**, *58*, 77–100. [CrossRef]
22. García de Prado, J.; De Tena, M.T.; Pro, C. Estudio Topográfico y prospección geoeléctrica de las áreas de depósito de una pequeña cuenca hidrográfica en la provincia de Cáceres (España). *Mapping* **2007**, *115*, 45–49.
23. Zambrano Nájera, J. Estimación de la Producción y Transporte de Sedimentos en Cuencas Urbanas Pequeñas a Escala de Evento Mediante un Modelo de Base Física Basado en SIG. Ph.D. Thesis, Universidad Politécnica de Cataluña, Barcelona, Spain, 2015.
24. De Alba, S.; Lacasta, C.; Benito, G.; Perez Gonzalez, A. *Influence of Soil Management on Water Erosion in a Mediterranean Semiarid Environment in Central Spain*; Garcia Torres, L., Benitez, J., Martinez Vilela, A., Eds.; Food and Agriculture Organization-European Conservation Agriculture Federation: Brussels, Belgium, 2001; pp. 173–177.
25. De Alba, S.; Alcázar, M.; Cermeño, F.I.; Barbero, F. *Erosión y Manejo del Suelo. Importancia del Laboreo Ante los Procesos Erosivos Naturales y Antrópicos. En: Agricultura Ecológica en Secano: Soluciones Sostenibles en Ambientes Mediterráneos*; Ministerio De Agricultura, Alimentación y Medio Ambiente: Madrid, Spain, 2011; pp. 13–38.
26. Lindstrom, M.J.; Nelson, W.W.; Schumacher, T.E. Quantifying tillage erosion rates due to moldboard plowing. *Soil Tillage Res.* **1992**, *24*, 243–255. [CrossRef]
27. Govers, G.; Lobb, D.A.; Quine, T.A. Tillage erosion and translocation: Emergence of a new paradigm in soil erosion research. *Soil Tillage Res.* **1999**, *51*, 167–174.
28. Torri, D.; Borselli, L.; Calzonari, C.; Yáñez, M.; Salvador-Sanchis, M.P. *Soil Erosion, Land Use, Soil Quality and Soil Functions: Effects of Erosion*; Rubio, J.L., Morgan, R.P.C., Asins, S., Andreu, V., Eds.; Geoforma Ediciones: Logroño, Spain, 2002; pp. 131–148.
29. De Alba, S. *Erosión y Redistribución Mecánica del Suelo (Tillage Erosion): Transformación de los Paisajes Agrícolas*; Cerdá, A., Ed.; Erosión y Degradación del suelo agrícola en España, Cátedra de Divulgación Científica, Universidad de Valencia: Valencia, Spain, 2008; pp. 149–182.
30. Gómez Gutierrez, A. Estudio de la Erosión en Cárcavas en Áreas con Aprovechamiento Silvopatoril. Ph.D. Thesis, Universidad de Extremadura, Cáceres, Spain, 2009.
31. Rodríguez Grajera, A. Una norma preliberal. El Real Decreto 28 de abril de 1793 y sus repercusiones en Extremadura. Josep Fontana. Historia I Projecte Social. Reconeixement a una trajectoria. *Crit. Barc.* **2004**, *2*, 212–228.
32. Melón Jiménez, M.A.; Rodriguez Grajera, A. *Extremadura*; La Historia: Badajoz, Spain, 1997.
33. Wysocka-Czubaszek, A.; Czubaszek, R. Tillage erosion: The principles, controlling factors and main implications for future research. *J. Ecol. Eng.* **2014**, *15*, 150–159. [CrossRef]

 *sustainability*

Article

Experimental Setup for Splash Erosion Monitoring—Study of Silty Loam Splash Characteristics

David Zumr [1,*], **Danilo Vítor Mützenberg** [1], **Martin Neumann** [1], **Jakub Jeřábek** [1], **Tomáš Laburda** [1], **Petr Kavka** [1], **Lisbeth Lolk Johannsen** [2], **Nives Zambon** [2], **Andreas Klik** [2], **Peter Strauss** [3] and **Tomáš Dostál** [1]

[1] Faculty of Civil Engineering, Czech Technical University in Prague, Thákurova 7,
166 29 Prague 6, Czech Republic; danilo.mutz@gmail.com (D.V.M.); martin.neumann@fsv.cvut.cz (M.N.);
jakub.jerabek@fsv.cvut.cz (J.J.); tomas.laburda@fsv.cvut.cz (T.L.); petr.kavka@fsv.cvut.cz (P.K.);
dostal@fsv.cvut.cz (T.D.)

[2] Institute for Soil Physics and Rural Water Management, University of Natural Resources and Life Sciences,
1190 Vienna, Austria; lisbeth.johannsen@boku.ac.at (L.L.J.); nives.zambon@boku.ac.at (N.Z.);
andreas.klik@boku.ac.at (A.K.)

[3] Federal Agency for Water Management, Institute for Land & Water Management Research, 3252
Petzenkirchen, Austria; peter.strauss@baw.at

[*] Correspondence: david.zumr@fsv.cvut.cz

Received: 21 November 2019; Accepted: 22 December 2019; Published: 24 December 2019

Abstract: An experimental laboratory setup was developed and evaluated in order to investigate detachment of soil particles by raindrop splash impact. The soil under investigation was a silty loam Cambisol, which is typical for agricultural fields in Central Europe. The setup consisted of a rainfall simulator and soil samples packed into splash cups (a plastic cylinder with a surface area of 78.5 cm^2) positioned in the center of sediment collectors with an outer diameter of 45 cm. A laboratory rainfall simulator was used to simulate rainfall with a prescribed intensity and kinetic energy. Photographs of the soil's surface before and after the experiments were taken to create digital models of relief and to calculate changes in surface roughness and the rate of soil compaction. The corresponding amount of splashed soil ranged between 10 and 1500 g m^{-2} h^{-1}. We observed a linear relationship between the rainfall kinetic energy and the amount of the detached soil particles. The threshold kinetic energy necessary to initiate the detachment process was 354 J m^{-2} h^{-1}. No significant relationship between rainfall kinetic energy and splashed sediment particle-size distribution was observed. The splash erosion process exhibited high variability within each repetition, suggesting a sensitivity of the process to the actual soil surface microtopography.

Keywords: splash erosion; rainfall simulator; splash cup; soil loss; soil detachment; disdrometer; rainfall kinetic energy

1. Introduction

The initial stage of the erosion process (splash erosion) occurs when raindrops with high kinetic energy hit bare soil, breaking down aggregates and detaching soil particles. Such particles are translocated a short distance from the raindrop's impact and they then settle on the soil's surface and block the interaggregate pores reducing the topsoil's infiltration capacity and accelerating the formation of surface runoff. Hence, understanding the relationships between various rainfall characteristics and splash erosion is important to be able to predict the dominant runoff mechanisms of unprotected soils and to determine the rainfall kinetic energy threshold for erosion initiation.

Various monitoring techniques have been developed over the years to measure the degree of soil detachment in relation to the kinetic energy of raindrops. Besides splash cups (which are used in this study), splash boards or tracers have also been used in previous studies (as reviewed by Fernández-Raga et al. [1]). When monitoring splash erosion, there are several considerations to take into account when developing study design:

(A) collection mechanism:

- detached soil is splashed into a collector located around the soil sample (e.g., [2], Figure 1a)
- detached soil is splashed into a collector surrounded by the soil material (e.g., [3–5], Figure 1b)

(B) sample preparation:

- disturbed soil sample (e.g., [6,7])
- in situ undisturbed soil (e.g., [2–4]).

Each approach has its advantages and disadvantages. One needs to estimate the contributing area of the surrounding soil (Figure 1b), otherwise it is not possible to calculate the detached soil amount per specific area. This problem occurs (to some extent) in the Morgan setup [2] because some soil particles are transported within the sampling area (Figure 1a). Therefore, the optimum sampling area is a tradeoff between underestimating the splash, representativeness of the collected sample, as well as ease of sample handling [3].

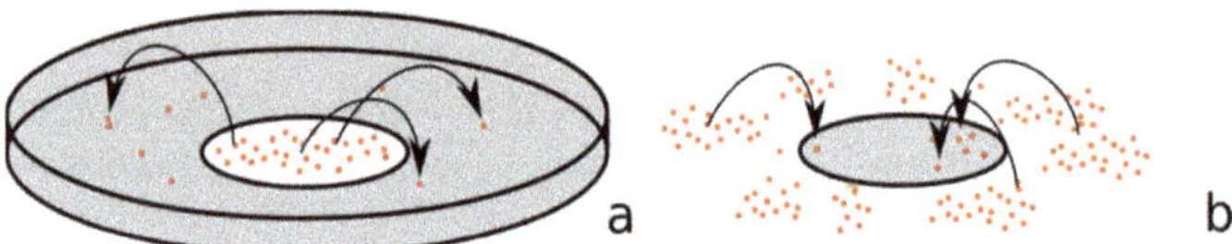

Figure 1. Two methods to collect detached soil particles: (**a**) collector is around the soil sample; (**b**) collector is surrounded by the sampled soil.

Wei et al. [8] showed that the splash erosion rate is dependent on soil sample water saturation and texture. Sandy soil was not sensitive to degree of saturation whereas samples with increasing clay content were, and similar results were reported by Khaledi Darvishan et al. [9]. To the contrary, Watung et al. [10] did not observe any significant difference in soil splash under variable saturated conditions for tropical soil (Oxisol). Utilizing disturbed soil samples allows for the control of soil sample conditions. In situ measurement on undisturbed samples does not allow for the control of soil conditions; on the other hand the measurement may be more representative for a given location since soil structure and soil surface characteristics are preserved. Therefore, there are a lot of factors to consider when designing an experimental setup.

The splash-collecting device needs to have sufficient dimensions to trap most of the detached particles. As Leogun et al. [11] and Marzen et al. [12] showed, the amount of detached soil decreases exponentially with increasing distance from the soil sample. The transport splash distance increases with decreasing soil particle size. Fu et al. [13] added that the splashed distance is not only related to particle or aggregate sizes, but also to rainfall kinetic energy.

Transport distances ranged between 10 cm and 20 cm for aggregates with diameters between 0.5 mm and 1 mm and up to 35 cm for soil aggregates with diameters from 0.05 mm to 0.5 mm in a study by Legout et al. [11]. Most of the splashed particles were observed within 20 cm from the soil sample in Fu et al. [14] and within 35 cm in [12].

The most common problem with splash erosion experiments is that it is difficult to compare results across studies due to each author's differing experimental setup [15]. Many different splash cup setups with various sizes and trapping principles have been used to measure splash erosion.

Pioneering studies were performed using single soil fractions and cylindrically shaped cups [16–18]. Kinnell [6] added a thin external ring around the soil sample to exclude surface runoff which may occur during ponding conditions. Scholten et al. [7] developed splash cups in which sample saturation was controlled, their setup maintained nearly constant water content of the sample, and allowed for the simultaneous draining of rainfall. Two types of splash cups (cups and funnels) were compared by Fernández-Raga et al. [4]. The funnels collected systematically more particles because they prevent particle transport back to the surrounding soil (backsplash). The most frequently used splash cup design is inspired by Morgan (1981) (for example, [9,19–22]). In all studies reported above, the sediment loss was determined by weighing the dry sample prior to and after the measurement period. The Morgan setup is also suitable for use in both indoor and outdoor conditions with disturbed or undisturbed soil samples.

In general, the splash erosion process is very complex and the reported results usually exhibit high variability. Angulo-Martinez et al. [23] evaluated the effects of rainfall characteristics, rainfall erosivity index and soil type with a linear mixed-effects model. The rainfall erosivity index explained 55% of the data variability but soil type did not have a statistically significant influence on erosion. Up to 74% of the variability within a single soil type was attributed to random effects. The role of slope (and upward/downward splash) and rainfall intensity were investigated. It was reported that slope altered the splashed particle-size distribution and the role of slope for total splashed material varied for various rainfall intensities [22]. The rainfall itself is also a very important factor, and authors emphasize the need of accurate drop-shape estimation in order to obtain adequate kinetic energy of the rainfall. Rainfall changes the surface microtopography [24] which may have further effects on water infiltration, surface water retention and surface runoff [25]. A common method for the analysis of surface relief changes is close-range photogrammetry [26]. It has been shown that especially loose soils are prone to a fast decrease in microrelief roughness, leading to accelerated soil erosion [27,28].

Rainfall kinetic energy (KE) is often estimated based on measured rainfall intensity (KE-I relationship) due to the lack of a direct rainfall kinetic energy measurement [29]. Lobo and Monilla [30] tested several KE-I relationships for various geographical locations and concluded that parameters of the KE-I relationships are site specific. Meshesa et al. [31] tested parametric relationships between rainfall intensity and rainfall kinetic energy using artificial rainfall, noting that artificial rainfall exhibits raindrops with different sizes than natural rainfall. Therefore, KE-I relationships derived under natural conditions should not be applied to rainfall simulators.

The relationship between the rainfall kinetic energy and the amount of splash erosion on a bare surface varies for different soil types and tillage practices. Most of the studies of splash erosion on real soils come from arid or semi-arid climates, such as the Mediterranean region, Loess Plateau of China or southern states of the USA [1]. In Central Europe, soil erosion processes have been studied extensively, but splash erosion has not usually been considered or evaluated. Rainfall in Central Europe often does not generate overland flow (due to low intensity and/or short duration), but soil detachment and resulting soil surface changes take place from the impact of first drops with sufficient kinetic energy [32]. The lack of knowledge of splash erosion rates on agriculturally cultivated Cambisols is the main motivation for the presented research.

In this study we present a splash erosion experimental setup which utilizes techniques from previously published works. We utilize the Morgan design and provide an open-source, easy to manufacture splash cup. The objective of these experiments is to determine the impact of rainfall kinetic energy on splash detachment for a typical agricultural soil in Central Europe. An associated aim is to evaluate the particle-size distribution of the eroded material and to analyze the effects of the rainfall kinetic energy on soil consolidation.

2. Materials and Methods

2.1. Splash Erosion Collection Device

The monitoring setup is designed for both indoor and outdoor measurements and is similar to the system proposed by Morgan [2].

The monitoring device includes (Figure 2): (1) the sediment collector, (2) the cylindric splash cup, (3) the photogrammetry reference targets, (4) the LED illumination ring, (5) the outlet for the sediment collection and (6) the splash cup holder. The sizes and proportions of the device components and the angle between the rim of the splash cup and the collector were adjusted to capture the maximum amount of splashed soil while allowing unchanged raindrop impact on the soil sample.

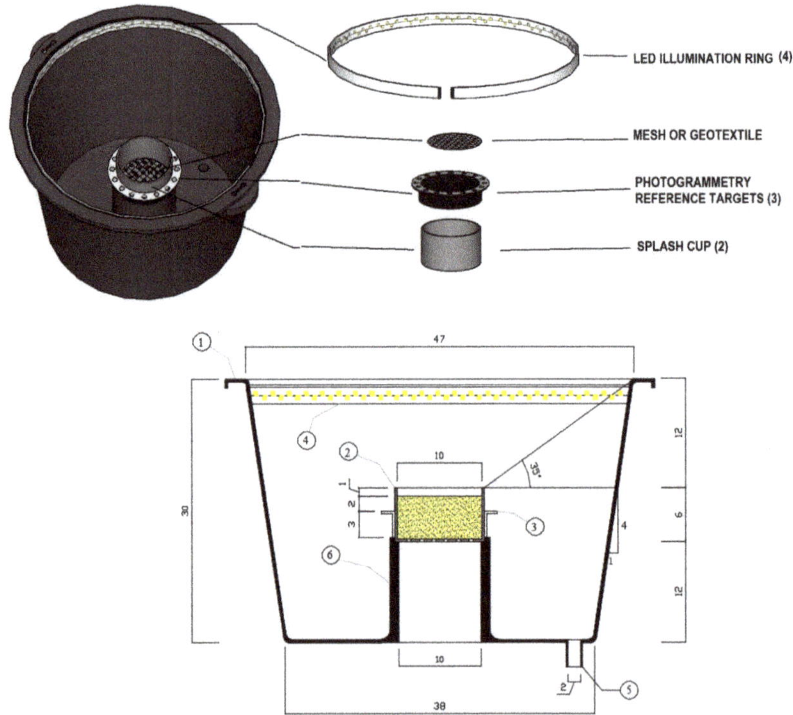

Figure 2. Splash erosion device and its parts and dimensions (units in centimeters).

The splash cup (2 in Figure 2) consists of a polypropylene cylinder with a height of 60 mm, inner diameter of 100 mm and a surface area of 7854 mm^2. The cylinder's wall is 2.7 mm thick and the upper rim of the splash cup is sharpened. The bottom of the cup is perforated in order to allow for draining of percolated water. A fine mesh is placed in the bottom of the cup to prevent soil loss during manipulation and water percolation. Each tested soil is filled to one centimeter below the rim of the splash cup to prevent runoff in case of ponding water on the sample's surface. The sediment collector's purpose is to capture eroded (splashed) soil particles. Its walls are angled in order to funnel the water with soil particles to the outlet (5 in Figure 2) which leads to a storage bucket under the sediment collector. In the center, there is a holder (6 in Figure 2) which is used to support the splash cup above the sediment collector's base so that the soil sample is protected from the backsplash of soil particles that accumulate inside the collector. The holder does not have a bottom, therefore, water percolating through the soil sample can freely drip out. This water is not usually collected. The collector is made from a commercially available polypropylene bucket, and all the firmly attached components (the holder and the outlet) are butt welded to the collector. The CAD drawings of the splash cup's components as well as mounting procedure are freely available here: rain.fsv.cvut.cz/splashcup.

After every experiment, the detached particles still attached to the collector's walls or settled on the collector's bottom were washed into the outlet and added to the remaining eroded particles. The suspension of the collected sediment and rainfall water was then filtered, oven dried and weighed.

The splash erosion device was then prepared for photogrammetrical analysis of soil sample surface changes due to rainfall impact. Typically, 10 to 15 referenced photographs from different angles were taken for the successful reconstruction of a digital surface model. Therefore, around the splash cup there is a white ring with the photogrammetry reference targets (3 in Figure 2). An LED illumination ring (4 in Figure 3) was attached to the sediment collector to provide adequate illumination to ensure that there were no shadows on the surface. The specification of the LED light strip was: chromaticity 4250 K, power 12 W/m, 60 LEDs/m, luminous flux 1050 lm/m.

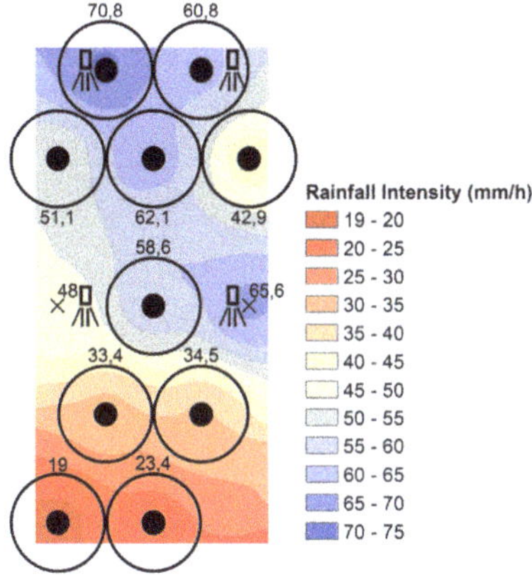

Figure 3. Sketch of the 10 fixed positions and the rainfall intensity distribution below the nozzles. The positions of four nozzles used in the experiment are depicted with a spray symbol, and the X marks denote additional positions where the rainfall intensity was also measured.

2.2. Soil Sample Preparation and Analysis

The soil used for testing of the splash collection system was taken from a topsoil horizon (upper 10 cm) of a cultivated field at the experimental site of Bykovice in Central Bohemia, Czech Republic. The soil type is classified as a Cambisol, and the texture corresponds to silty loam according to the World Reference Base (WRB) classification [33] (12.7% sand, 76.6% silt, 10.7% clay), $CaCO_3$ content is <0.92%, pH 6.9, and total organic carbon 1.7%.

Soil was collected in April 2017 during seedbed conditions. The soil was transported to the laboratory, stripped of large organic residues (stems, roots), large clods and stones, and then air dried.

Collected soil was sieved to remove particles and aggregates larger than 10 mm before filling the splash cup. A piece of permeable geotextile was placed inside the cup to prevent the soil from passing through the splash cup's perforations. The splash cup was then loosely packed with the same amount of prepared soil to reach a similar bulk density to that of seedbed conditions (0.83 g cm^{-3}). The soil sample was not compacted; we only distributed soil aggregates equally along the sample surface and removed any remaining organic residues. Then, the filled splash cups were placed inside the sediment collectors so that the splash cup's surface was level.

After each rainfall simulation the eroded soil particles were carefully washed out from the sediment collector, and the suspension of rain water and eroded sediment was transferred to the laboratory. The obtained sample was filtered on a paper filter with a mesh size of 5 mm, oven dried (at 40 °C) and weighed.

The dried soil was further analyzed using a laser diffraction particle-size analyzer (Mastersizer 3000, Malvern Panalytical Ltd., UK) to determine soil texture. We mixed the splashed material from all the repetitions to obtain enough soil for this analysis. Each soil sample was dispersed in distilled water and placed into an ultrasonic bath for 320 s to disaggregate the soil. Then the sample was analyzed by the laser diffractometer. Measurements were repeated 25 times for every sample. The procedure is described in detail by Kubínová [34].

2.3. Rainfall Simulation

A laboratory Norton Ladder type rainfall simulator was used to generate rainfall. The simulator had an experimental area of 0.9 × 4 m. The rainfall was produced by eight oscillating nozzles, type Veejet 80100, which were mounted in two parallel sections at 2.6 m above the soil samples. Tap water was used, water pressure was set to 32 kPa and rainfall intensity was controlled by the nozzle oscillating frequency. The average raindrop diameter generated by the simulator was 2.3 mm according to monitoring with disdrometers [35].

In this experiment we took advantage of the fact that rainfall intensity spatially varies over the experimental plot. Eleven positions with the rainfall intensity between 20 and 70 mm h^{-1} were chosen for further testing. Figure 3 shows the spatial distribution of rainfall intensity. The pattern is based on the intensity measurements of the splash cup positions and inverse distance weighted interpolation. The rainfall kinetic energy was measured with the Laser Precipitation Monitor (LPM) by Thies Clima® and the KE-I relationship for the given rainfall simulator was established in advance of the splash erosion simulations. The rainfall simulation lasted 15 min, then the soil samples were collected and the splashed amount was analyzed. The whole procedure was repeated five times (totaling 55 samples analyzed).

3. Results and Discussion

The measured relationship between rainfall intensity and rainfall kinetic energy is shown in Figure 4. The observed trend of the KE-I is linear with the slope of 18.69. Compared to the published relationships for natural rainfall (e.g., [36–39]) the KE of the simulated rainfall is lower by approximately 35%. Similar results show that underestimation of the simulated rainfall kinetic energy were also obtained by Petrů and Kalibová [40]. The KE-I relationship is strongly dependent on the rainfall simulator design, nozzle types and water pressure (e.g., simulators generating larger raindrops than the natural rainfall overestimate KE) [31].

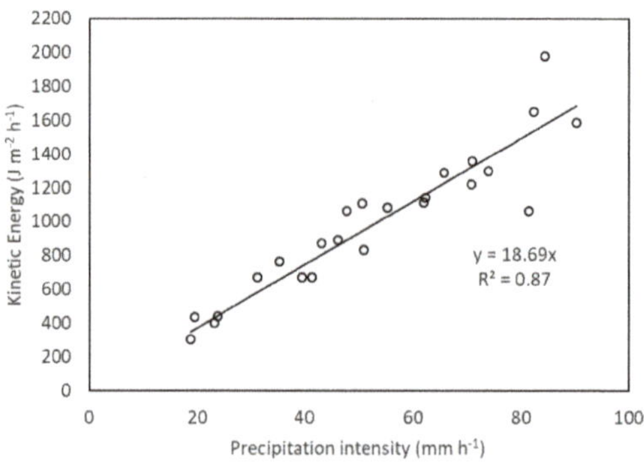

Figure 4. Measured relationship between simulated rainfall intensity and kinetic energy.

The measured soil splash rate ranged between 10 and 2012 g m^{-2} h^{-1} for rainfall kinetic energy between 380 and 1450 J m^2 h^{-1}. The threshold kinetic energy needed to initiate the detachment process was identified by extrapolation to be 354 J m^{-2} h^{-1}. The recorded mass of the detached particles exhibits large variability across the five replicates at each position (Figure 5). The variability was higher for the positions where higher soil erosion was recorded (positions with higher rainfall intensity and kinetic energy). Similar variability during comparable experiments was reported in literature [23]. The variability could be explained by very complex soil erosion behavior which is influenced by size distribution and arrangement of the soil particles and aggregates on the sample's surface (random roughness).

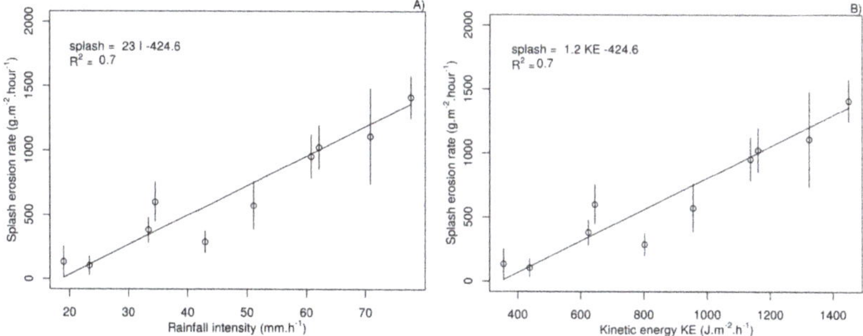

Figure 5. Relationships between rainfall intensity (**A**) and rainfall kinetic energy (**B**) and splash erosion rate. The bars stand for standard deviation, the marker denotes the mean value.

The overall relationship between the mass of the eroded particles and the rainfall kinetic energy has a linear trend (Figure 5). The coefficient of determination is 0.7 which is higher than or in a similar range as presented in comparable studies (e.g., [6,15,41]). Table 1 summarizes selected splash erosion experiments from the literature to show how the experiments vary in setup. They show similar, most often linear, relationships between the amount of detached soil particles and various rainfall characteristics. The slope of the linear trendline differs in each study because of differing experimental setups (rainfall duration, sample preparation) and soil properties [6,15,42]. Bisal [17] and Mazurak and Mosher [18] already experimentally showed that the splashed amount is linearly dependent on drop size and velocity. Surprisingly, in contrast to the more recently published studies, Bisal [17] did not find a significant relationship between rainfall intensity and the amount of sand splashed as long as no ponding occurred on the sand's surface.

Table 1. Comparison of the results with the published splash erosion studies.

Reference	Soil	Sample Preparation, Experiment Specifications	Rainfall Intensity (mm h^{-1})	Rainfall Kinetic Energy (J m^{-2} mm^{-1})	Splash–Rainfall Relationship
Bisal 1960 [17,18]	Sand	Leveled with the rim	76–152	-	No significant relationship (R^2 = 0.31)
Angulo-Martinez et al. 2012 [23]	Silty soil	Leveled 25 mm below the rim, under natural rainfall	12–93	2–12	Linear function
Geissler et al., 2012 [41]	Fine sand	Leveled with the rim; measured under forest vegetation where the throughfall's KE is reported as 2.53 times higher	1–45	-	Linear function (R^2 = 0.74)
Boroghani et al., 2012 [43]	Silt-clay-loam	Not known, only three datapoints measured	69–120	-	Linear function (R^2 = 0.91)
Wu et al., 2019 [44]	Silty loam, seedbed conditions	Leveled with the rim	48–150	4–7	Polynomial function
Fernández-Raga et al., 2019 [15]	Fine sand	Leveled to the rim	38–160	26–29	Linear function (R^2 = 0.18)
This study	Silty loam, seed bed conditions (Cambisol)	Leveled 10 mm below the rim	19–78	11	Linear function (R^2 = 0.70)

The observed splash–KE trendline is strongly influenced by duration of the rainfall experiment. Splash erosion varies over time, especially in the case of structured soils with developed aggregates that are initially broken down into smaller fractions by raindrops. It has been observed that the splash increases with decreasing aggregate size [45,46] and increasing event duration [47,48]. The fact that splash erosion is strongly dependent on surface microtopography is another reason why it is difficult to compare results across studies. Artificial samples filled with smooth, fine-grained sand produce different erosion than natural soils with higher surface roughness and particle cohesion. This is, as noted above, due to the smaller size of individual particles, but is also due to microrelief variation. The effect of surface roughness on splash erosion is not a straightforward process [49]. Some authors found decreasing erosion with increasing roughness [50], and some the opposite trend [49]. The changes in surface microrelief in one sample (applied kinetic energy of 1150 J m^{-2} h^{-1}, recorded average soil surface consolidation of 0.75 mm) is shown in Figure 6. The soil surface consolidated due to rainfall and splash erosion. Figure 7 shows the linear relationship between rainfall kinetic energy and soil consolidation even though the measured soil settling is very heterogeneous and the coefficient of determination is low.

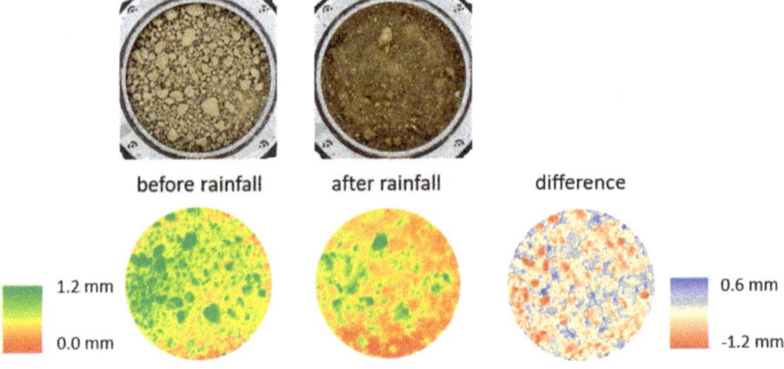

Figure 6. Soil sample surface before and after simulated rainfall. The lower photos represent digital surface models (DSM) in which green areas have higher elevation than the red areas. The soil surface consolidated in the average by 0.75 mm.

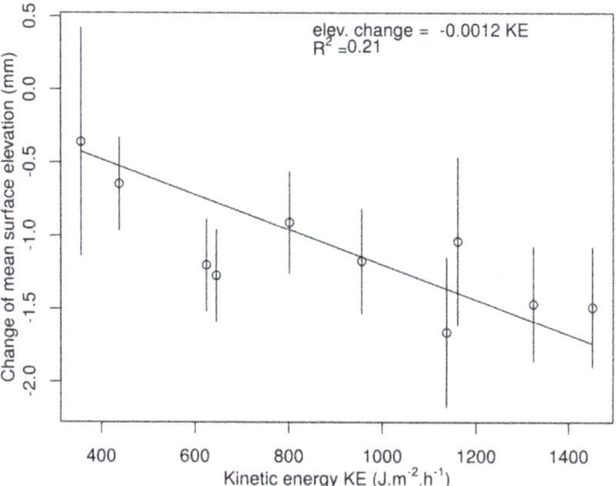

Figure 7. Relationship between rainfall kinetic energy and soil surface settling.

Analysis of the splashed material particle-size distribution (PSD) did not show a significant relationship between rainfall kinetic energy (KE) and detached sediment texture (Figure 8). The Pearson's correlation coefficient of KE versus clay content was 0.36 (p-value 0.39, for a = 0.05), KE versus silt −0.22 (p-value 0.61) and KE versus sand −0.24 (p-value 0.56). The PSD of the detached sediment is not significantly different from the texture of the original soil sample (see the horizontal dashed lines on Figure 7). Therefore, all particle fractions are detached uniformly with no preference toward fine or coarse fractions, no matter the kinetic energy applied. It is important to note that the splashed sediment is usually detached in the form of aggregates and therefore the aggregate size distribution should be evaluated. We have not done this analysis as we were not able to collect the undisturbed splashed soil aggregates. For example, Fu et al. [13] show that especially the fine particle and aggregate (<0.053 mm) ratios change with variable rainfall KE. The KE per mm of rainfall is the same for all the measured points shown in Figure 8, which is due to the design of the rainfall simulator. The results may change if different types of the rainfall simulators (with various drop size distribution or drop velocities) are applied.

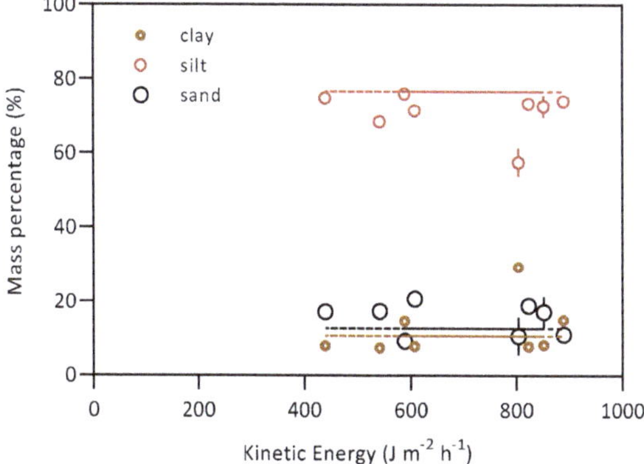

Figure 8. Relationship between rainfall kinetic energy and textural classes (clay, silt, sand) of detached sediment. The dashed lines represent the texture of the Bykovice soil. No significant difference between the soil sample and splashed sediment was observed.

Fernández-Raga et al. [15] demonstrated that splash erosion estimation is strongly dependent on the splash collection setup. Poesen and Torri [3] found that the area of the splash cup is the main influencer defining the amount splashed and therefore coefficients for different splash cup areas should be used. Even the trendline between the amount splashed and the rainfall kinetic energy differs based on the methodology of collection. The experimental design proposed in this study, a modified version of Morgan's splash cup, proved to be reliable, practical and easy to handle. As the splash cups are compact, light, and robust, they can be easily mounted to any support mechanism either in the laboratory and in the terrain. Due to the materials used, the device is durable and can be used for several seasons under field conditions. The soil sample can be packed separately from the collection tray and fixed to its position just before the rainfall experiment.

4. Conclusions

A splash cup methodology was presented and used to analyze splash erosion of a silty loam agricultural topsoil with simulated rainfall across various kinetic energies.

The splash cup, which consists of commercially available components, proved to be a versatile and practical tool for the monitoring of splash erosion. The design follows the dimensions proposed by Morgan, therefore, the splash cup can be used for comparison with other studies in which Morgan's device was employed. Even though the soil particle detachment process is very sensitive to factors other than experimental design, standardization or harmonization of splash cup designs would be a beneficial step forward in the complex research area of the splash erosion process. Therefore, we provide a detailed description of the splash erosion setup, including technical drawings, assembly manual and description of sample preparation and collection on the website rain.fsv.cvut.cz/splashcup.

The results of the presented splash erosion experiment show similar results to previously published studies. The relationship between rainfall kinetic energy and splashed soil amount is linear and there is a kinetic energy threshold to initiate erosion. Even under controlled experimental conditions, when the soil samples were prepared the same way and rainfall characteristics remained constant during the experiment, the eroded soil amount varies across each replicate. This reinforces that splash erosion is a very complex process and the resulting erosion is sensitive to small changes in soil properties and soil surface relief which is problematic and remains an open question that needs to be studied further.

Author Contributions: Conceptualization, D.Z. and T.D.; Formal analysis, D.V.M., J.J., M.N., T.L., L.L.J. and N.Z.; Methodology, D.Z. and P.K.; Project administration, T.D. and A.K.; Supervision, A.K., P.S. and T.D.; Visualization, D.V.M., T.L. and J.J.; Writing—original draft, D.Z.; Writing—review & editing, All. All authors have read and agreed to the published version of the manuscript.

Funding: This research was performed within the project "Kinetic energy of rainfall as a driving force of soil detachment and transport". Financial support was provided through the Czech Science Foundation (GACR): GF17-33751L and the Austrian Science Fund (FWF): I 3049-N29.

Conflicts of Interest: The authors declare no conflict of interest.

References

1. Fernández-Raga, M.; Palencia, C.; Keesstra, S.; Jordán, A.; Fraile, R.; Angulo-Martínez, M.; Cerdà, A. Splash erosion: A review with unanswered questions. *Earth-Sci. Rev.* **2017**, *171*, 463–477. [CrossRef]
2. Morgan, R.P.C. Field measurement of splash erosion. *Int. Assoc. Sci. Hydrol. Publ.* **1981**, *133*, 373–382.
3. Poesen, J.; Torri, D. The effect of cup size on splash detachment and transport measurements. Part I. Field measurements. *Geomorphic Process. Environ. Strong Seas. Contrasts Vol. I Hillslope Process. Catena Suppl.* **1988**, *12*, 113–126.
4. Fernández-Raga, M.; Fraile, R.; Keizer, J.J.; Teijeiro, M.E.V.; Castro, A.; Palencia, C.; Calvo, A.I.; Koenders, J.; Marques, R.L.D.C. The kinetic energy of rain measured with an optical disdrometer: An application to splash erosion. *Atmos. Res.* **2010**, *96*, 225–240. [CrossRef]
5. Truman, C.C.; Bradford, J.M. Soil Science Society of America journal. *Soil Sci. Soc. Am. J.* **1976**, *59*, 519–526. [CrossRef]

6. Kinnell, P.I.A. Splash Erosion: Some Observations on the Splash-Cup Technique 1. *Soil Sci. Soc. Am. J.* **1974**, *38*, 657–660. [CrossRef]
7. Scholten, T.; Geißler, C.; Goc, J.; Kühn, P.; Wiegand, C. A new splash cup to measure the kinetic energy of rainfall. *J. Plant Nutr. Soil Sci.* **2011**, *174*, 596–601. [CrossRef]
8. Wei, Y.; Wu, X.; Cai, C. Splash erosion of clay–sand mixtures and its relationship with soil physical properties: The effects of particle size distribution on soil structure. *Catena* **2015**, *135*, 254–262. [CrossRef]
9. Khaledi Darvishan, A.; Sadeghi, S.H.; Homaee, M.; Arabkhedri, M. Measuring sheet erosion using synthetic color-contrast aggregates. *Hydrol. Process.* **2014**, *28*, 4463–4471. [CrossRef]
10. Watung, R.L.; Sutherland, R.A.; El-Swaify, S.A. Influence of rainfall energy flux density and antecedent soil moisture content on splash transport and aggregate enrichment ratios for a Hawaiian Oxisol. *Soil Technol.* **1996**, *9*, 251–272. [CrossRef]
11. Legout, C.; Leguedois, S.; Le Bissonnais, Y.; Issa, O.M. Splash distance and size distributions for various soils. *Geoderma* **2005**, *124*, 279–292. [CrossRef]
12. Marzen, M.; Iserloh, T.; Casper, M.C.; Ries, J.B. Quantification of particle detachment by rain splash and wind-driven rain splash. *CATENA* **2015**, *127*, 135–141. [CrossRef]
13. Fu, Y.; Li, G.; Wang, D.; Zheng, T.; Yang, M. Raindrop Energy Impact on the Distribution Characteristics of Splash Aggregates of Cultivated Dark Loessial Cores. *Water* **2019**, *11*, 1514. [CrossRef]
14. Fu, Y.; Li, G.; Zheng, T.; Li, B.; Zhang, T. Splash detachment and transport of loess aggregate fragments by raindrop action. *Catena* **2017**, *150*, 154–160. [CrossRef]
15. Fernández-Raga, M.; Campo, J.; Rodrigo-Comino, J.; Keesstra, S.D. Comparative Analysis of Splash Erosion Devices for Rainfall Simulation Experiments: A Laboratory Study. *Water* **2019**, *11*, 1228. [CrossRef]
16. Bisal, F. Calibration of splash cup for soil erosion studies. *Agric. Eng.* **1950**, *31*, 621–622.
17. Bisal, F. The effect of raindrop size and impact velocity on sand-splash. *Can. J. Soil Sci.* **1960**, *40*, 242–245. [CrossRef]
18. Mazurak, A.P.; Mosher, P.N. Detachment of Soil Particles in Simulated Rainfall 1. *Soil Sci. Soc. Am. J.* **1968**, *32*, 716–719. [CrossRef]
19. Vigiak, O.; Okoba, B.O.; Sterk, G.; Groenenberg, S. Modelling catchment-scale erosion patterns in the East African Highlands. *Earth Surf. Process. Landf.* **2005**, *30*, 183–196. [CrossRef]
20. Vigiak, O.; Sterk, G.; Romanowicz, R.J.; Beven, K.J. A semi-empirical model to assess uncertainty of spatial patterns of erosion. *Catena* **2006**, *66*, 198–210. [CrossRef]
21. Moghadam, B.K.; Jabarifar, M.; Bagheri, M.; Shahbazi, E. Effects of land use change on soil splash erosion in the semi-arid region of Iran. *Geoderma* **2015**, *241*, 210–220. [CrossRef]
22. Sadeghi, S.H.; Harchegani, M.K.; Asadi, H. Variability of particle size distributions of upward/downward splashed materials in different rainfall intensities and slopes. *Geoderma* **2017**, *290*, 100–106. [CrossRef]
23. Angulo-Martinez, M.; Beguería, S.; Navas, A.; Machin, J. Splash erosion under natural rainfall on three soil types in NE Spain. *Geomorphology* **2012**, *175*, 38–44. [CrossRef]
24. Dexter, A.R. Effect of rainfall on the surface micro-relief of tilled soil. *J. Terramechanics* **1977**, *14*, 11–22. [CrossRef]
25. Assouline, S.; Mualem, Y. Modeling the dynamics of seal formation and its effect on infiltration as related to soil and rainfall characteristics. *Water Resour. Res.* **1997**, *33*, 1527–1536. [CrossRef]
26. Bretar, F.; Arab-Sedze, M.; Champion, J.; Pierrot-Deseilligny, M.; Heggy, E.; Jacquemoud, S. An advanced photogrammetric method to measure surface roughness: Application to volcanic terrains in the Piton de la Fournaise, Reunion Island. *Remote Sens. Environ.* **2013**, *135*, 1–11. [CrossRef]
27. de Oro, L.A.; Buschiazzo, D.E. Degradation of the soil surface roughness by rainfall in two loess soils. *Geoderma* **2011**, *164*, 46–53. [CrossRef]
28. Luo, J.; Zheng, Z.; Li, T.; He, S. Spatial heterogeneity of microtopography and its influence on the flow convergence of slopes under different rainfall patterns. *J. Hydrol.* **2017**, *545*, 88–99. [CrossRef]
29. Angulo-Martínez, M.; Beguería, S.; Kyselý, J. Use of disdrometer data to evaluate the relationship of rainfall kinetic energy and intensity (KE-I). *Sci. Total Environ.* **2016**, *568*, 83–94. [CrossRef]
30. Lobo, G.P.; Bonilla, C.A. Sensitivity analysis of kinetic energy-intensity relationships and maximum rainfall intensities on rainfall erosivity using a long-term precipitation dataset. *J. Hydrol.* **2015**, *527*, 788–793. [CrossRef]

31. Meshesha, D.T.; Tsunekawa, A.; Tsubo, M.; Haregeweyn, N.; Tegegne, F. Evaluation of kinetic energy and erosivity potential of simulated rainfall using Laser Precipitation Monitor. *CATENA* **2016**, *137*, 237–243. [CrossRef]
32. Bauer, B. Soil splash as an important agent of erosion. *Geogr. Pol.* **1990**, *58*, 99–106.
33. Chesworth, W.; Camps Arbestain, M.; Macías, F.; Spaargaren, O.; Spaargaren, O.; Mualem, Y.; Morel-Seytoux, H.J.; Horwath, W.R.; Almendros, G.; Chesworth, W.; et al. *Classification of Soils: World Reference Base (WRB) for Soil Resources*; Springer: Dordrecht, The Netherlands, 2008; pp. 120–122.
34. Kubínová, R. Grain Size Distribution of Eroded Soil. Master's Thesis, Faculty of Civil Engineering, Czech Techical University in Prague, Prague, Czech Republic, 2019.
35. Kavka, P.; Neumann, M.; Laburda, T.; Zumr, D. Developing of the laboratory rainfall simulator for testing the technical soil surface protection measures and droplets impact. In Proceedings of the XVII ECSMGE-2019 European Conference on Soil Mechanics and Geotechnical Engineering, Reykjavik, Iceland, 1–6 September 2019.
36. Van Dijk, A.I.J.; Bruijnzeel, L.; Rosewell, C. Rainfall intensity–kinetic energy relationships: A critical literature appraisal. *J. Hydrol.* **2002**, *261*, 1–23. [CrossRef]
37. Steiner, M.; Smith, J.A.; Steiner, M.; Smith, J.A. Reflectivity, Rain Rate, and Kinetic Energy Flux Relationships Based on Raindrop Spectra. *J. Appl. Meteorol.* **2000**, *39*, 1923–1940. [CrossRef]
38. Wischmeier, W.H.; Smith, D.D. *Predicting Rainfall Erosion Losses—A Guide to Conservation Planning*; Science and Education Administration United States Department of Agriculture: Hyattsville, MD, USA, 1978; p. 58.
39. Brown, L.C.; Foster, G.R. Storm Erosivity Using Idealized Intensity Distributions. *Trans. ASABE* **1987**, *30*, 0379–0386. [CrossRef]
40. Petrů, J.; Kalibová, J. Measurement and computation of kinetic energy of simulated rainfall in comparison with natural rainfall. *Soil Water Res.* **2018**, *13*, 226–233.
41. Geißler, C.; Kühn, P.; Böhnke, M.; Bruelheide, H.; Shi, X.; Scholten, T. Splash erosion potential under tree canopies in subtropical SE China. *CATENA* **2012**, *91*, 85–93. [CrossRef]
42. Angulo-Martínez, M.; Beguería, S.; Latorre, B.; Fernández-Raga, M. Comparison of precipitation measurements by OTT Parsivel^2 and Thies LPM optical disdrometers. *Hydrol. Earth Syst. Sci.* **2018**, *22*, 2811–2837. [CrossRef]
43. Boroghani, M.; Hayavi, F.; Noor, H. Affectability of splash erosion by polyacrylamide application and rainfall intensity. *Soil Water Res.* **2012**, *7*, 159–165. [CrossRef]
44. Wu, B.; Wang, Z.; Zhang, Q.; Shen, N.; Liu, J. Evaluating and modelling splash detachment capacity based on laboratory experiments. *CATENA* **2019**, *176*, 189–196. [CrossRef]
45. Farres, P.J. The dynamics of rainsplash erosion and the role of soil aggregate stability. *CATENA* **1987**, *14*, 119–130. [CrossRef]
46. Ekwue, E.I. The effects of soil organic matter content, rainfall duration and aggregate size on soil detachment. *Soil Technol.* **1991**, *4*, 197–207. [CrossRef]
47. Ma, R.-M.; Li, Z.-X.; Cai, C.-F.; Wang, J.-G. The dynamic response of splash erosion to aggregate mechanical breakdown through rainfall simulation events in Ultisols (subtropical China). *CATENA* **2014**, *121*, 279–287. [CrossRef]
48. Rezaei Arshad, R.; Mahmoodabadi, M.; Farpoor, M.H.; Fekri, M. Experimental investigation of rain-induced splash and wash processes under wind-driven rain. *Geoderma* **2019**, *337*, 1164–1174. [CrossRef]
49. Luo, J.; Zheng, Z.; Li, T.; He, S. Assessing the impacts of microtopography on soil erosion under simulated rainfall, using a multifractal approach. *Hydrol. Process.* **2018**, *32*, 2543–2556. [CrossRef]
50. Helming, K.; Roth, C.H.; Wolf, R.; Diestel, H. Characterization of rainfall-microrelief interactions with runoff using parameters derived from digital elevation models (DEMs). *Soil Technol.* **1993**, *6*, 273–286. [CrossRef]

Article

Using Machine Learning-Based Algorithms to Analyze Erosion Rates of a Watershed in Northern Taiwan

Kieu Anh Nguyen [1], Walter Chen [1,*], Bor-Shiun Lin [2] and Uma Seeboonruang [3,*]

[1] Dept. of Civil Engineering, National Taipei University of Technology, Taipei 10608, Taiwan; t106429401@ntut.edu.tw

[2] Disaster Prevention Technology Research Center, Sinotech Engineering Consultants, Taipei 11494, Taiwan; bosch.lin@sinotech.org.tw

[3] Faculty of Engineering, King Mongkut's Institute of Technology Ladkrabang, Bangkok 10520, Thailand

* Correspondence: waltchen@ntut.edu.tw (W.C.); uma.se@kmitl.ac.th (U.S.);
Tel.: +886-2-27712171 (ext. 2628) (W.C.); +66-2329-8334 (U.S.)

Received: 14 February 2020; Accepted: 4 March 2020; Published: 6 March 2020

Abstract: This study continues a previous study with further analysis of watershed-scale erosion pin measurements. Three machine learning (ML) algorithms—Support Vector Machine (SVM), Adaptive Neuro-Fuzzy Inference System (ANFIS), and Artificial Neural Network (ANN)—were used to analyze depth of erosion of a watershed (Shihmen reservoir) in northern Taiwan. In addition to three previously used statistical indexes (Mean Absolute Error, Root Mean Square of Error, and R-squared), Nash–Sutcliffe Efficiency (NSE) was calculated to compare the predictive performances of the three models. To see if there was a statistical difference between the three models, the Wilcoxon signed-rank test was used. The research utilized 14 environmental attributes as the input predictors of the ML algorithms. They are distance to river, distance to road, type of slope, sub-watershed, slope direction, elevation, slope class, rainfall, epoch, lithology, and the amount of organic content, clay, sand, and silt in the soil. Additionally, measurements of a total of 550 erosion pins installed on 55 slopes were used as the target variable of the model prediction. The dataset was divided into a training set (70%) and a testing set (30%) using the stratified random sampling with sub-watershed as the stratification variable. The results showed that the ANFIS model outperforms the other two algorithms in predicting the erosion rates of the study area. The average RMSE of the test data is 2.05 mm/yr for ANFIS, compared to 2.36 mm/yr and 2.61 mm/yr for ANN and SVM, respectively. Finally, the results of this study (ANN, ANFIS, and SVM) were compared with the previous study (Random Forest, Decision Tree, and multiple regression). It was found that Random Forest remains the best predictive model, and ANFIS is the second-best among the six ML algorithms.

Keywords: Erosion rate; ANFIS; ANN; SVM; Shihmen Reservoir watershed

---

1. Background and Introduction

Soil erosion is of major concern to agriculture and has had a detrimental long-term effect on both soil productivity and the sustainability of agriculture in particular. Soil erosion can lead to water pollution, increased flooding, and sedimentation, which damage the environment [1]. This has influenced the introduction of erosion control practices and policies as a necessity in almost every country of the world and under virtually every type of land use. Soil erosion causes both on-site and off-site consequences [2]. The on-site effects, such as soil losses from a field and depleted organic matter or nutrients of the soil, are particularly relevant on agricultural land. Off-site, downstream sedimentation decreases the capacity of rivers and reservoirs, increasing the threat from flooding. Many hydroelectricity and irrigation projects have been ruined as a consequence of soil erosion [3].

In Taiwan, 74% of the land area is slopes. Moreover, the average annual rainfall amount is 2500 mm, mainly occurred from May to August. As a result of climate change, and the concentration of rainfall duration and increasing rainfall intensity, year-round water availability has become a critical issue in Taiwan. Most of the soil erosion in Taiwan has been the result of unsuitable agricultural behaviors and overuse of slope lands. For example, about 65% of the hillslope lands are for crop production [4]. Coupled with the abundant rainfall, severe soil erosion occurs. Brought into service in 1964, the Shihmen Reservoir ranks third among all reservoirs in the country in terms of storage capacity. However, typhoons cause serious sediment problems and soil erosion [5]. Therefore, a thorough evaluation of the causative factors (predictors) of soil erosion in the watershed is important to the people of Taiwan.

Lo [6] applied the Agricultural Non-Point Source Pollution (AGNPS) model, which simulates water erosion and the transport of sediments to predict sedimentation in the watershed of the Shihmen reservoir. The results showed that the depth of sedimentation in the watershed averages around 2.5 mm/yr. Soil erosion in the watershed was also evaluated using USLE on different Digital Elevation Models (DEMs), and the average annual soil erosion rate varied greatly depending on the DEM [7]. Chiu et al. [8] used 137Cs radionuclide collected from 60 hillslope sampling sites of the basin of the Shihmen reservoir to determine the soil erosion rate and found it to be one or two orders of magnitude lower than predicted by USLE. Recently, Liu et al. [9] improved the analysis of the Shihmen reservoir watershed using the slope unit method in addition to the common grid cell approach. The results were verified with erosion pin measurements.

Erosion pins are a conventional method of measuring soil erosion. Erosion pins have been employed to measure ground-lowering rates and compare erosion rates of rill and interrill areas, plots of different sizes, slope areas, and many other regions in need of study. For example, Sirvent et al. [10] used erosion pins to measure the ground lowering in two plots every six months. The result demonstrated that erosion rates in the rill areas were 25%–50% higher than those in the interrill areas. Edeso et al. [11] used 29 erosion pins to evaluate soil erosion of different plots in northern Spain and showed that soil erosion in all plots was increasing over time. In Taiwan, Lin et al. [12] concluded that the soil erosion depth measured by erosion pins sharply increased when the accumulated rainfall exceeded 200 mm in a rainfall event.

An Artificial Neural Network (ANN) is an artificial intelligence model that was used to process information and was inspired by the human brain [13]. An Adaptive Neuro-Fuzzy Inference System (ANFIS) is an artificial intelligence model that combines neural networks with fuzzy inference systems [14]. The unique structure of ANFIS incorporates the ability of fuzzy inference systems (FIS) to improve the precision of prediction. Finally, SVM (Support Vector Machine) is a machine learning (ML) method that is very suited to intricate classification problems. In recent years, these artificial intelligence techniques have been applied to many different fields [15]. For example, Quej et al. [16] used ANN, ANFIS, and SVM to predict the daily global solar radiation in the Yucatan Peninsula, Mexico. Model performance was evaluated by the Mean Absolute Error (MAE), Root Mean Squared of Error (RMSE), and R-squared (R^2). The result indicated that SVM has a better performance than the other techniques. Zhou et al. [17] compared the accuracy of four models including ANN, SVM, WANN (Wavelet preprocessed ANN), and WSVM (Wavelet preprocessed SVM) to predict groundwater depth. The results showed that the models are ranked as follows: WSVM > WANN > SVM > ANN. Angelaki et al. [18] also used ANFIS, ANN, and SVM to predict the cumulative infiltration of soil. The results of model performance based on RMSE and Correlation Coefficient (CC) suggest that ANFIS works better than both SVM and ANN.

In the field of soil erosion modeling, there have been many studies carried out to estimate soil erosion using models such as USLE [9,19], SWAT [20,21], WEPP [22,23], WaTEM/SEDEM [24,25], and EUROSEM [26]. However, there has been no application of ML algorithms to estimate soil erosion except for in our previous work [27], in which we used Decision Tree (DT), Random Forest (RF), and Multiple Regression (MR) to create ML models to predict the soil erosion depths (rates). Therefore, this study aims to extend the investigation to the application of other ML techniques. Our main goal is to further our understanding of soil erosion rates in the study area.

2. Dataset and Research Method

The research design involves the use of site-specific data collected in the study area. They are described in the sections below.

2.1. Area of Study

The Shihmen reservoir dam is situated in the northern part of Taiwan on the banks of the Tahan River. Its watershed has an area of approximately 759.53 km^2 (Figure 1), and elevation rises towards the south. Hills extend over most areas, and for more than 60% of the watershed, the slope is more than 55% [28]. The yearly average temperature is 19 °C and average humidity, 82%. The typical rainy season is from May to October and the dry season from November to April. The average annual precipitation is approximately 2500 mm/yr [28].

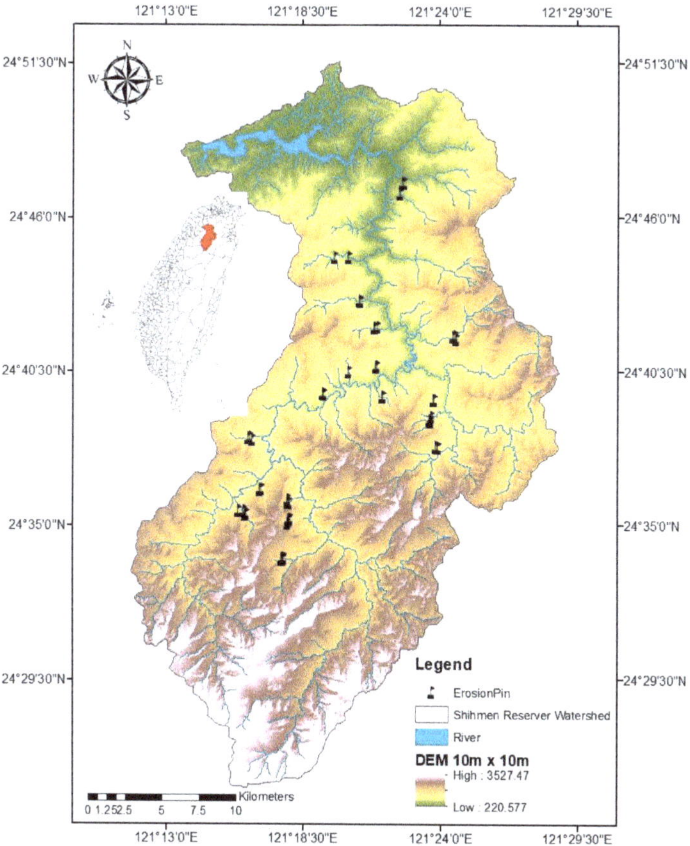

Figure 1. The study area is Northern Taiwan's Shihmen reservoir.

2.2. Data Preparation

We collected various data about the locations of the installed erosion pins as well as the measurements of the erosion pins themselves, as described in the following sections.

2.2.1. Predictors

A total of 14 environmental factors were utilized as the predictors (independent factors or input variables) in the model, namely distance to river, distance to road, type of slope, sub-watershed, slope direction, elevation, slope class, rainfall, epoch, lithology, and the amount of organic content, clay, sand, and silt in the soil. These factors have been gathered from different sources and become a geospatial database, as described in Nguyen et al. [27].

2.2.2. Target

The specification and installation of the erosion pins were described in Lin et al. [12]. Within the boundary of the Shihmen reservoir watershed, a total of 550 pins were installed on 55 slopes (10 pins per slope). The measurement data were collected from 8 September 2008 to 10 October 2011. The annual erosion depths were averaged at each slope, and the value ranges from 2.17 to 13.03 mm/yr. The metal rods (pins) used in this analysis were mounted on slopes without any signs of a landslide, collapse, or gully erosion. Therefore, our findings cannot be generalized beyond sheet and rill erosion.

2.3. Model Configuration

In this study, ML algorithms were used to predict the erosion rates of sheet erosion and rill erosion. The overall framework of the study consisted of three main parts (Figure 2), as summarized below.

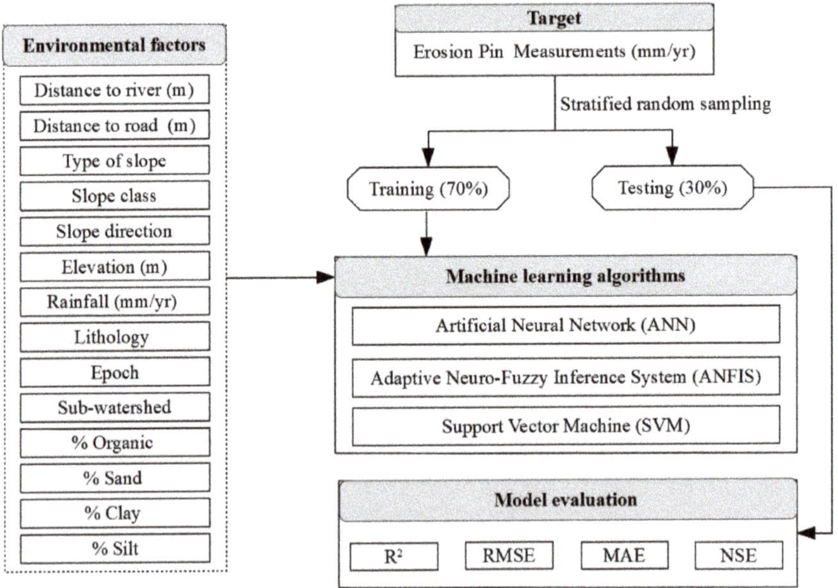

Figure 2. The research framework of this study.

First, the entire dataset, which includes 14 predictors and one target, was divided into a training set (70% or 38 samples) and a test set (30% or 17 samples) as is commonly done in the literature [29–32]. This step was repeated three times (i.e., Grouping #1, Grouping #2, and Grouping #3) to reduce the data variability to sampling. Because stratified random sampling has been shown to produce better outcomes than the simple random sampling [27], only stratified random sampling was used in this study to ensure the proper representation of the population. In stratified random sampling, the dataset was divided into several strata, and each stratum was sampled proportionately (70/30). Sub-watershed was selected as the stratification variable for this study, because it had been shown that the erosion depths in different watersheds were statistically different [33]. Hence, using stratified random sampling with sub-watershed as the stratification variable can provide a better estimate of the sample statistics and therefore create a better ML model.

Second, the ANN, ANFIS, and SVM models were built using the 70% training data, and the resulting models were tested using the test data.

Third, the performance metrics of R^2, NSE (Nash–Sutcliffe Efficiency), RMSE, and MAE were calculated on the training and test data. The Wilcoxon signed-rank test was conducted to determine if a statistically significant difference exists between the three models. Finally, the results of ANN, ANFIS, and SVM were tabulated and compared with results from our previous research [27].

2.4. Machine Learning Algorithms

Three widely used and potentially applicable ML algorithms are used in this study. They are described below.

2.4.1. Artificial Neural Network

An Artificial Neural Network (ANN) mimics how the human brain processes information. The purpose of the ANN model is to predict a target outcome by using input data through a back-propagation learning algorithm [34]. A typical ANN model has a multi-layer feed-forward structure that is connected by nodes with three main layers, namely the input layer, the hidden layer(s), and the output layer. The ANN determines the weight for each node and builds its results through training. In this study, the ANN model was created using the 'nntool' in the MATLAB 2016 software. A three layers feed-forward back-propagation network type was used. It consists of an input layer (14 neurons representing 14 environmental factors), one hidden layer (29 neurons), and one output layer (erosion rate), as shown in Figure 3. The number of neurons of the hidden layer is determined based on the following equation [35]:

$$N = 2x + 1 \tag{1}$$

where N is the number of hidden nodes, and x is the number of input nodes.

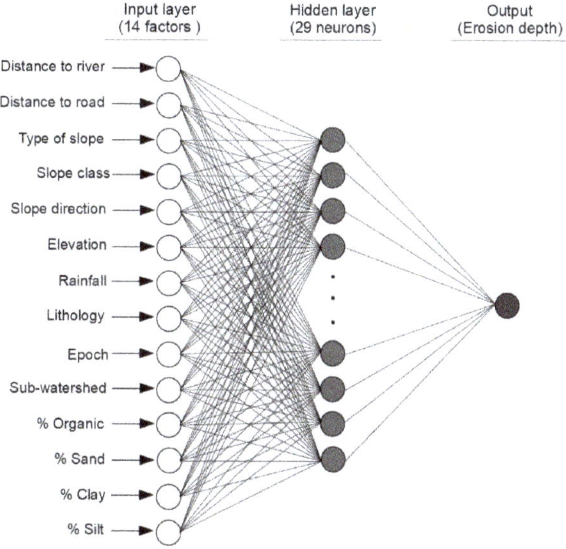

Figure 3. The model of ANN used in this study.

2.4.2. Adaptive Neuro-Fuzzy Inference System

An Adaptive Neuro-Fuzzy Inference System (ANFIS) is a combination of ANN and fuzzy logic, which utilizes the strengths of both techniques. Jang [14] introduced the concept of ANFIS in 1993. ANFIS contains five layers connected by directed links. These five layers are the Fuzzification layer, the Product layer, the Normalized layer, the Defuzzification layer, and the Output layer. The main purpose of ANFIS is to define the optimum parameter values of an equivalent fuzzy inference system by applying a learning algorithm. ANFIS can be constructed using two different methods, namely Genfis 1 (grid partitioning) and Genfis 2 (subtractive clustering). Genfis 1 has a limitation of six input variables. Since we have 14 factors, we used Genfis 2 in our analysis instead of Genfis 1. In this research, the ANFIS model was constructed using the "anfisedit" tool in the MATLAB 2016 software following the steps of loading data, generating FIS (sub clustering), and testing.

2.4.3. Support Vector Machine

A Support Vector Machine (SVM) is a supervised learning model developed by Schölkopf et al. [36] that can be used for regression and classification. The SVM model algorithm creates a line or a hyperplane that divides a dataset into two classes. The distance from the hyperplane to the nearest data points on both sides is defined as the margin. The purpose is to select a hyperplane with the greatest possible margin, thus giving a greater chance of new data being classified correctly into these two classes. In this study, the SVM model was implemented using custom codes and the 'fitrsvm' package on the MATLAB 2016 software.

2.5. Evaluation Criteria of Model Performance

Model evaluation is an indispensable part of developing a useful model. It supports the discovery and selection of a good model that can be used in the future. There are several statistical indices commonly used to estimate and calculate the performance and the validity of the ML algorithms. Here, the statistical parameters employed to measure the errors between the predicted and the observed values are R^2, NSE, RMSE, and MAE [37–39].

The R^2 value indicates the consistency with which the predicted values versus the measured values following a regression line [27]. It ranges from zero to one. If the value is equal to one, the predictive model is considered "perfect." The definition of R^2 is as follows:

$$R^2 = 1 - \frac{SSE}{SST} = 1 - \frac{\sum(Y - Y_1)^2}{\sum(Y - \overline{Y})^2} \tag{2}$$

where SST represents the total sum of squares, SSE is the error sum of squares, Y is the prediction of the model, Y_1 is the prediction of the regression line, and $\overline{Y}$ is the average of predicted values (Figure 4).

It is worth noting that although R^2 has been widely used for model evaluation, the statistics are highly sensitive to extreme values and are insensitive to additive and proportional differences between model and measured data [40]. More importantly, R^2 is calculated against the regression line (Figure 4), not the 1:1 line. A high R^2 only means a good fit to the regression line, not necessarily a set of good predictions concerning the observations (see the distinction made between the regression line and the 1:1 line in Figure 4). Therefore, we retain R^2 in this study only for completeness. Instead of relying on R^2, we computed RMSE, MAE, and NSE to compare the model performance. They are more appropriate than R^2.

RMSE and MAE are statistical parameters that are 'dimensioned.' They express the average errors of the model in the unit of the output variable (Equations 3 and 4). The RMSE is of particular importance, because it is one of the most commonly reported parameters in the climatic and environmental literature [41]. Smaller values of RMSE and MAE suggest nearer approximation of observed values by the models. The RSME and MAE are widely used basic metrics for assessing the performance of predictive models [42]. They are defined as follows:

$$RMSE = \sqrt{\frac{\sum (Y - Y_2)^2}{n}}$$

(3)

$$MAE = \frac{1}{n} \sum |Y - Y_2|$$

(4)

where Y_2 is the predicted value of the 1:1 line, and n is the number of samples.

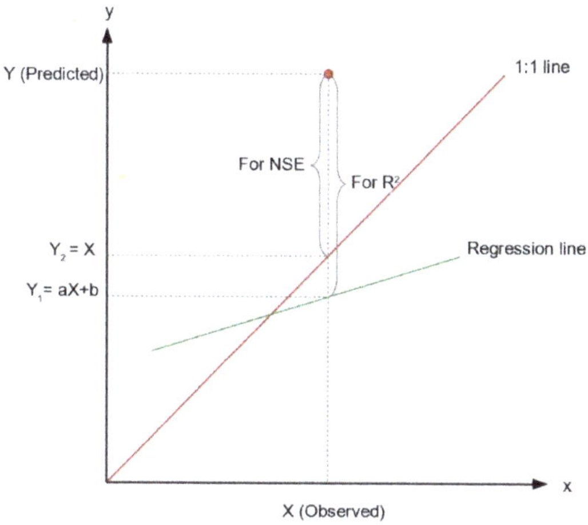

Figure 4. Illustration of the symbols used in the Equations (2)–(5).

In addition to R^2, RMSE, and MAE, which all have been used in our previous study [27], an additional parameter (NSE) is included in this study. As shown in Equation (5), NSE is a normalized statistical parameter that defines the relative magnitude of the residual variance compared to the measured data variance [43]. It shows how well the plot of the predicted values versus the observed values fits the 1:1 line. The value of NSE ranges from $-\infty$ to one, with NSE = 1 being the optimal value. If the value of NSE is between zero and one, the model is said to have acceptable performance (the higher, the better). On the other hand, if the value of NSE is smaller than 0.0, it will indicate that the average of observed values is better than the predicted value, and the model performance is not acceptable [43].

$$NSE = 1 - \frac{\sum (Y - Y_2)^2}{\sum \left(X - \overline{X}\right)^2}$$

(5)

where X is the observed value and $\overline{X}$ is the average of observed values.

2.6. Wilcoxon Signed-Rank Test

In addition to using NSE, RMSE, and MAE to evaluate the effectiveness of the models, it is necessary to examine if the differences in NSE, RMSE, and MAE are statistically significant. In this study, we used the Wilcoxon signed-rank test to compare the errors generated by different predictive models because the Wilcoxon signed-rank test is non-parametric (distribution-free). The steps of the Wilcoxon signed-rank test are as follows: Let Y denote the observed value, M_1 denote the predicted value of model 1, and M_2 denote the predictive value of model 2. The absolute error of each prediction is measured (Equations (6) and (7)) by:

$$E_1 = |Y - M_1| \tag{6}$$

$$E_2 = |Y - M_2| \tag{7}$$

To determine whether one model predicts more accurately than the others do, we perform the one-tailed hypothesis test. The null hypothesis (H_0) is that "Two models have the same predictive error" $(E_1 = E_2)$. The alternative hypothesis (H_a) is that "The first model has a smaller error than the second model" $(E_1 < E_2)$. In the Wilcoxon signed-rank test, we chose a significance level of 0.05. Then, the decision to whether to reject the null hypothesis or not is based on the resulting p-value. If the p-value is greater than 0.05, it will fail to reject the null hypothesis. Otherwise, if the p-value is less than 0.05, the null hypothesis will be rejected at a confidence level of 95%. In this case, the Wilcoxon signed-rank test is run using the R command wilcox.test.

3. Results and Discussions

We conducted all analyses using the dataset described in Section 2. Our findings are presented in the following sections.

3.1. Evaluation of Predictive Models

This study used quantitative criteria to assess the performance of predictive models. Table 1 shows the computed R^2, NSE, RMSE, and MAE statistics for ANFIS, ANN, and SVM under three different stratified random samplings. Among the four parameters, R^2 and NSE are statistics used to evaluate the predictive performance of the models under consideration. The higher the R^2 and NSE, the better the models simulate the results. However, as shown in Figure 4, R^2 is a measure of the goodness-of-fit of the regression line, not the 1:1 line. Therefore, a high R^2 does not necessarily mean a good performance of the model for predicting the soil erosion rate. In this regard, NSE is a much more suitable index to use because it is based on the differences between the predicted and observed values (1:1 line). Therefore, in the following, we will restrict our discussion to NSE only and ignore R^2.

On the other hand, MAE and RMSE are evaluation metrics commonly used to gauge the performance of the models that output continuous numbers. Both RMSE and MAE produce average errors of the models in units of the model output variables. However, it is worth noting that RMSE and MSE could be calculated against the 1:1 line or the regression line [27], and different results could be obtained. In this study, we computed both RMSE and MSE based on the 1:1 line to reflect the differences between the predicted and observed values.

As can be seen from Table 1, the average NSE of the training data for the ANN, ANFIS, and SVM models is 0.62, 1.00, and 0.51, respectively. Theoretically, the value of NSE ranges from $-\infty$ to one, and the higher the NSE, the better the model performs. Thus, it can be inferred from these numbers that ANFIS is the best model with a perfect NSE value of 1.00, and it outperforms the other two models substantially. However, the average NSE deteriorates significantly when the test data are used to judge the true performance of the models. They are 0.32, 0.49, and 0.17 for the ANN, ANFIS, and SVM models, respectively. The ANFIS model remains the best performing model, while the ANN model still beats the SVM. It is worth noting that a negative NSE was obtained for grouping #2 of SVM. This means that the model did not contribute to the improvement of the prediction, and the average of the observed values is better than the predicted value [44].

According to Table 1, there is also a significant difference in the RMSE values across the three models. The average RMSE of the training data for the ANN, ANFIS, and SVM models is 1.23, 0.01, and 1.43 mm/yr, respectively. Again, the ANFIS model outperforms the other two models by a substantial margin. The prediction of the ANFIS model was so accurate that there was almost no error in calculating the differences between the predicted and the observed values. However, the model performance falls rapidly when test data were used. The average RMSE of the test data for the

ANN, ANFIS, and SVM models increases to 2.36 mm/yr, 2.05 mm/yr, and 2.61 mm/yr, respectively. The disparity in model performance between the training data and the test data indicates that an over-fitting problem has occurred in the training of the ANFIS model. The other two models do not show signs of overtraining, but their predictive performances are inferior to that of the ANFIS model.

Lastly, if we compare the MAE results of the three models, a largely similar conclusion to the observation made above will be reached. The average MAE of the training data for the ANN, ANFIS, and SVM models is 0.75 mm/yr, 0.01 mm/yr, and 0.99 mm/yr, respectively. The ANFIS model performs much better than the ANN and SVM models. Moreover, the average MAE of the test data for the ANN, ANFIS, and SVM models increases to 1.85, 1.67, and 2.14 mm/yr, respectively. The errors of the test data are bigger than the errors in the training data. This again shows that the ANFIS model has been over-fitted. Overall, a general picture emerging from the analysis of NSE, RMSE, and MAE is that ANFIS is the best model, followed by ANN and SVM. They are ranked as follows, from best to worst: ANFIS, ANN, SVM.

Table 1. The performance metrics of the ANFIS, ANN, and SVM models.

		ANN		ANFIS		SVM	
		Train	Test	Train	Test	Train	Test
Grouping #1	R^2	0.59	0.49	1.00	0.66	0.37	0.33
	NSE	0.53	0.46	1.00	0.54	0.37	0.33
	RMSE (mm/yr)	1.38	2.20	0.02	2.04	1.60	2.46
	MAE (mm/yr)	0.71	1.71	0.01	1.73	1.03	2.11
Grouping #2	R^2	0.62	0.40	1.00	0.54	0.59	0.28
	NSE	0.45	0.17	1.00	0.35	0.58	-0.05
	RMSE (mm/yr)	1.61	2.46	0.01	2.17	1.40	2.76
	MAE (mm/yr)	1.18	1.81	0.01	1.61	1.03	2.18
Grouping #3	R^2	0.90	0.39	1.00	0.60	0.61	0.24
	NSE	0.88	0.33	1.00	0.57	0.59	0.23
	RMSE (mm/yr)	0.69	2.43	0.01	1.95	1.29	2.60
	MAE (mm/yr)	0.36	2.02	0.01	1.67	0.90	2.13
Average	R^2	**0.70**	**0.43**	**1.00**	**0.60**	**0.52**	**0.28**
	NSE	**0.62**	**0.32**	**1.00**	**0.49**	**0.51**	**0.17**
	RMSE (mm/yr)	**1.23**	**2.36**	**0.01**	**2.05**	**1.43**	**2.61**
	MAE (mm/yr)	**0.75**	**1.85**	**0.01**	**1.67**	**0.99**	**2.14**

3.2. Visual Comparison of Models

The data in Table 1 provide convincing evidence that ANFIS is the best performing model. We further plotted the predicted values versus the observed values with the regression line and the 1:1 line in the scatter plots of Figure 5. The left-side figures show the training datasets, while the right-side figures are the test datasets. Three different sampling results (groupings) are presented in Figure 5. The first row is grouping #1 (5a and 5b). The second row is grouping #2 (5c and 5d). Finally, the bottom-most row is grouping #3 (5e and 5f). Figure 5 shows that during the training stage, ANFIS successfully tunes its parameters to minimize the errors associated with its predictions. Therefore, the regression line of ANFIS (red) almost coincides with the 1:1 line, and it resulted in an average NSE of 1.00. Coincidentally, the regression line of ANN (blue) is also very close to the 1:1 line in all three cases. However, the data points (blue) are much more scattered around the 1:1 line than those of ANFIS (red) are. Therefore, ANN has a much lower average NSE of 0.62.

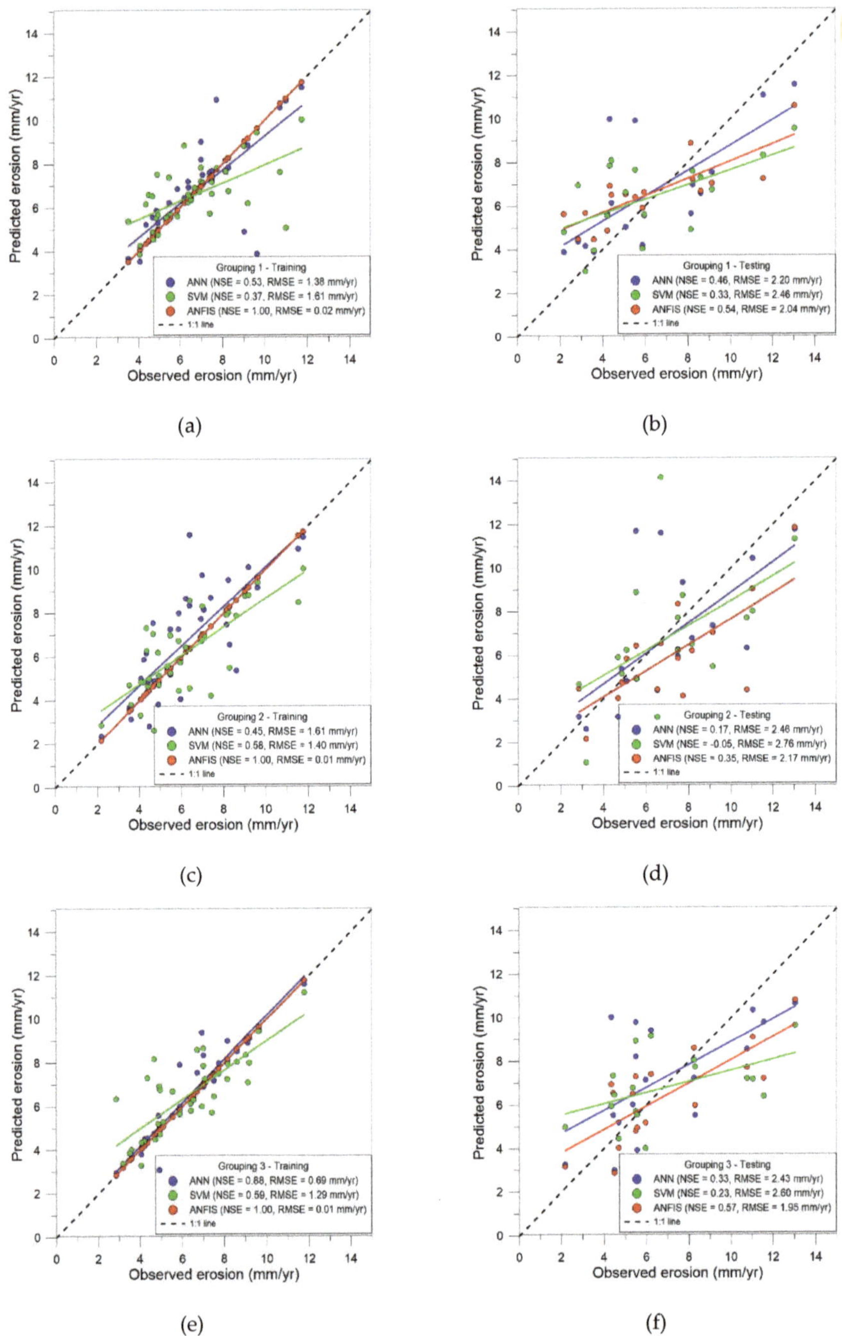

Figure 5. X-Y plots of predicted values vs. observed values: (**a**) Grouping #1—Training data; (**b**) Grouping #1—Test data; (**c**) Grouping #2—Training data; (**d**) Grouping #2—Test data; (**e**) Grouping #3—Training data; (**f**) Grouping #3—Test data.

As for the test data on the right-hand side of Figure 5, all three models showed substantially more scatter than that of the training data, which indicates substantially bigger errors between the predictions and the observations. The regression line of ANFIS is no longer the closest to the 1:1 line. However, the ANFIS model still has the highest average NSE (0.49) and the lowest average RMSE (2.05 mm/yr). It appears that ANFIS gained its performance advantage due to its hybrid learning approach of combining the ANN and the Fuzzy Inference System.

3.3. Results of Wilcoxon Signed-Rank Test

This study was undertaken to determine the best ML model using NSE, RMSE, and MAE. To determine if the differences between different model predictions were statistically significant, we used the Wilcoxon signed-rank test to compare the errors between different models. With the absolute error of each model being E_{ANFIS}, E_{ANN}, and E_{SVM}, respectively, the results of the Wilcoxon signed-rank test on the training data for each combination of the three models are summarized in Table 2. The results of the test data were shown in Table 3.

It can be seen from Table 2 that for the training data, the p-values between ANFIS and ANN are all very small and less than the threshold value of 0.05. The same can be said for the p-values between ANFIS and SVM. Taken together, the data presented here provide evidence that ANFIS was a statistically better model than both ANN and SVM when the models were trained with the training data. By contrast, the p-values between ANN and SVM are not small enough to reject the null hypothesis. Therefore, it is inconclusive whether a statistical difference exists in the predictive performance between the two models.

In terms of the test data, however, no statistically significant difference exists between the three models. As shown in Table 3, the p-values are all higher than the threshold value of 0.05. Therefore, we cannot reject the null hypothesis in favor of the alternative hypothesis.

Table 2. The results of the Wilcoxon signed-rank test (training data).

P-value (Training)	ANFIS-ANN	ANFIS-SVM	ANN-SVM
H_0: Null hypothesis H_a: Alternative hypothesis	H_0: $E_{ANFIS} = E_{ANN}$ H_a: $E_{ANFIS} < E_{ANN}$	H_0: $E_{ANFIS} = E_{SVM}$ H_a: $E_{ANFIS} < E_{SVM}$	H_0: $E_{ANN} = E_{SVM}$ H_a: $E_{ANN} < E_{SVM}$
Grouping #1	< 0.001	< 0.001	0.046
Grouping #2	< 0.001	< 0.001	0.777
Grouping #3	< 0.001	< 0.001	< 0.001

Table 3. The results of the Wilcoxon signed-rank test (test data).

P-value (Test)	ANFIS-ANN	ANFIS-SVM	ANN-SVM
H_0: Null hypothesis H_a: Alternative hypothesis	H_0: $E_{ANFIS} = E_{ANN}$ H_a: $E_{ANFIS} < E_{ANN}$	H_0: $E_{ANFIS} = E_{SVM}$ H_a: $E_{ANFIS} < E_{SVM}$	H_0: $E_{ANN} = E_{SVM}$ H_a: $E_{ANN} < E_{SVM}$
Grouping #1	0.6267	0.1317	0.1530
Grouping #2	0.6944	0.1030	0.1421
Grouping #3	0.2293	0.0540	0.3221

3.4. Comparison with Other ML Algorithms

In our previous research, DT, RF, and MR were used to predict the depths of erosion at the Shihmen reservoir watershed study site [27]. The result showed that RF outperforms other ML algorithms. In this study, we further compared the performance of ANN, ANFIS, and SVM to predict soil erosion depths (rates). The results of the above six models are shown together in Figure 6.

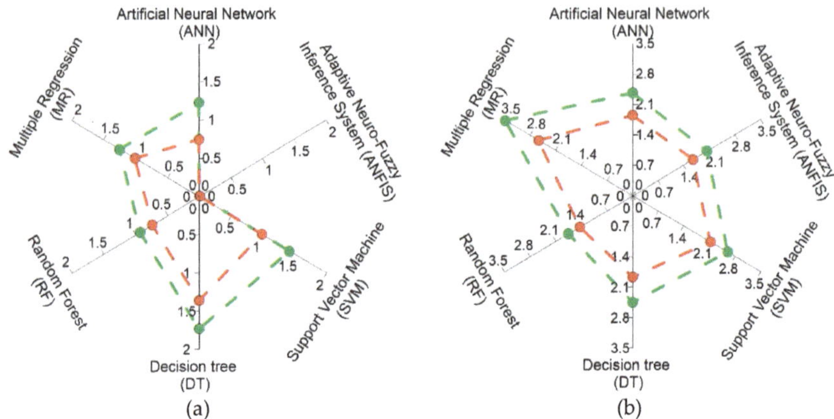

Figure 6. Radar plots of six ML models: (**a**) Training data; (**b**) Test data.

Figure 6 shows the comparison between six ML models in terms of the average RMSE (green points) and the average MAE (red points). The sub-Figure 6(a) is based on the training data, and 6(b) is based on the test data. If the points are closer to the center of the radar web, the model will have a better predictive performance. It can be seen from Figure 6a that in terms of RMSE starting from the best to the worst, the six models can be ranked as follows: ANFIS (0.01 mm/yr) < RF (0.93 mm/yr) < ANN (1.23 mm/yr) < MR (1.25 mm/yr) < SVM (1.43 mm/yr) < DT (1.73 mm/yr). It is obvious that ANFIS is the best model for training data. However, because models can perform very well during training but perform poorly in the testing against new data (over-fitted), it is the results of test data that distinguish a good model from a poor one. To evaluate model performance, we should focus on the metrics of test data. It can be seen from Figure 6 that although ANFIS is the best model among the three ML models compared in this study, it is still not quite as good as RF in the previous study. If we rank the six ML models again using the average RMSE of the test data, we will obtain a different rank (starting from the best to the worst): RF (1.75 mm/yr) < ANFIS (2.05 mm/yr) < ANN (2.36 mm/yr) < DT (2.45 mm/yr) < SVM (2.61 mm/yr) < MR (3.47 mm/yr). In other words, RF replaces ANFIS and becomes the favored choice. It is also possible to draw similar conclusions from the MAE values, also in Figure 6.

The results of test data are critical and are emphasized because evaluating the predictive performance of models by training data could lead to over-optimistic and overfitting models. As a result, although ANFIS performs better in training, RF is still considered the best model for predicting the soil erosion rate in the study area. Hannan et al. [45] and Barenboim et al. [46] also compared the performance of RF and ANFIS in their studies. Both studies indicated that the ANFIS and RF models were effective; however, Hannan et al. [45] preferred RF to ANFIS and ANN, and Barenboim et al. [46] recommended both RF and ANFIS.

Figure 7 shows the interpolated distribution of erosion rates (mm/yr) in the study watershed using the Inverse Distance Weighting (IDW) method. Figure 7a was obtained from erosion pin measurements, whereas Figure 7b,c were predicted by ANFIS and RF, respectively. It can be seen from the figures that the same distribution pattern is observed. The erosion rate is the highest on the east side of the study area, the lowest on the west side, and the north and south sides have an in-between erosion rate.

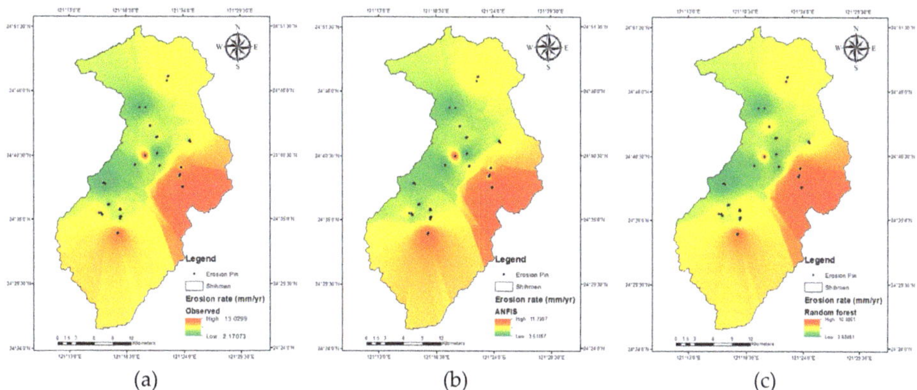

Figure 7. Distribution of erosion rates in the study area: (**a**) Observed; (**b**) Predicted by ANFIS; (**c**) Predicted by RF.

4. Conclusions

In this study, the ANN, ANFIS, and SVM algorithms were used to create predictive models of soil erosion rates in the study area of the Shihmen reservoir. The soil erosion rates were measured by 550 erosion pins installed on 55 slopes, and the results of the measurements reflect the sheet and rill erosion that took place within the study area. After dividing the dataset by a 70/30 ratio into training and test datasets using stratified random sampling, ANN, ANFIS, and SVM were used to generate respective models based on the 14 types of factors included in the training data. Then the models were applied to the test data, and the discrepancies from the real measurements were evaluated by R^2, NSE, RMSE, and MAE.

Without making an ex-ante choice of soil erosion model, the ex-post outcomes of ML models were quite satisfactory. The average RMSE of the training data ranges from mere 0.01 to 1.43 mm/yr. Among the three models, the performance of ANFIS is considerably higher than those of ANN and SVM, as indicated by its RMSE of 0.01 mm/yr. However, the performance of all three models degraded when they are applied to the test data. Results showed that the average RMSE of the test data varies from 2.05 to 2.61 mm/yr, with ANFIS still the best among the three models. To examine if the difference in prediction is statistically significant, the Wilcoxon signed-rank test was used to conduct pairwise comparisons of the three models. The results indicate that the ANFIS model is better than both the ANN and SVM models for the training data. However, no statistically significant difference exists between the three models when the models are applied to test data.

Moreover, the advantage of ANFIS disappeared when it was compared with the ML models (DT, RF, and MR) developed in our previous study. Although the average RMSE of ANFIS on training data is still unmatched, the average RMSE of ANFIS on test data was worse than that of RF. This shows that ANFIS may have been over-trained, and RF is still considered the best model for predicting the soil erosion rate in the study area.

In this and previous studies, we have made a substantial effort and progress in applying ML algorithms to the prediction of soil erosion rates without resorting to any soil erosion models. Although the effort was made, there is still no shortage of ML algorithms that promise better results than what has been obtained to date. It remains to be seen if ML algorithms are truly viable alternatives to traditional soil erosion models. Future research will have to address this issue in more detail.

Finally, because of the easy installation and wide availability of erosion pins, we believe that the approach presented here is generally applicable to other regions of the world. It would be desirable to obtain such measurements and carry out similar analyses for comparison.

Author Contributions: Conceptualization, W.C.; Data curation, K.A.N. and B.-S.L.; Formal analysis, K.A.N.; Funding acquisition, W.C.; Investigation, B.-S.L.; Methodology, W.C.; Project administration, W.C.; Software, K.A.N.; Supervision, W.C. and U.S.; Writing—original draft, K.A.N.; Writing—review and editing, W.C., B.-S.L. and U.S. All authors have read and agreed to the published version of the manuscript

Funding: This study was partially supported by the National Taipei University of Technology-King Mongkut's Institute of Technology Ladkrabang Joint Research Program (grant number NTUT-KMITL-108-01) and the Ministry of Science and Technology (Taiwan) Research Project (grant number MOST 108-2621-M-027-001).

Acknowledgments: We thank Kent Thomas for proofreading an early draft of this manuscript. We also thank the anonymous reviewers for their careful reading of our manuscript and insightful suggestions to improve the paper.

Conflicts of Interest: The authors declare no conflict of interests.

References

1. Jose, J.; District, P.C.C.; Roosevelt, F.D. *Erosion and Sedimentation*; University of Connecticut: Haddam, UK, 1993.

2. Zeneli, G.; Loca, S.; Diku, A.; Lila, A. On-Site and Off-Site Effects of Land Degradation in Albania. *Ecopersia* **2017**, *5*, 1787–1797.

3. Hagans, D.K.; Weaver, W.E.; Madej, M.A. *Long-Term on-Site and off-Site Effects of Logging and Erosion in the Redwood Creek Basin, Northern California*; Paper Presented at the American Geophysical Union Meeting on Cumulative Effects; Technical Bulletin; National Council of the Paper Industry for Air and Stream Improvement: New York, NY, USA, 1986.

4. Chen, Z.-S. *Establishment of Potential Soil Erosion Map in Western Taiwan and Its Best Management Strategies*; Paper presented at the Asia-EC JRC Joint Conference 2017 on "All That Soil Erosion the Global Task to Conserve Our Soil Resource"; Research Center of Surface Soil Resources Inventory and Integration (SSORii): Seoul, South Korea, 2017.

5. Lin, B.-S.; Chen, C.-K.; Thomas, K.; Hsu, C.-K.; Ho, H.-C. Improvement of the K-Factor of USLE and Soil Erosion Estimation in Shihmen Reservoir Watershed. *Sustainability* **2019**, *11*, 355. [CrossRef]

6. Lo, K.F.A. Quantifying soil erosion for the Shihmen reservoir watershed, Taiwan. *Agric. Syst.* **1994**, *45*, 105–116. [CrossRef]

7. Chen, W.; Li, D.-H.; Yang, K.-J.; Tsai, F.; Seeboonruang, U. Identifying and comparing relatively high soil erosion sites with four DEMs. *Ecol. Eng.* **2018**, *120*, 449–463. [CrossRef]

8. Chiu, Y.J.; Chang, K.T.; Chen, Y.C.; Chao, J.H.; Lee, H.Y. Estimation of soil erosion rates in a subtropical mountain watershed using 137 Cs radionuclide. *Nat. Hazards* **2011**, *59*, 271–284. [CrossRef]

9. Liu, Y.-H.; Li, D.-H.; Chen, W.; Lin, B.-S.; Seeboonruang, U.; Tsai, F. Soil erosion modeling and comparison using slope units and grid cells in Shihmen reservoir watershed in northern Taiwan. *Water* **2018**, *10*, 1387. [CrossRef]

10. Sirvent, J.; Desir, G.; Gutierrez, M.; Sancho, C.; Benito, G. Erosion rates in badland areas recorded by collectors, erosion pins and profilometer techniques (Ebro Basin, NE-Spain). *Geomorphology* **1997**, *18*, 61–75. [CrossRef]

11. Edeso, J.; Merino, A.; Gonzalez, M.; Marauri, P. Soil erosion under different harvesting managements in steep forestlands from northern Spain. *Land Degrad. Dev.* **1999**, *10*, 79–88. [CrossRef]

12. Lin, B.; Thomas, K.; Chen, C.; Ho, H. Evaluation of soil erosion risk for watershed management in Shenmu watershed, central Taiwan using USLE model parameters. *Paddy Water Environ.* **2016**, *14*, 19–43. [CrossRef]

13. Pradhan, B.; Lee, S. Landslide susceptibility assessment and factor effect analysis: Backpropagation artificial neural networks and their comparison with frequency ratio and bivariate logistic regression modelling. *Environ. Model. Softw.* **2010**, *25*, 747–759. [CrossRef]

14. Jang, J.-S. ANFIS: Adaptive-network-based fuzzy inference system. *IEEE Trans. Syst. ManCybern.* **1993**, *23*, 665–685. [CrossRef]

15. Mokhtarzad, M.; Eskandari, F.; Vanjani, N.J.; Arabasadi, A. Drought forecasting by ANN, ANFIS, and SVM and comparison of the models. *Environ. Earth Sci.* **2017**, *76*, 729. [CrossRef]

16. Quej, V.H.; Almorox, J.; Arnaldo, J.A.; Saito, L. ANFIS, SVM and ANN soft-computing techniques to estimate daily global solar radiation in a warm sub-humid environment. *J. Atmos. Sol. Terr. Phys.* **2017**, *155*, 62–70. [CrossRef]

17. Zhou, T.; Wang, F.; Yang, Z. Comparative analysis of ANN and SVM models combined with wavelet preprocess for groundwater depth prediction. *Water* **2017**, *9*, 781. [CrossRef]

18. Angelaki, A.; Singh Nain, S.; Singh, V.; Sihag, P. Estimation of models for cumulative infiltration of soil using machine learning methods. *ISH J. Hydraul. Eng.* **2018**, 1–8. [CrossRef]

19. Sousa, A.A.R.; Barandica, J.M.; Rescia, A. Ecological and economic sustainability in olive groves with different irrigation management and levels of erosion: A case study. *Sustainability (Switzerland)* **2019**, *11*, 4681. [CrossRef]

20. Abdelwahab, O.M.M.; Ricci, G.F.; De Girolamo, A.M.; Gentile, F. Modelling soil erosion in a Mediterranean watershed: Comparison between SWAT and AnnAGNPS models. *Environ. Res.* **2018**, *166*, 363–376. [CrossRef]

21. Panagopoulos, Y.; Dimitriou, E.; Skoulikidis, N. Vulnerability of a northeast Mediterranean island to soil loss. Can grazing management mitigate erosion? *Water (Switzerland)* **2019**, *11*, 1491. [CrossRef]

22. Srivastava, A.; Brooks, E.S.; Dobre, M.; Elliot, W.J.; Wu, J.Q.; Flanagan, D.C.; Gravelle, J.A.; Link, T.E. Modeling forest management effects on water and sediment yield from nested, paired watersheds in the interior Pacific Northwest, USA using WEPP. *Sci. Total Environ.* **2020**, *701*, 134877. [CrossRef]

23. Kinnell, P.I.A.; Wang, J.; Zheng, F. Comparison of the abilities of WEPP and the USLE-M to predict event soil loss on steep loessal slopes in China. *Catena* **2018**, *171*, 99–106. [CrossRef]

24. Krasa, J.; Dostal, T.; Jachymova, B.; Bauer, M.; Devaty, J. Soil erosion as a source of sediment and phosphorus in rivers and reservoirs – Watershed analyses using WaTEM/SEDEM. *Environ. Res.* **2019**, *171*, 470–483. [CrossRef] [PubMed]

25. Liu, Y.; Fu, B. Assessing sedimentological connectivity using WATEM/SEDEM model in a hilly and gully watershed of the Loess Plateau. *Ecol. Indic.* **2016**, *66*, 259–268. [CrossRef]

26. Morgan, R.; Quinton, J.; Smith, R.; Govers, G.; Poesen, J.; Auerswald, K.; Chisci, G.; Torri, D.; Styczen, M. The European Soil Erosion Model (EUROSEM): A dynamic approach for predicting sediment transport from fields and small catchments. Earth Surf. Process. *Landf. J. Br. Geomorphol. Res. Group* **1998**, *23*, 527–544. [CrossRef]

27. Nguyen, K.A.; Chen, W.; Lin, B.S.; Seeboonruang, U.; Thomas, K. Predicting Sheet and Rill Erosion of Shihmen Reservoir Watershed in Taiwan Using Machine Learning. *Sustainability* **2019**, *11*, 3615. [CrossRef]

28. Tsai, F.; Lai, J.-S.; Chen, W.W.; Lin, T.-H. Analysis of topographic and vegetative factors with data mining for landslide verification. *Ecol. Eng.* **2013**, *61*, 669–677. [CrossRef]

29. Chen, W.; Panahi, M.; Pourghasemi, H.R. Performance evaluation of GIS-based new ensemble data mining techniques of adaptive neuro-fuzzy inference system (ANFIS) with genetic algorithm (GA), differential evolution (DE), and particle swarm optimization (PSO) for landslide spatial modelling. *Catena* **2017**, *157*, 310–324. [CrossRef]

30. Hong, H.; Panahi, M.; Shirzadi, A.; Ma, T.; Liu, J.; Zhu, A.X.; Kazakis, N. Flood susceptibility assessment in Hengfeng area coupling adaptive neuro-fuzzy inference system with genetic algorithm and differential evolution. *Sci. Total Environ.* **2018**, *621*, 1124–1141. [CrossRef]

31. Lee, S.; Kim, Y.-S.; Oh, H.-J. Application of a weights-of-evidence method and GIS to regional groundwater productivity potential mapping. *J. Environ. Manag.* **2012**, *96*, 91–105. [CrossRef]

32. Riaz, M.T.; Basharat, M.; Hameed, N.; Shafique, M.; Luo, J. A Data-Driven Approach to Landslide-Susceptibility Mapping in Mountainous Terrain: Case Study from the Northwest Himalayas, Pakistan. *Nat. Hazards Rev.* **2018**, *19*, 05018007. [CrossRef]

33. Chen, W.; Chen, A. *A Statistical Test of Erosion Pin Measurements*; Paper Presented at the 39th Asian Conference on Remote Sensing (ACRS 2018); Asian Association of Remote Sensing (AARS): Kuala Lumpur, Malaysia, 2018.

34. Lee, S.; Ryu, J.-H.; Lee, M.-J.; Won, J.-S. Use of an artificial neural network for analysis of the susceptibility to landslides at Boun, Korea. *Environ. Geol.* **2003**, *44*, 820–833. [CrossRef]

35. Hecht-Nielsen, R. Kolmogorov's mapping neural network existence theorem. In *Proceedings of the International Conference on Neural Networks*; IEEE Press: New York, NY, USA, 1987; Volume 3.

36. Schölkopf, B.; Burges, C.J.; Smola, A.J. *Advances in Kernel Methods: Support Vector Learning*; MIT Press: Cambridge, MA, USA, 1999.

37. Despotovic, M.; Nedic, V.; Despotovic, D.; Cvetanovic, S. Evaluation of empirical models for predicting monthly mean horizontal diffuse solar radiation. *Renew. Sustain. Energy Rev.* **2016**, *56*, 246–260. [CrossRef]

38.	Moriasi, D.N.; Arnold, J.G.; Van Liew, M.W.; Bingner, R.L.; Harmel, R.D.; Veith, T.L. Model evaluation guidelines for systematic quantification of accuracy in watershed simulations. *Trans. ASABE* **2007**, *50*, 885–900. [CrossRef]

39.	Verstraeten, G.; Prosser, I.P.; Fogarty, P. Predicting the spatial patterns of hillslope sediment delivery to river channels in the Murrumbidgee catchment, Australia. *J. Hydrol.* **2007**, *334*, 440–454. [CrossRef]

40.	Gupta, S.K.; Goyal, M.R. *Soil Salinity Management in Agriculture: Technological Advances and Applications*; Apple Academic Press Inc.: Waretown, NJ, USA, 2017.

41.	Willmott, C.J.; Matsuura, K. Advantages of the mean absolute error (MAE) over the root mean square error (RMSE) in assessing average model performance. *Clim. Res.* **2005**, *30*, 79–82. [CrossRef]

42.	Teke, A.; Yıldırım, H.B.; Çelik, Ö. Evaluation and performance comparison of different models for the estimation of solar radiation. *Renew. Sustain. Energy Rev.* **2015**, *50*, 1097–1107. [CrossRef]

43.	Nash, J.E.; Sutcliffe, J.V. River flow forecasting through conceptual models part I—A discussion of principles. *J. Hydrol.* **1970**, *10*, 282–290. [CrossRef]

44.	Krause, P.; Boyle, D.; Bäse, F. Comparison of different efficiency criteria for hydrological model assessment. *Adv. Geosci.* **2005**, *5*, 89–97. [CrossRef]

45.	Hannan, M.; Ali, J.A.; Mohamed, A.; Uddin, M.N. A random forest regression based space vector PWM inverter controller for the induction motor drive. *IEEE Trans. Ind. Electron.* **2017**, *64*, 2689–2699. [CrossRef]

46.	Barenboim, M.; Masso, M.; Vaisman, I.I.; Jamison, D.C. Statistical geometry based prediction of nonsynonymous SNP functional effects using random forest and neuro-fuzzy classifiers. *Proteins Struct. Funct. Bioinform.* **2008**, *71*, 1930–1939. [CrossRef]

 *sustainability*

Article

Evaluation of the SEdiment Delivery Distributed (SEDD) Model in the Shihmen Reservoir Watershed

Kent Thomas [1], Walter Chen [1,*], Bor-Shiun Lin [2] and Uma Seeboonruang [3,*]

[1] Department of Civil Engineering, National Taipei University of Technology, Taipei 10608, Taiwan; t107429401@ntut.edu.tw

[2] Disaster Prevention Technology Research Center, Sinotech Engineering Consultants, Taipei 11494, Taiwan; bosch.lin@sinotech.org.tw

[3] Faculty of Engineering, King Mongkut's Institute of Technology Ladkrabang, Bangkok 10520, Thailand

* Correspondence: waltchen@ntut.edu.tw (W.C.); uma.se@kmitl.ac.th (U.S.); Tel.: +886-2-27712171 (ext. 2628) (W.C.); +66-2329-8334 (U.S.)

Received: 3 June 2020; Accepted: 30 July 2020; Published: 2 August 2020

Abstract: The sediment delivery ratio (SDR) connects the weight of sediments eroded and transported from slopes of a watershed to the weight that eventually enters streams and rivers ending at the watershed outlet. For watershed management agencies, the estimation of annual sediment yield (SY) and the sediment delivery has been a top priority due to the influence that sedimentation has on the holding capacity of reservoirs and the annual economic cost of sediment-related disasters. This study establishes the SEdiment Delivery Distributed (SEDD) model for the Shihmen Reservoir watershed using watershed-wide SDR_w and determines the geospatial distribution of individual SDR_i and SY in its sub-watersheds. Furthermore, this research considers the statistical and geospatial distribution of SDR_i across the two discretizations of sub-watersheds in the study area. It shows the probability density function (PDF) of the SDR_i. The watershed-specific coefficient (β) of SDR_i is 0.00515 for the Shihmen Reservoir watershed using the recursive method. The SY mean of the entire watershed was determined to be 42.08 t/ha/year. Moreover, maps of the mean SY by 25 and 93 sub-watersheds were proposed for watershed prioritization for future research and remedial works. The outcomes of this study can ameliorate future watershed remediation planning and sediment control by the implementation of geospatial SDR_w/SDR_i and the inclusion of the sub-watershed prioritization in decision-making. Finally, it is essential to note that the sediment yield modeling can be improved by increased on-site validation and the use of aerial photogrammetry to deliver more updated data to better understand the field situations.

Keywords: sediment delivery distributed model; sediment yield; SEDD; sediment delivery ratio; β coefficient; Shihmen Reservoir watershed

---

1. Introduction

Land degradation has been confirmed as a threat to agricultural productivity worldwide [1]. The gradual dissipation of quality topsoil transported as sediments along hill slopes and deposited into rivers and reservoirs has impacted multiple agricultural areas across the world. Sediment delivery by soil erosion and landslides has also led to the decreased longevity and failure of reservoirs and other water infrastructure. Moreover, due to the importance of reservoirs and dams to the water security of large populations in the developed world, the impact of sedimentation has been highlighted as a major issue in need of research worldwide over the past half a century and into the future. These processes influence many of the United Nations' (UN) Sustainable Development Goals (SDGs) [1–3]. Heavy sustained sedimentation leads to a severe influence on industries and economic growth for agriculture-based communities (SDG 8, 9, 10), the sustainability of cities and communities (SDG 11), and the availability of clean water downstream (SDG 6). These processes also pollute rivers, streams, and the sea by the transport of chemicals with sediments damaging marine flora and fauna and influencing the sustainability of marine environments (SDG 14).

Southeast Asia presents an even more advanced dilemma as the environmental and geomorphological characteristics across the region displays a strong influence on the rate of soil erosion, and the region's population and economies are already suffering from the impact [1]. After Typhoon Gloria impacted Taiwan, the Shihmen reservoir watershed was one of the many watersheds heavily inundated by sediments from soil erosion and landslides. This event was an example of typhoon events affecting Taiwan due to its tropical monsoon climate. The impact of these events creates frequent and high magnitude sediment disasters and long-term sediment inundation. The effect of extreme meteorological events on Shihmen and other reservoirs and the threat to urban and agricultural water supply quality throughout the nation has encouraged the Government of Taiwan to facilitate soil conservation research and countermeasures within the watershed through the introduction of soil conservation policies and programs. Taiwan's geographical location places it in a precarious position that geomorphologically has a high frequency of highly erodible soil materials and, additionally, its tropical monsoon climate brings frequent extreme rainfall or typhoon events which have led to massive sediment loads, dissociated by mass movements, and soil erosion, to be transported from slopes of watersheds into rivers and streams, affecting water supply in many communities. These events increase the siltation of reservoirs, such as the Shihmen Reservoir in Northern Taiwan, which diminishes the country's capacity to sustainably manage its water supply and poses a threat to the long-term sustainability of the dam and the populations it serves [4,5].

Therefore, sediment modeling and monitoring in Taiwan is of paramount importance to improve watershed conservation and sustainable water management by understanding how sediments are dissociated, translocated, and transported down the slopes of watersheds, entering streams as massive sediment loads. These sediments deposit along the waterway with detrimental impacts to river flow and along slopes, increasing future sediment disaster hazards and clogging reservoirs and dams, affecting water supplies to nearby populated districts and farms.

1.1. Sediment Delivery Ratio (SDR)

Researchers have established that the sediment yield (SY) and siltation found in rivers and dams are not equivalent to the gross sediments that have dissociated from the slopes through land degradation processes. The sediment delivery ratio (SDR) was established as the relationship between the SY and the gross erosion (A, in some research denoted as Soil Loss, SL). The SDR acts as a reducer to equate the maximum amount of soil erosion to the sediments delivered to the outlet. SDR is a continuous variable with a range from zero to one. In the initial stages of soil erosion research, SDR was evaluated for an entire watershed and a single outlet, such as a reservoir, denoted as SDR_w. However, as computational power has improved and research methods have become more developed, SDR is discretized to sub-watersheds, micro watersheds, and even more minute hydrological units within a watershed. This has been denoted as a geospatial SDR_i and is defined as the fraction of the gross soil loss (SL) from the hydrological unit, i, that reaches a flow pathway [6].

Many methods for determining the SDR have been developed over the past six decades. Long-term monitoring of SY in test plots and different regions were the starting point in SDR research and established multiple geophysical and hydrological factors that affect SDR. Wu et al. [7] stated that there are three prominent methods of estimating SDR, which are: (1) the ratio between soil erosion and sediment yield, (2) empirical formula relationships between SDR and its contributing factors, and (3) spatial modeling and numerical modeling using hydraulic and hydrological theories. These include models for the calculation and classification of SDR based on regional and local observed data.

Li [8] determined from the comparison of the USLE results and the outlet sediments at the Shihmen Reservoir that the SDR was 0.49 from the annual siltation amount of 71.2 t/ha/year (between 2011 and 2015), which includes sheet and rill erosion, gully erosion, and landslides. There have been other studies of the SDR in the Shihmen Reservoir watershed but these have mainly focused on the post-landslide erosion SDR or event-based SDR. Tsai et al. [9] defined the SDR of the landslide areas in the watershed from 0.42 to 0.78 for eroded areas and zero for areas where aggradation occurred. Moreover, Tsai et al. [10] employed the model of Bathurst et al. [11] of the sediment delivery of landslides. They determined that the sediment delivery of landslides in the Shihmen Reservoir watershed ranged from 0.75 to 0.89 (1986–1998), 0.48 to 0.64 (1998–2004), and 0.40 to 0.58 for typhoon Aere (2004). Lin et al. [12] explored the impact of different factors (such as curve number and Manning's) on the SDR in the Shihmen Reservoir watershed.

SWCB [13] previously employed DEMs of difference (DoD) method to determine a correlation between SDR and the entire watershed (SDR_w) in the Shihmen Reservoir watershed and concluded the relationship to be $SDR = 41.23W_{Area}^{-0.017}$ where W_{Area} is the watershed area, and the SDR_w was 0.368. Chen et al. [14] used the Grid-based Sediment Production and Transport Model (GSPTM) to determine SDR_w to be 0.299 (rounded to 0.30) for a simulation of the impact of the Typhoon Morakot on the Yufeng and Xiuluan sub-watersheds, and Chiu et al. [15] extended the GSPTM to identify unstable stream reaches. Additionally, Liu [16] determined that the total soil loss (also commonly-termed gross soil erosion or the total amount of soil erosion) was 90.6 t/ha/year by converting measured average erosion depth of erosion pins [17,18] to the amount of soil erosion. This study emphasizes the SDR of recent years and assumes $SDR_w = 0.49$.

1.2. Spatial Discretization

This research compares the implementation of the SEDD with grid–cell unit analysis across the watershed and two discretization levels of sub-watersheds (25 and 93 SW). Spatial discretization of watersheds in hydrological models has been a critical issue for many researchers within the remote sensing, physical modeling, hydrology, and mathematical computations fields [19]. Wood et al. [20] elaborated that watersheds are areas made of infinite minute points of related hydrological processes (for example, evaporation, infiltration, and runoff). Watersheds can be further divided into sub-watersheds or sub-basins that are continuous assemblages or spatial objects that drain towards a specific point, contains similar vegetation layers and/or linked slope, or elevation characteristics, and are commonly used in the study of landslides, soil erosion, hydrology, and other natural processes and are specific areas within a watershed. There are many methods of defining watersheds, sub-watersheds, and even more minute forms of discretization, such as hydrological units depending on the characteristics of the areas [19–22].

Contextually, at the time of the development of many soil erosion models and concepts from the 1970s to the 1990s, computational cost, time, and power were limitations that hindered the discretization of watersheds to more elements [19]. Therefore, many older models were developed with sub-watershed as the smallest hydrological unit. This has continued even to the 2000s with the SWAT (Soil and Water Assessment Tool) model, and research has supported the object area discretization in soil erosion, LULC (Land Use and Land Cover), and sedimentation models and has criticized the use of areal discretization. SWAT is a comprehensive soil erosion model that is popular in the field of hydrology. In this study, we utilized the first stage in the SWAT analysis, the discretization of a study area into sub-basins/sub-watersheds, to develop 25-subwatershed divisions. The two inputs used in this analysis was the surface topography and the land use type [23]. Previously in the Shihmen Reservoir watershed, grid cells and slope units have also been employed in soil erosion research [16].

Researchers implementing the SEdiment Distributed Delivery (SEDD) model have mainly focused on the hydrological or morphological unit [6,24–29] but Di Stefano and Ferro [30] utilized grid cells in the SEDD model.

1.3. Study Objectives

Geomorphological changes, climate change, and land degradation processes are highly influential to soil erosion studies. Therefore, it is important to highlight the time scale. The prioritization of watersheds or sub-watersheds for soil erosion, landslide, and conservation works has been a long-standing topic in geomorphology, hydrology, disaster management, and environmental planning research. In the past, the gross soil loss (often derived by the USLE model) and its contributing factors have been the primary focus for prioritization of soil conservation planning. This study explored the sediment yield value to the prioritization of sub-watersheds for remediation and conservation.

This research incorporated and considered the implementation of the SEDD model in the Shihmen Reservoir watershed and the research objectives, methods, and assumptions are as follows:

1. The main focus of this study was on the advancement of the soil erosion research in the Shihmen Reservoir watershed to include determination of the watershed-specific coefficient, β, using known SDR_w values, and developing the SDR_i relationship and SEDD model for the watershed, applying the findings to the assessment of the sediment yield (SY) by sheet and rill erosion in the watershed.
2. In this study, the non-point source sheet and rill erosion calculated by the USLE is the main target land degradation process. Landslides, gully erosion, and channel type erosions are not considered.
3. The datasets employed are assumed to be from the same period and are the most recent available data for this study area.
4. This study, additionally, explores the statistical properties of the SDR, SL, and SY of the watershed using two sub-watershed discretizations (25 and 93) of the Shihmen Reservoir watershed and compares the results.
5. This study determines the watershed prioritization of soil conservation using the SY determined by the SEDD.

This study is a theoretical and statistical analysis of the SDR distribution based upon the SEDD model using real data from the Shihmen Reservoir watershed.

2. Study Area

The Shihmen Reservoir watershed (elevation map shown in Figure 1a) is a 759.53 km^2 watershed found in Northern Taiwan between latitudes 24.426° N and 24.861° N, and longitudes 121.192° E and 121.479° E. Local authorities use the reservoir to regulate 23% of the water supply for the nearby communities, industries, and agriculture. The effective storage volume of the reservoir is 207 million m^3. Still, it has been affected by heavy sedimentation, and some of the additional check dams established to diminish sedimentation have been affected or damaged in past years. The established subdivision of the watershed by local government is into five sub-watersheds, namely Ku-Chu, Yu-Feng, San-Kuang, Pai-Shih, Tai-Kang sub-watersheds. However, in this study, the entire watershed is delineated into 25 sub-watersheds (Figure 1b) and 93 sub-watersheds (Figure 1c), as explained in Section 3.1. The watershed is traversed by the Tahan River and its irregular pattern of tributaries that spread across the steep, high slopes of the sub-watersheds (angle 20° to 85° with an elevation between 220 m and 3527 m). The subtropical monsoon climate and the annual rainfall between 2200 mm and 2800 mm in the watershed encourages a densely vegetated state, which is mostly natural and artificial coniferous and broadleaved trees, and the trend in rainfall pattern has increased over the past decades to create concerns of soil erosion (Figure 1d) [9,31].

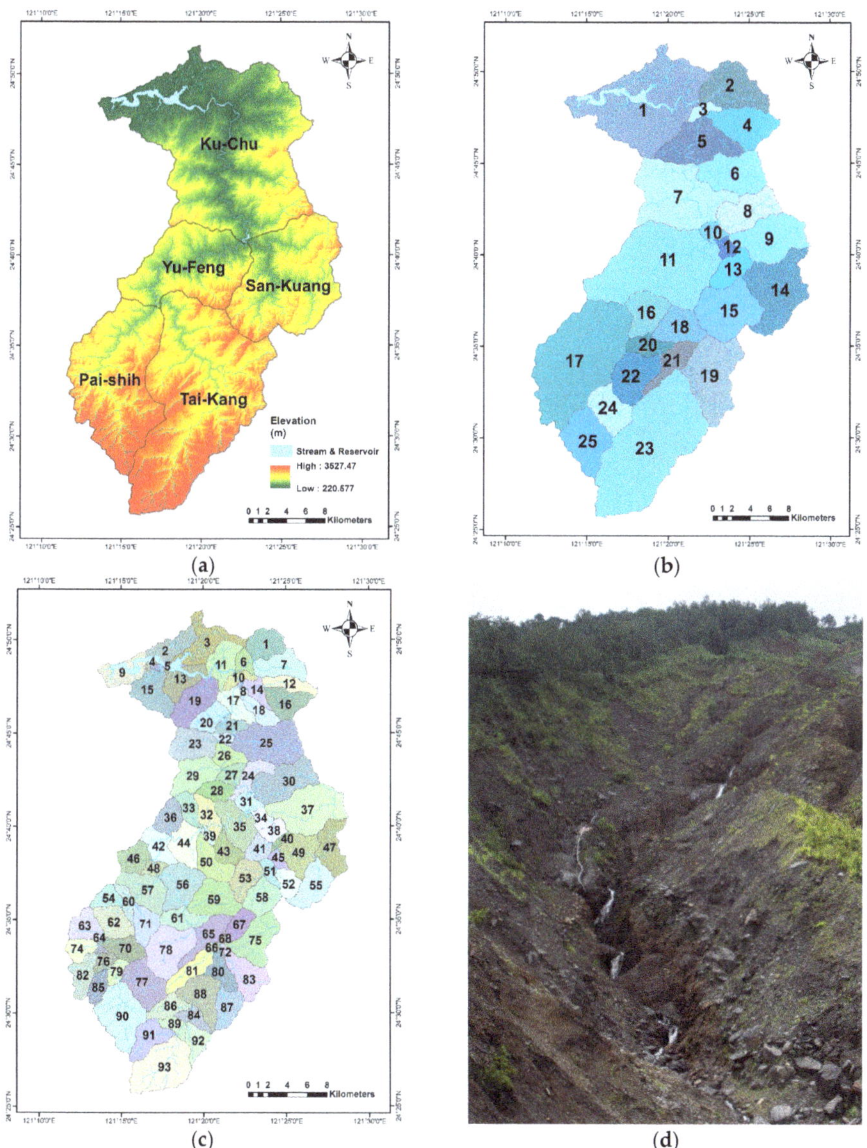

Figure 1. Maps of the Shihmen Reservoir watershed: (**a**) elevation, (**b**) 25 sub-watersheds, and (**c**) 93 sub-watersheds. (**d**) An illustrative picture of soil erosion.

3. Methodology

The SEDD model developed by Ferro and Minacapilli [24] analyzes the spatially distributed SY by determining the parameters of SDR and SL found using any soil loss model. The SDR is found by a relationship between the travel time ($t_{p,i}$) and a watershed-specific coefficient, β. β cannot be determined analytically but is determined iteratively, by balancing the relationship between a known SY, SL, and SDR. The SY is evaluated for the entire watershed and the two discretizations, 25 and 93 sub-watershed models, to compare the impact of aggregation on the model outputs.

The flow chart of the SEDD model implemented in this study is distributed among four stages as shown in Figure 2:

Stage 1 (1.1):

1. Determine the study area and input the DEM.
2. Determine the slope, flow direction, and river/stream map from a derived depression-less DEM.
3. Use the flow direction and river/stream map as input to determine the flow distance, which is then used with the slope to determine the travel time.

Stage 2:

4. Find the six parameters of the USLE equation and estimate SL.
5. Determine the SDR using measured outlet SY and estimated SL.

Back to stage 1.2:

6. Determine β from the SDR, SL, and the travel time found.

Stage 3:

7. Return to the DEM and discretize the study area into sub-watersheds/hydrological units or any other applicable unit of analysis.

Stage 4:

8. Re-estimate the SDR for the parent watershed and each sub-watershed.
9. Estimate the sediment yield (SY) and perform any further analysis necessary.

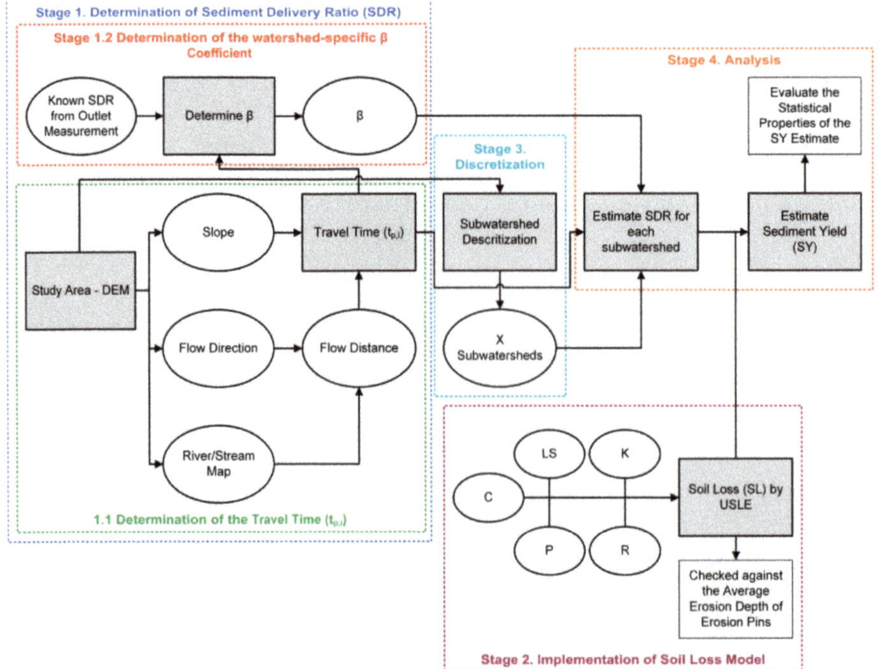

Figure 2. Flow chart of the SEDD model in this study.

SDR evaluation for SEDD modeling requires the development of three (3) main parameters: the slope, hydraulic length, and the β coefficient. The "flow distance" module of the ArcGIS software was employed to develop the horizontal flow paths and required the use of the "flow accumulation" and "flow direction" modules. The slope map and other component datasets of the USLE/SEDD model were developed in the ArcGIS platform.

This study relied upon the R programming language. Specifically, the "raster," "e1071," and "LaplacesDemon" libraries were employed for statistical analysis, conversion of data from raster to point datasets, and other analysis, while the "ggplot2" libraries of R were employed for data visualization [32–35].

The USLE model in this study was developed in the ArcGIS model builder framework updating the model originally developed by Jhan [36], Yang [37], Li [8], and Liu [38].

The USLE model was developed by Wischmeier and Smith [39,40] and is designed to predict the average annual SL by sheet and rill erosion. It has been continuously improved upon and localized by many researchers over the past decades. The USLE model can be denoted as follows:

$$A_m = R_m K_m LSCP \tag{1}$$

where A_m is the computed soil loss per unit area (t/ha/year), R_m is the rainfall and runoff factor (MJ-mm ha^{-1} h^{-1} year^{-1}), K_m is the soil erodibility factor (t-h MJ^{-1} mm^{-1}), L is the slope-length factor (dimensionless), S is the slope-steepness factor (dimensionless), C is the cover and management factor, and P is the support practice factor.

3.1. Sediment Distributed Delivery Model (SEDD)

The SEDD utilizes the calculation of the flow length for SDR developed by Ferro and Minacapilli [24], which includes considerations for slope and the roughness/runoff coefficient. SY, the sediment yield by the annual scale, for a basin is discretized into N_i morphological units and is measured in tonnes (t):

$$\sum_{i=1}^{N_i} SY_i = \sum_{i=1}^{N_i} A_{mi} SDR_i SU_i \tag{2}$$

where A_{mi} is the computed soil loss per unit area (t/ha/year) for the morphological unit i, SU_i is the area of the morphological unit i (ha), and N_i is the number of morphological units over the watershed area (dimensionless).

Ferro and Minacapilli [24] concluded that the SDR_w can be physically defined using watershed morphological data without the consideration of the soil loss model. SDR_i in Equation (2) and the relationship between the SDR_w and SDR_i was defined as follows (Equations (3) and (4)):

$$SDR_i = e^{-\beta t_{p,i}} = \exp\left[-\beta \frac{l_{p,i}}{\sqrt{S_{p,i}}}\right] = \exp\left[-\beta \sum_{j=1}^{N_p} \frac{\lambda_{i,j}}{\sqrt{S_{i,j}}}\right] \tag{3}$$

$$SDR_w = \frac{\sum_{i=1}^{N_i} \exp\left[-\beta \frac{l_{p,i}}{\sqrt{S_{p,i}}}\right] A_{mi} SU_i}{\sum_{i=1}^{N_i} A_{mi} SU_i} \tag{4}$$

where $t_{p,i}$ is the travel time from ith morphological unit to the nearest stream reach (m), β is the roughness and runoff coefficient for the watershed (m^{-1}), $s_{p,i}$ is the slope of the hydraulic flow path (m/m), $l_{p,i}$ is the length of the hydraulic flow path (m), $\lambda_{i,j}$ and $S_{i,j}$ are the length and slope of each morphological unit i localized along the hydraulic path j, and N_p is the number of morphological units localized along the hydraulic path j [6,24,26–29].

In this study, the parent watershed (the Shihmen Reservoir watershed) was discretized into two discrete sets of children sub-watersheds, specifically 25 and 93 sub-watersheds. The 25 sub-watersheds were delineated using the SWAT model with the input of the DEM and river course of the Shihmen

Reservoir watershed. The output was 25 hydrologically connected sub-watersheds, as shown in Figure 1b. The 93 sub-watersheds were delineated using the ArcGIS Watershed module utilizing an input of the flow direction and a stream map derived from flow accumulation. The output was 93 hydrologically connected sub-watersheds, as shown in Figure 1c.

The SEDD model was implemented as 10 m grid data, with over seven million data points and aggregated into sub-watersheds. This study derived the grid-based SEDD model and aggregated the dataset into sub-watersheds to examine the impact this can have on the output SY data. A grid–cell-based analysis is more computationally expensive, and larger watersheds can be quite time consuming or hardware-expensive but may provide alternative results. On the other hand, the hydrological unit analysis is less time consuming to produce, has a lower computational cost, and can garner results that are aggregated by specific areas. However, there can be discrepancies due to oversimplification, some vast.

3.2. Sediment Yield Determination

Flow direction (FD) algorithms estimate the flow of a specified material from a source cell (S1) continuously to the next neighboring cells (N1) until a stopping point or limitation point (river or stream in Figure 3) has been reached, such as a stream, river, or dam. The SDR model considers the hydraulic flow of sediments through each morphological/hydrological unit to the trunk river or its tributaries (in this scenario, the Tahan river, and its tributaries) using the single flow direction algorithm, deterministic D8. The distance traveled by particles has been termed the "flow path length" and has varying definitions and methods of calculation. The flow distance module in ArcGIS was used to define the horizontal flow distance to the river using the D8 flow direction.

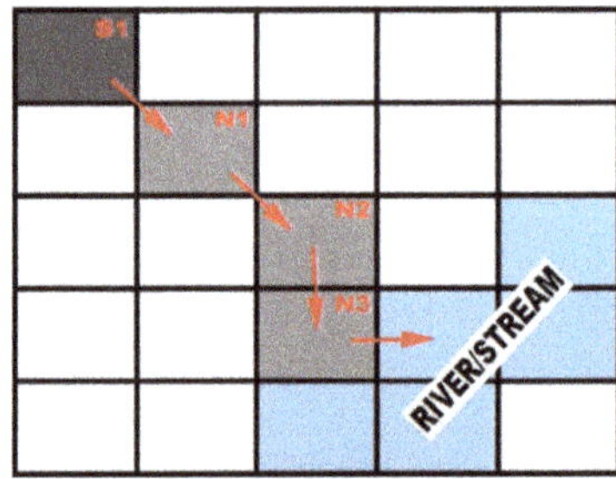

Figure 3. A flow path showing the single flow D8 algorithm where S = source and N = Node (re-drawn from [41]).

The relative time taken to reach the sink from the source has been termed the "travel time." Moreover, Jain and Kothyari [27] state that the travel time from a cell i is equivalent to the total travel time through each of the N_p cells along its flow path until terminating at a channel. The flow path length and the slope in conjunction with a watershed-wide β coefficient are used to derive, firstly, the travel time and, secondly, the SDR within the SEDD model.

The SY was first derived using a watershed-wide grid of the USLE and SDR_i, and then compared against the discretization of the 25 sub-watersheds model and the 93 sub-watersheds model using the mean, median, and mode for each sub-watershed.

4. Results

The results of SEDD modeling in the Shihmen Reservoir watershed and its sub-watersheds are shown in the following sections.

4.1. The β Coefficient

Using GIS, we could calculate the flow lengths, slope, and travel time needed in Equations (3) and (4) for the Shihmen Reservoir watershed. At the same time, soil loss could be computed by USLE, and the SDR_w is known to be 0.49 previously determined from the annual siltation of the reservoir [8]. Therefore, the only unknown is Equation (4) was the β coefficient. By assuming an initial value of β, we can solve for the β coefficient iteratively to satisfy the Equation (4). We found the β coefficient to be 0.00515. This is the first result of applying the SEDD model to the Shihmen Reservoir watershed using the iterative method, and a number like this has never been obtained before. The determined β coefficient is watershed specific.

4.2. Grid-Based Travel Time, SDR, SL, and SY

The following Figures 4–7 depict the travel time, sediment delivery ratio (SDR), soil loss (SL), and sediment yield (SY) determined using the SEDD model.

The probability density functions (PDF) of the grid–cell-based SDR, SL, and SY of the Shihmen Reservoir watershed have asymmetric, right-skewed distributions (which will be shown later in Section 4.3 with two representative sub-watersheds). This study determined in the Shihmen Reservoir watershed for SDR: the mean value of the distribution was 0.474 (not the same as SDR_w), the median was 0.459, and the mode was 0.317. For the SL of the entire watershed, the mean was estimated as 87.07 t/ha/year, and the median was 71.74 t/ha/year. Lastly, the mean SY for the entire watershed was 42.08 t/ha/year, and the median was 29.80 t/ha/year.

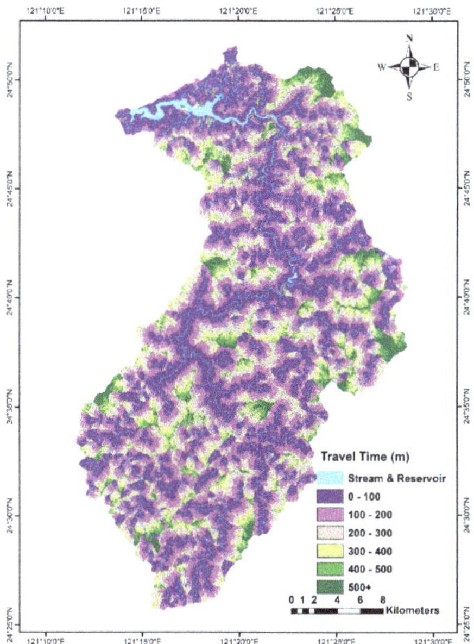

Figure 4. Travel Time of the Shihmen Reservoir watershed.

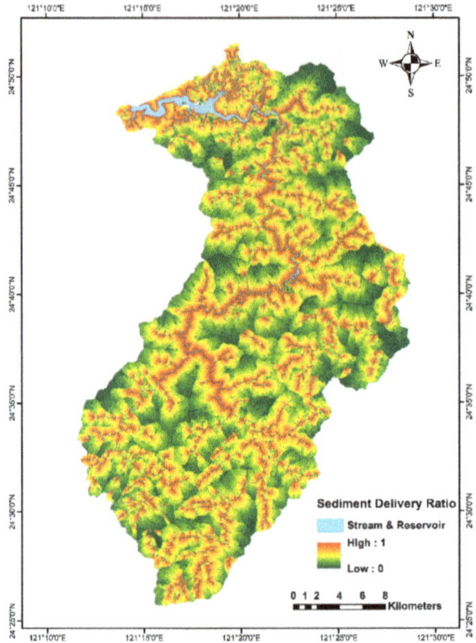

Figure 5. Sediment Delivery Ratio of the Shihmen Reservoir watershed.

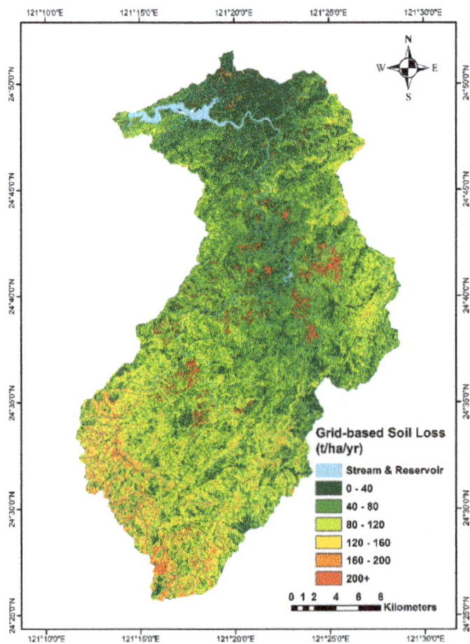

Figure 6. Soil Erosion/Soil Loss Distribution of the Shihmen Reservoir watershed.

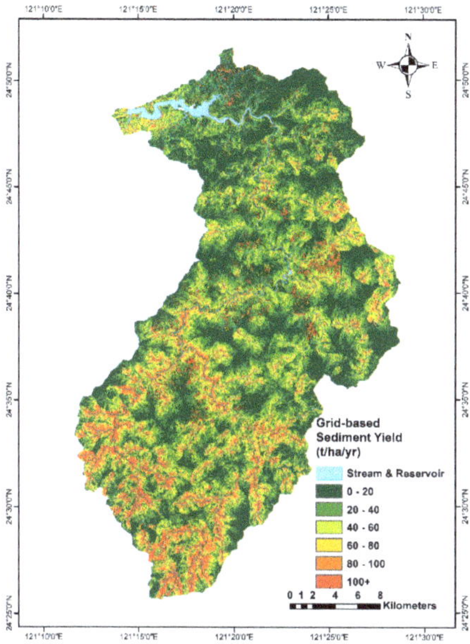

Figure 7. Sediment Yield of the Shihmen Reservoir watershed.

4.3. Comparison by Sub-Watershed Discretization on SDR, SL, and SY

This study considered the distribution of SDR, SL, and SY in the Shihmen Reservoir watershed discretized into 93 sub-watersheds. From Figure 8, it is evident that both the median and the mean of SDR by 93 SW had very similar distributions with a central tendency centered around 0.500.

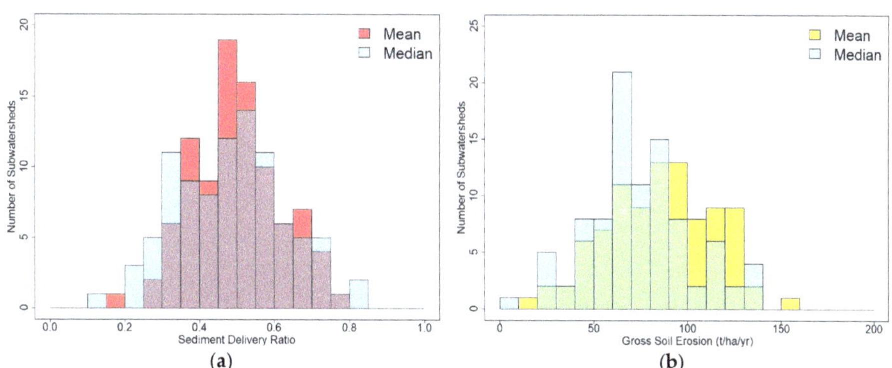

Figure 8. Histograms of the distributions of means and medians: (**a**) sediment delivery ratio (SDR) and (**b**) soil loss of the Shihmen Reservoir sub-watersheds.

The range of mean SL by sub-watershed was between 15.60 t/ha/year (SW #8) and 151.21 t/ha/year (SW #45) and the range of the median of SL was 1.15 t/ha/year (SW #5) and 136.36 t/ha/year (SW #69). The evaluation of the median of the SL for 93 SW showed that the median tended to be between 60 and 90 t/ha/year, and over 20 sub-watersheds had a median between 60–70 t/ha/year. In contrast, the mean SL by sub-watersheds were more widely spread and had a number of peaks between 60 and 120 t/ha/year with over 20 sub-watersheds valued at 80 to 100 t/ha/year.

The range of the mean SY by sub-watershed in the 93 sub-watershed discretization was between 10.28 t/ha/year (SW#1) and 92.97 t/ha/year (SW#73) while the range of the median was 0.38 t/ha/year (SW#8) and 88.11 t/ha/year (SW #73). For the 25 SW, the mean SY was between 10.52 t/ha/year (SW #3) and 73.92 t/ha/year (SW #8) and the range of the median was between 5.50 t/ha/year (SW #3) and 49.43 t/ha/year (SW #16).

To investigate the mean, median, and mode values by sub-watershed discretization (25 SW and 93 SW) and the correlation of each aspect of sediment yield analysis (SDR, SL, and SY), this study created the boxplots displayed in Figure 9a–e. This study additionally developed the PDF plot of each sub-watershed to investigate similarities and differences between the sub-watersheds—a sample of these plots from the 93 SW are displayed in Figure 10. The variance in the shapes of the probability density functions of SDR, SL, and SY was evident.

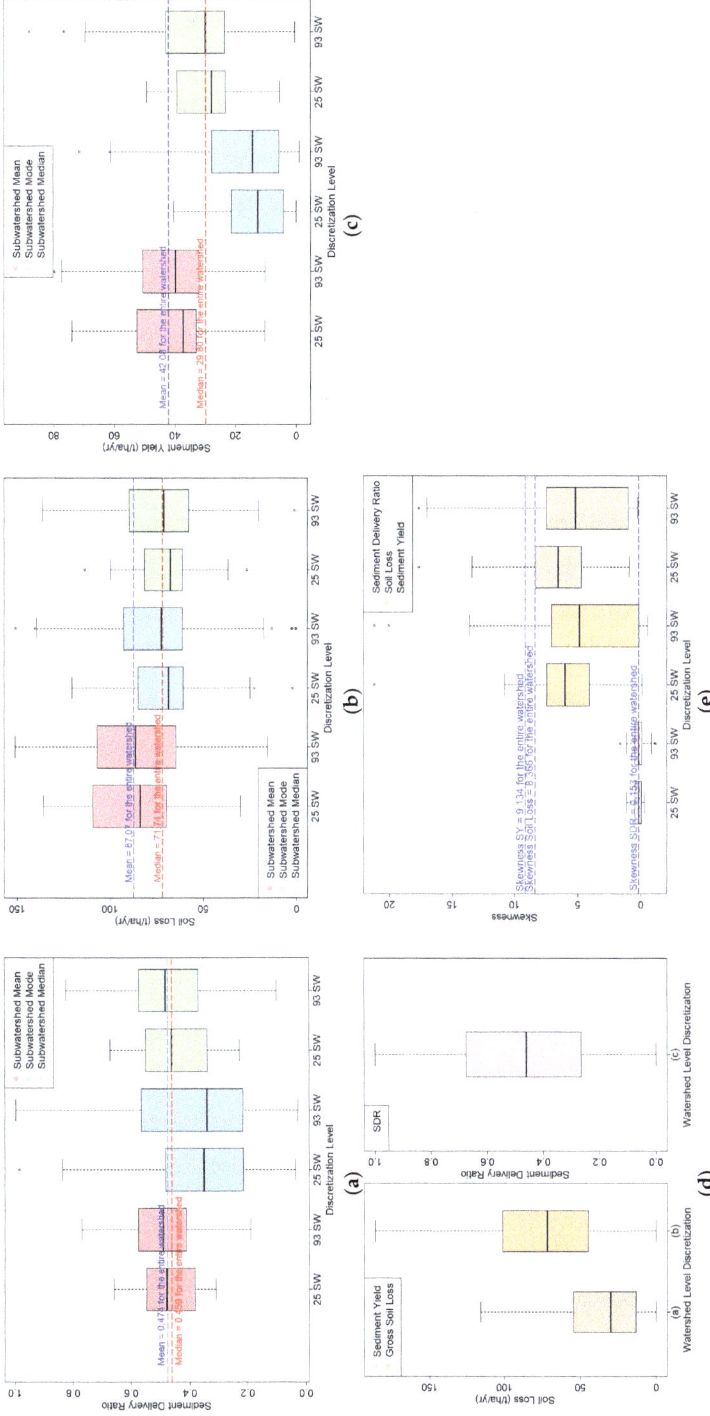

Figure 9. Boxplots of: (**a**) the distributions of means, medians, and modes of sub-watershed SDR, (**b**) sub-watershed soil loss (SL), and (**c**) sub-watershed sediment yield (SY); (**d**) the distributions of the Shihmen Reservoir watershed SDR, SL, and SY, and (**e**) the distributions of skewness of sub-watershed SDR, SL, and SY.

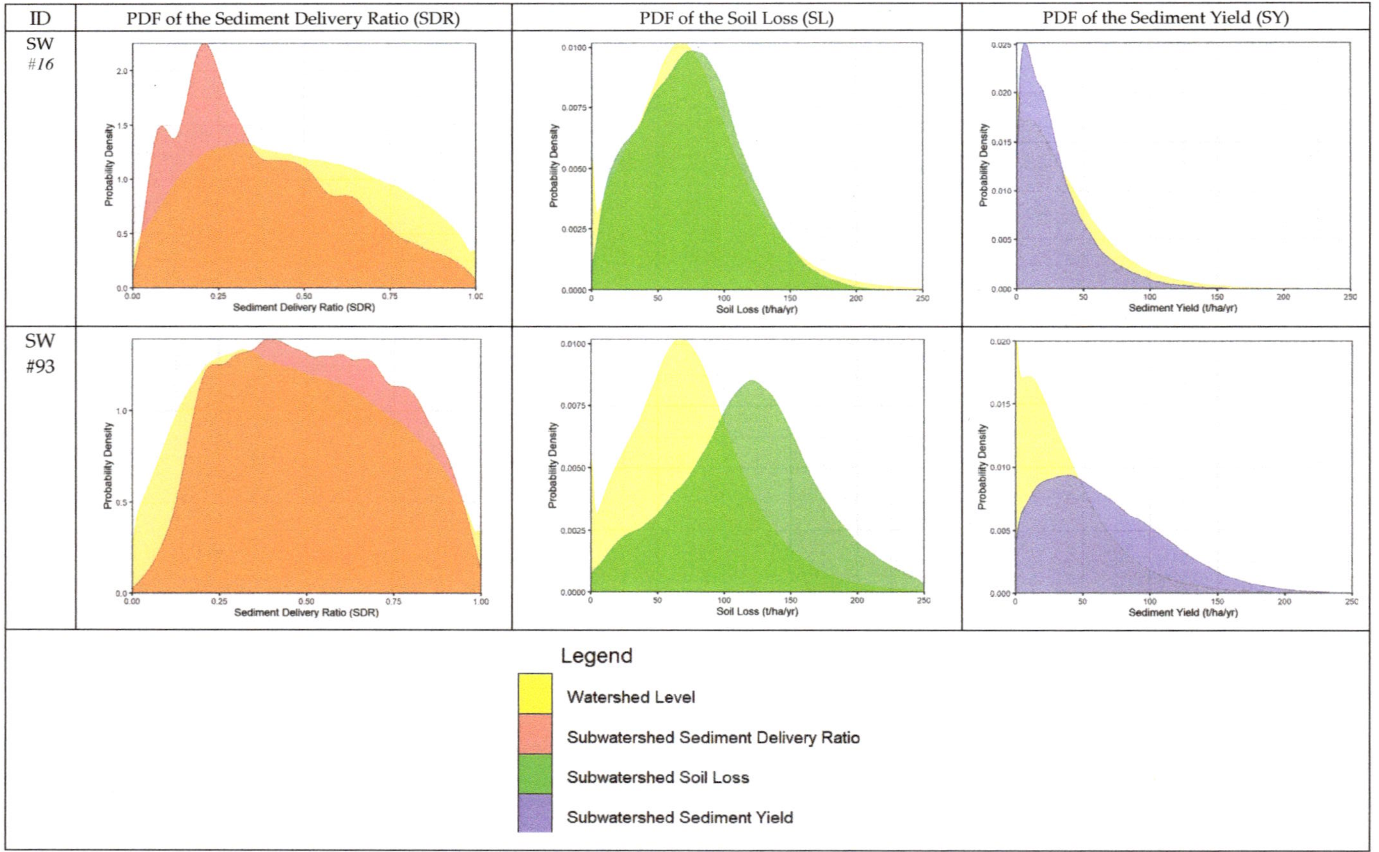

Figure 10. Example probability density functions of SDR, SL, and SY of two sub-watersheds and the Shihmen Reservoir watershed.

5. Discussion

Past studies have shown the distribution of soil erosion of the Shihmen Reservoir watershed, and this paper estimates the geospatially distributed sediment delivery ratio (SDR_i) under a beta determined by the recursive method of the SEdiment Delivery Distributed (SEDD) model and a known watershed-wide SDR_w.

5.1. The Importance of SDR

The SDR plays an essential role in soil erosion research as an additional parameter when considering sediment yield, slope conservation/remediation, and sediment control projects for engineers and decision-makers. An SDR of almost 100% denotes an area where the dislodged sediments have a near-perfect chance to reach a nearby river or stream. In comparison, an SDR of 0% within a highly erodible soil may mean that the soil has a high likelihood of depositing before reaching the river or stream. The SDR consideration creates a twofold issue for engineers: the classification of sediment loads entering the stream and the classification of the deposition quantities on the slopes that may lead to future sediment related disasters.

The traditional method of conveying the SDR across a region has historically been studies of small watersheds or empirical equations. The SDR values obtained from these methods were extrapolated to the entire watershed and utilized to calculate the SY from the gross erosion model's output SL. With the advancement of GIS, it is now possible to use morphological units or grid cells to compute the SDR for each location in the watershed. However, unlike SL or SY, it is essential to note that the individual SDR (SDR_i) cannot be averaged directly. The average of the SDR_i is not the SDR_w because of the definition of SDR. To calculate the SDR_w, individual SY has to be computed from the SDR_i first. SDR_w is then the ratio of the sum of SY divided by the total soil loss. In this study, we calculated the mean of SDR_i purely for the sake of statistical analysis.

The determined SDR can be used to determine SY, which is then used as a basis for decision-making, countermeasure designs, and other uses. Therefore, the accuracy of these evaluations is critical to the management of a watershed, such as the Shihmen Reservoir watershed, where significant soil erosion risk is apparent.

5.2. The β Coefficient in the Model

The SDR calculation within the SEDD model utilizes one unique parameter requiring expert opinion or field experiments, β. The β coefficient in the model is calculated using three methodologies, the recursive method, the trial-and-error method, and the field experiments method, using Cs^{137} [6,42–44]. The β coefficient is a watershed-wide constant, but in some cases, this parameter is evaluated for each spatial discretization unit (for sub-watersheds). Additionally, some researchers have subjectively set a value using known published values of β [45]. The "recursive approach" or "recursive fitting approach" as used in this study defines β using known values of SDR_w. When β is determined, SDR_i can be determined from Equation (3). Sometimes, SDR_w is obtained from empirical equations. For example, Vanoni's method [46] was used to estimate β as a SEDD parameter [6]. Burguet et al. [44] evaluated β over two olive orchard catchments using different annual C factors and R factors, determining the median β from multiple events and improving the assessment of the β parameter for their two study catchments. Other studies have utilized an additional factor, the vegetation or land-cover parameter, α [27,47], but Lopez-Vicente and Navas [29] concluded that the SDR in the SEDD model has limited sensitivity to the land use type. The slope and length of the flow paths have more influence on the model.

Additionally, Lai [48] explored the use of the SEDD in the Shihmen Reservoir watershed but determined the β coefficient by setting it to values between 0 and 200 using the Jain and Kothyari's variant of the original model which introduces the coefficient a_i (also denoted as k_i) to consider different land uses based on expert opinions from a table introduced by Haan [49]. However, this study utilizes the original model developed by Ferro and Minacapilli [24] because the use of the other model introduces high spatial variance in the a_i coefficient [27] and increases uncertainty to the SDR model. This is because the expert opinions have to be utilized to compare Haan's table to land use types not considered originally or are geographically dissimilar. This study determines the $\beta = 0.00515$ using the known SDR_w value, while Lai [48] determined $\beta = 8.5$. For these reasons, this study has been distinguished as a more accurate implementation of the SEDD model.

The SDR_w (0.49) used in this study was derived by Li [8], which was calculated from reservoir sedimentation, and the soil loss included sheet and rill erosion, gully erosion, and landslides sediments. This recursive method is discussed thoroughly by Porto and Walling [42] and is employed in regions lacking data to obtain reliable values for β. The recursive method of the SEDD model in our study was developed within the model builder of the ArcGIS software. It extended the works of the model originally developed by Jhan et al. [50] and Chen et al. [51].

5.3. SEDD Model Using the Grid–Cell-Based Analysis

The SEDD model determines the sediment yield at the outlet from each hydrological unit or grid cell unit. First, the USLE equation is applied to the study area, and the gross erosion derived is then reduced by the SDR to the equivalent sediment yield at the outlet. The SDR equation within the SEDD model uses physical and hydrological parameters (slope and flow distance, respectively) to determine the sediment delivery to channels and eventually to the reservoir outlet.

The correlation between the SY, SL, and the ratio SDR is evident in the previously shown Figures 5–7. A visual comparison between Figures 4 and 5 can easily distinguish that the SDR values are inversely related to the travel time. As the travel time or distance to a river channel increases, the probability of eroded particles entering the stream decreases and consequently SDR decreases.

In Figure 5, the SDR of the watershed ranged from 0 to 1, and the areas where sediment delivery has a higher likelihood to reach channels or the outlet is focused mainly around the rivers and streams distributed throughout the watershed. The soil loss evaluation by USLE has been previously discussed in Chen et al. [51] and Liu et al. [16]. This study builds upon the model of Liu et al. [16] by increasing the number of rainfall stations in the analysis of R_m to 41 stations in and around the watershed [38]. Significantly, the SL is 0–40 t/ha/year at the lower elevation in the northern areas of the watershed that surround the reservoir. In contrast, in the southern reaches of the watershed of higher elevations, the SL increases to 80 t/ha/year or more and often exceeding 100 t/ha/year. The delivery to the channel is limited by the SDR, as shown in Figure 5. Therefore, the results of SY in these areas follow the pattern of the SDR with increases around the river channels. The influence of the SDR on SY is evident from visual inspection of the maps.

5.4. Mean, Median, and Mode of 25/93 Sub-Watersheds

This study determined the SDR and SY of the SEDD model. It analyzed the probability density function of the SDR, SL, and SY for a discretization of the watershed into 25 sub-watersheds and 93 sub-watersheds. The distributions of SDR, SL, and SY for the watershed and sub-watersheds were found to be all non-normal distributions.

For each sub-watershed, we calculated its mean, median, and mode of SDR, SL, and SY from its distributions. The results were then aggregated into Figure 9 using boxplots. Additionally shown in Figure 9a–c are the mean and median of the entire watershed (horizontal dash lines). Comparing the 25 sub-watersheds with the 93 sub-watersheds, we can see that the range (between the ends of whiskers) of 25 SW is always smaller than that of the 93 SW, no matter what the distribution type is (SDR, SL, or SY). This means that with finer discretization, there is more variation, and it is more likely to obtain extreme values. Interestingly, the interquartile ranges (IQR) of the means of SDR, SL, and SY of 25 SW and 93 SW are about the same. It can also be noted that although the mean values of SDR and SL of 93 SW have broader ranges than those of the 25 SW, the resulting mean values of SY for both the 25 and 93 SW have about the same range. Figure 9d shows the distribution of the watershed level

SDR, SL, and SY. Compared with Figure 9b,c, the ranges of the watershed SL and SY are larger than those of both sub-watershed discretizations.

Figure 9e shows the skewness of the SDR, SL, and SY datasets of the 25 SW and 93 SW in comparison with the skewness of the grid–cell-based analysis of the entire watershed. The difference in the skewness of the SDR between all three discretizations (Shihmen as a whole, 25 SW, and 93 SW) is small. Their range (SDR) is much smaller than that of the SL and SY. Additionally, the skewness of the SL and SY are similarly distributed. The lower quartile range is more dominant in the 93 SW, while the 25 SW shows a generally even distribution between the upper and low quartiles. The skewness of the entire watershed SDR was 0.153, for SL 8.359 and for sediment yield 9.134. The upper whiskers of the 93 SW for SL and SY are significantly longer than the lower whiskers. Sixty-two of the 93 sub-watersheds (66.7%) for SDR were positively skewed (right-skewed), meaning the majority of SDR were below the 0.500 while for the 25 sub-watersheds, 17 were positively skewed (68.8%). For the SL, 24 of 25 sub-watersheds (96%) were positively skewed (right-skewed), and 81 of 93 sub-watersheds (87%) were, too. All of the 25 sub-watersheds (100%) and all 93 sub-watersheds were positively skewed for the PDF of SY. The similar positive relationship between the skewness of SL and SY shows that the SL is more influential on the SY than the SDR is. Both example cases in Figure 10 support this. SW #16 has a skewness of 0.625 (SDR), 4.568 (SL), and 5.463 (SY), and SW #93 has a skewness 0.064 (SDR), 0.357 (SL), and 0.827 (SY).

Applying Equation (4) to each of the 25 or 93 sub-watersheds, we also calculated the sub-watershed SDR_w, as shown in Figure 11 (not averaging of SDR_i). It can be seen that the discretization of 93 sub-watershed shows a broader range of SDR_w values than the 25 sub-watersheds. These SDR_w values are essential if a sub-watershed level study is to assess detailed SY features for each sub-watershed. The SDR_w can also serve as one of the criteria to select the most critical sub-watersheds to monitor.

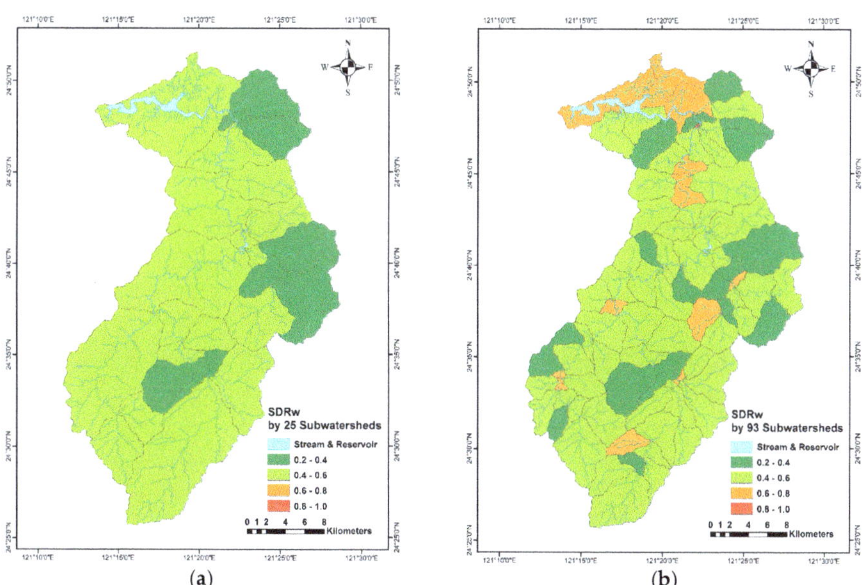

(a)   (b)

Figure 11. Sub-watershed SDR_w of (**a**) 25 sub-watersheds and (**b**) 93 sub-watersheds.

5.5. Sub-Watershed Prioritization

Soil Conservation and countermeasures against sedimentation of river waterways are a costly endeavor and involve significant budgetary concerns for nations such as Taiwan. Climate change and its effects, such as the increase of global land surface temperatures and changing precipitation patterns, pose increased concerns for water security for large population centers such as Taipei City and Taoyuan. These budgetary concerns reinforce the need for sustainable watershed management and evidence-based decision making for targeted soil conversation and soil erosion mitigation programs. Therefore, sub-watershed prioritization can improve the effectiveness of these interventions while controlling budgets. In this study, the three measurements of central tendency (mean, median, and mode) of SDR, SL, and SY were explored. The SEDD model introduces a physically-based methodology for determining the SDR that is useful in understanding what probability of soil will enter the stream or outlet while modeling changing rainfall or land cover levels.

The prioritization by mean SY for 25 and 93 sub-watersheds is shown in Figure 12, each with five levels of prioritization from 0 to 20 t/ha/year (low) to 80 to 100 t/ha/year (high) derived from the ranges previously discussed. It is evident that there are striking similarities between both of these prioritization schemes. Generally speaking, the SY at the outlet is low, but in the eastern and southern extremes of the watershed, there are specific sub-watersheds with medium to high sediment yield values.

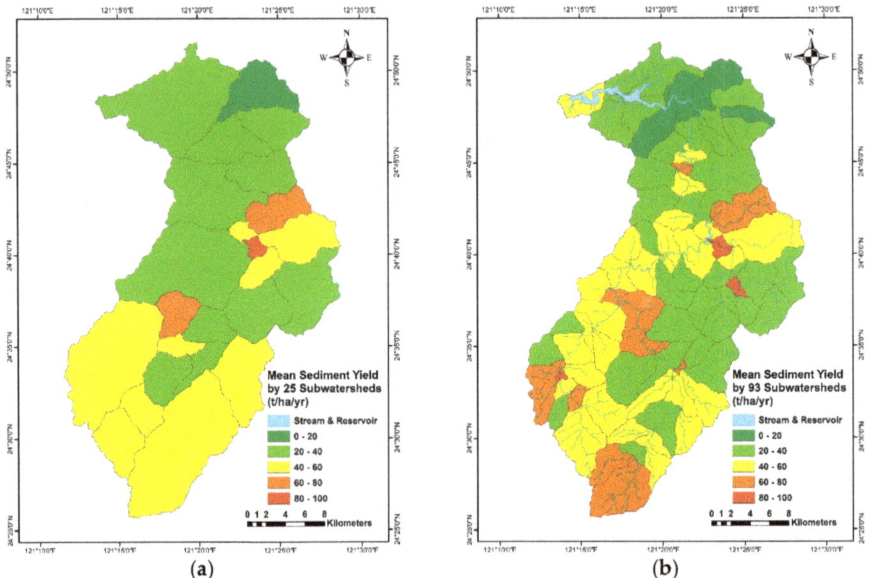

Figure 12. Sub-watershed prioritization by sediment yield of (**a**) 25 sub-watersheds and (**b**) 93 sub-watersheds.

6. Conclusions

This study applied the SEdiment Delivery Distributed (SEDD) model to the Shihmen Reservoir watershed in Taiwan. The main merit of the study lies in the first attempt to derive the sediment yield by soil erosion for the entire Shihmen Reservoir watershed and its sub-watersheds. By using the recursive method, this study determined the SEDD β coefficient to be 0.00515 and predicted the spatial distributions (maps) of travel time, SDR_i, soil erosion (soil loss), and sediment yield of the Shihmen Reservoir watershed. The resulting average of SY for the Shihmen Reservoir watershed was 42.08 t/ha/year and by sub-watershed 10.52–73.92 t/ha/year for 25 sub-watersheds and 10.28–92.97 t/ha/year for 93 sub-watersheds.

This study also recommended the use of mean SY by sub-watershed in sub-watershed prioritization for future soil conservation decision-making. The model presented takes advantage of all currently available data for the Shihmen reservoir watershed. However, it is essential to note that the sediment yield modeling can be improved by increased on-site validation and the use of aerial photogrammetry to deliver more updated data to better understand the field situations. Considering the implications of climate change, there is also a great need for further research on the sediment yield by soil erosion and other land degradation sources and the impacts of changing precipitation regimes. Using geospatial models of sediment yield as guidance can help the local government to better implement engineering and ecological solutions for soil conservation to achieve sustainable land management (SLM), thereby reducing sediment yield risks and creating a well-balanced solution for all stakeholders.

Author Contributions: Conceptualization, W.C.; Data curation, K.T.; Formal analysis, K.T.; Funding acquisition, W.C. and U.S.; Investigation, B.-S.L. and U.S.; Methodology, W.C.; Project administration, W.C.; Resources, B.-S.L.; Software, K.T.; Supervision, W.C.; Visualization, K.T.; Writing—original draft, K.T. and W.C.; Writing—review & editing, W.C., B.-S.L. and U.S. All authors have read and agreed to the published version of the manuscript.

Funding: This study was partially supported by the National Taipei University of Technology-King Mongkut's Institute of Technology Ladkrabang Joint Research Program (Grant Numbers NTUT-KMITL-106-01, NTUT-KMITL-107-02, and NTUT-KMITL-108-01) and the Ministry of Science and Technology (Taiwan) Research Project (Grant Number MOST 108-2621-M-027-001).

Acknowledgments: We thank the anonymous reviewers for their careful reading of our manuscript and insightful suggestions to improve the paper.

Conflicts of Interest: The authors declare no conflict of interest.

References

1. Young, A. *Land Degradation in South Asia: Its Severity, Causes and Effects upon the People*; Final Report for the Economic and Social Council of the United Nations; Food and Agriculture Organization of the United Nations, United Nations Development Programme, and United Nations Environment Programme: Rome, Italy, 1993; Available online: http://www.fao.org/3/v4360e/V4360E00.htm (accessed on 2 August 2020).

2. IPCC. *2007: Climate Change 2007: Synthesis Report*; Pachauri, R.K., Reisinger, A., Eds.; Contribution of Working Groups I, II and III to the Fourth Assessment Report of the Intergovernmental Panel on Climate Change; IPCC: Geneva, Switzerland, 2007; 104p.

3. Keesstra, S.; Mol, G.; de Leeuw, J.; Okx, J.; Molenaar, C.; de Cleen, M.; Visser, S. Soil-related sustainable development goals: Four concepts to make land degradation neutrality and restoration work. *Land* **2018**, *7*, 133. [CrossRef]

4. Hsiao, C.Y.; Lin, B.S.; Chen, C.K.; Chang, D.W. Application of Airborne Lidar Technology in Analyzing Sediment-Related Disasters and Effectiveness of Conservation Management in Shihmen Watershed. *J. Geoengin.* **2014**, *9*, 55–73. [CrossRef]

5. Wang, H.W.; Kondolf, M.; Tullos, D.; Kuo, W.C. Sediment Management in Taiwan's Reservoirs and Barriers to Implementation. *Water* **2018**, *10*, 1034. [CrossRef]

6. Fernandez, C.; Wu, J.Q.; McCool, D.K.; Stöckle, C.O. Estimating Water Erosion and Sediment Yield with GIS, RUSLE, and SEDD. *J. Soil Water Conserv.* **2003**, *58*, 128–136.

7. Wu, L.; Liu, X.; Ma, X.-Y. Research Progress on the Watershed Sediment Delivery Ratio. *Int. J. Environ. Stud.* **2018**, *75*, 565–579. [CrossRef]

8. Li, D.-H. Analyzing Soil Erosion of Shihmen Reservoir Watershed Using Slope Units. Master's Thesis, National Taipei University of Technology, Taipei, Taiwan, 2017. (In Chinese).

9. Tsai, Z.X.; You, G.J.Y.; Lee, H.Y.; Chiu, Y.J. Use of a Total Station to Monitor Post-Failure Sediment Yields in Landslide Sites of the Shihmen Reservoir Watershed, Taiwan. *Geomorphology* **2012**, *139*, 438–451. [CrossRef]

10. Tsai, Z.X.; You, G.J.Y.; Lee, H.Y.; Chiu, Y.J. Modeling the Sediment Yield from Landslides in the Shihmen Reservoir Watershed, Taiwan. *Earth Surf. Process. Landf.* **2013**, *38*, 661–674. [CrossRef]

11. Bathurst, J.C.; Burton, A.; Ward, T.J. Debris Flow Run-out and Landslide Sediment Delivery Model Tests. *J. Hydraul. Eng.* **1997**, *123*, 410–419. [CrossRef]

12. Lin, L.-L.; Feng, M.-C.; Tu, Y.-T. Applying AGNPS to Investigate Sediment Delivery Ratio for Different Watershed. *J. Soil Water Conserv.* **2006**, *38*, 373–386. (In Chinese)

13. Soil and Water Conservation Bureau (SWCB). *Evaluation on Sediment Environment Change and Soil and Water Conservation Plan for Shihmen Reservoir Watershed*; SWCB Report; Soil and Water Conservation Bureau (SWCB): Taipei, Taiwan, 2018. (In Chinese)

14. Chen, Y.C.; Wu, Y.H.; Shen, C.W.; Chiu, Y.J. Dynamic Modeling of Sediment Budget in Shihmen Reservoir Watershed in Taiwan. *Water* **2018**, *10*, 1808. [CrossRef]
15. Chiu, Y.-J.; Lee, H.-Y.; Wang, T.-L.; Yu, J.; Lin, Y.-T.; Yuan, Y. Modeling sediment yields and stream stability due to sediment-related disaster in Shihmen reservoir watershed in Taiwan. *Water* **2019**, *11*, 332. [CrossRef]
16. Liu, Y.H.; Li, D.H.; Chen, W.; Lin, B.S.; Seeboonruang, U.; Tsai, F. Soil Erosion Modeling and Comparison Using Slope Units and Grid Cells in Shihmen Reservoir Watershed in Northern Taiwan. *Water* **2018**, *10*, 1387. [CrossRef]
17. Nguyen, K.A.; Chen, W.; Lin, B.-S.; Seeboonruang, U.; Thomas, K. Predicting Sheet and Rill Erosion of Shihmen Reservoir Watershed in Taiwan Using Machine Learning. *Sustainability* **2019**, *11*, 3615. [CrossRef]
18. Nguyen, K.A.; Chen, W.; Lin, B.-S.; Seeboonruang, U. Using Machine Learning-Based Algorithms to Analyze Erosion Rates of a Watershed in Northern Taiwan. *Sustainability* **2020**, *12*, 2022. [CrossRef]
19. Wei, Z.-Z.; Wang, G.-H.; Tan, Z.-L.; Dong, X.-M.; Jia, C.-L. Delineation of Hydrological Response Unit Based on Remote Sensing Data. *DEStech Trans. Eng. Technol. Res.* **2017**, 362–368. [CrossRef]
20. Wood, E.F.; Sivapalan, M.; Beven, K.; Band, L. Effects of Spatial Variability and Scale with Implications to Hydrologic Modeling. *J. Hydrol.* **1988**, *102*, 29–47. [CrossRef]
21. Bear, J. Dynamics of Fluids in Porous Media. *Soil Sci.* **1975**, *120*, 162–163. [CrossRef]
22. Savvidou, E. A Study of Alternative Hydrological Response Units (HRU) Configurations in the Context of Geographical Information Systems (GIS)—Based Distributed Hydrological Modeling. Ph.D. Thesis, Cyprus University of Technology, Limassol, Cyprus, 2018.
23. Arnold, J.G.; Kiniry, J.R.; Srinivasan, R.; Williams, J.R.; Haney, E.B.; Neitsch, S.L. CHAPTER 1—SWAT Input Data: Overview. In *SWAT Input/Output File Doc, Version 2012*; Texas Water Resources Institute: College Station, TX, USA, 2012; pp. 1–30.
24. Ferro, V.; Minacapilli, M. Sediment Delivery Processes at Basin Scale. *Hydrol. Sci. J.* **1995**, *40*, 703–717. [CrossRef]
25. Ferro, V.; Porto, P.; Tusa, G. Testing a Distributed Approach for Modelling Sediment Delivery. *Hydrol. Sci. J.* **1998**, *43*, 425–442. [CrossRef]
26. Ferro, V.; Porto, P. Sediment Delivery Distributed (SEDD) Model. *J. Hydrol. Eng.* **2000**, *5*, 411–422. [CrossRef]
27. Jain, M.K.; Kothyari, U.C. Estimation of Soil Erosion and Sediment Yield Using GIS. *Hydrol. Sci. J.* **2000**, *45*, 771–786. [CrossRef]
28. Fu, G.; Chen, S.; McCool, D.K. Modeling the Impacts of No-till Practice on Soil Erosion and Sediment Yield with RUSLE, SEDD, and ArcView GIS. *Soil Tillage Res.* **2006**, *85*, 38–49. [CrossRef]
29. López-Vicente, M.; Navas, A. Relating Soil Erosion and Sediment Yield to Geomorphic Features and Erosion Processes at the Catchment Scale in the Spanish Pre-Pyrenees. *Environ. Earth Sci.* **2010**, *61*, 143–158. [CrossRef]
30. Di Stefano, C.; Ferro, V. Assessing Sediment Connectivity in Dendritic and Parallel Calanchi Systems. *Catena* **2019**, *172*, 647–654. [CrossRef]
31. Lin, B.-S.; Chen, C.-K.; Thomas, K.; Hsu, C.-K.; Ho, H.-C. Improvement of the K-Factor of USLE and Soil Erosion Estimation in Shihmen Reservoir Watershed. *Sustainability* **2019**, *11*, 355. [CrossRef]
32. Wickham, H. Ggplot2 Book. *Media* **2009**, *35*, 211. [CrossRef]
33. R Core Team. R: A language and environment for statistical computing. R Foundation for Statistical Computing: Vienna, Austria, 2018. Available online: https://www.R-project.org/ (accessed on 2 August 2020).
34. Meyer, D.; Dimitriadou, E.; Hornik, K.; Weingessel, A.; Leisch, F.; Chang, C.-C.; Lin, C.-C. *Misc Functions of the Department of Statistics, Probability Theory Group (Formerly: E1071), TU Wien*; rdrr.io: Vienna, Austria, 2019.
35. Hijmans, R.J. Raster: Geographic Data Analysis and Modeling. R Package Version 3.0-12. 2020. Available online: https://CRAN.R-project.org/package=raster (accessed on 2 August 2020).
36. Jhan, Y.-K. The Analysis of Soil Erosion of Shihmen Reservoir Watershed. Master's Thesis, National Taipei University of Technology, Taipei, Taiwan, 2014. (In Chinese).
37. Yang, K.-J. Terrain Factor Analysis of Soil Erosion in Shihmen Reservoir Watershed. Master's Thesis, National Taipei University of Technology, Taipei, Taiwan, 2016. (In Chinese).
38. Liu, Y.-H. Analysis of Soil Erosion of Shihmen Reservoir Watershed in Taiwan and Lam Phra Ploeng Basin in Thailand. Master's Thesis, National Taipei University of Technology, Taipei, Taiwan, 2019. (In Chinese).

39. Wischmeier, W.H.; Smith, D.D. *Rainfall-Erosion Losses from Cropland East of the Rocky Mountains*; Agriculture Handbook No. 282; Agricultural Research Service, U.S. Department of Agriculture: Washington, DC, USA, 1965.

40. Wischmeier, W.H.; Smith, D.D. *Predicting Rainfall Erosion Losses*; Agriculture Handbook No. 537; USDA Science and Education Administration: Washington, DC, USA, 1978.

41. Ferro, V.; Di Stefano, C.; Minacapilli, M.; Santoro, M. Calibrating the SEDD Model for Sicilian Ungauged Basins. *IAHS-AISH Publ.* **2003**, *279*, 151–161.

42. Porto, P.; Walling, D.E. Use of Caesium-137 Measurements and Long-Term Records of Sediment Load to Calibrate the Sediment Delivery Component of the SEDD Model and Explore Scale Effect: Examples from Southern Italy. *J. Hydrol. Eng.* **2015**, *20*, 1–12. [CrossRef]

43. Di Stefano, C.; Ferro, V.; Porto, P.; Rizzo, S. Testing a Spatially Distributed Sediment Delivery Model (SEDD) in a Forested Basin by Cesium-137 Technique. *J. Soil Water Conserv.* **2005**, *60*, 148–157.

44. Burguet, M.; Taguas, E.V.; Gómez, J.A. Exploring Calibration Strategies of the SEDD Model in Two Olive Orchard Catchments. *Geomorphology* **2017**, *290*, 17–28. [CrossRef]

45. López-Vicente, M.; Lana-Renault, N.; García-Ruiz, J.M.; Navas, A. Assessing the Potential Effect of Different Land Cover Management Practices on Sediment Yield from an Abandoned Farmland Catchment in the Spanish Pyrenees. *J. Soils Sediments* **2011**, *11*, 1440–1455. [CrossRef]

46. Vanoni, V.A. *Sedimentation Engineering, Manuals and Reports on Engineering Practice*; No. 54; ASCE: New York, NY, USA, 1975.

47. Yang, M.; Li, X.; Hu, Y.; He, X. Assessing Effects of Landscape Pattern on Sediment Yield Using Sediment Delivery Distributed Model and a Landscape Indicator. *Ecol. Indic.* **2012**, *22*, 38–52. [CrossRef]

48. Lai, D.-H. Evaluation of Sediment Delivery Ratio and Completeness Ratio of the Reservoir Watershed. Ph.D. Thesis, National Chung Hsing University, Taichung, Taiwan, 2011. (In Chinese).

49. Haan, C.T.; Barfield, B.J.; Hayes, J.C. *Design Hydrology and Sedimentology for Small Catchments*; Academic Press Inc.: Cambridge, MA, USA, 1994.

50. Jhan, Y.-K.; Shen, Z.-P.; Chen, W.W.; Tsai, F. Analysis of soil erosion of Shihmen reservoir watershed. In Proceedings of the 34th Asian Conference on Remote Sensing 2013, ACRS, Bali, Indonesia, 20–24 October 2013; pp. 3034–3039.

51. Chen, W.; Li, D.H.; Yang, K.J.; Tsai, F.; Seeboonruang, U. Identifying and Comparing Relatively High Soil Erosion Sites with Four DEMs. *Ecol. Eng.* **2018**, *120*, 449–463. [CrossRef]

sustainability

Article

Effects of Land Use and Topographic Position on Soil Organic Carbon and Total Nitrogen Stocks in Different Agro-Ecosystems of the Upper Blue Nile Basin

Getu Abebe [1,2,*], Atsushi Tsunekawa [3], Nigussie Haregeweyn [4], Taniguchi Takeshi [3], Menale Wondie [2], Enyew Adgo [5], Tsugiyuki Masunaga [6], Mitsuru Tsubo [3], Kindiye Ebabu [1,5], Mulatu Liyew Berihun [1,7] and Asaminew Tassew [5]

[1] The United Graduate School of Agricultural Sciences, Tottori University, 1390 Hamasaka, Tottori 680-8553, Japan; kebabu2@gmail.com (K.E.); mulatuliyew@yahoo.com (M.L.B.)

[2] Amhara Agricultural Research Institute, Forestry Research Department, P.O. Box 527 Bahir Dar, Ethiopia; menalewondie@yahoo.com

[3] Arid Land Research Center, Tottori University, 1390 Hamasaka, Tottori 680-0001, Japan; tsunekawa@tottori-u.ac.jp (A.T.); takeshi@alrc.tottori-u.ac.jp (T.T.); tsubo@tottori-u.ac.jp (M.T.)

[4] International Platform for Dryland Research and Education, Tottori University, 1390 Hamasaka, Tottori 680-0001, Japan; nigussie_haregeweyn@yahoo.com

[5] College of Agriculture and Environmental Sciences, Bahir Dar University, P.O. Box 1289 Bahir Dar, Ethiopia; enyewadgo@gmail.com (E.A.); atassew2005@yahoo.com (A.T.)

[6] Faculty of Life and Environmental Science Shimane University, Shimane Matsue 690-0823, Japan; masunaga@life.shimane-u.ac.jp

[7] Faculty of Civil and Water Resource Engineering, Bahir Dar Institute of Technology, Bahir Dar University, P.O. Box 26 Bahir Dar, Ethiopia

[*] Correspondence: gabebe233@gmail.com

Received: 24 January 2020; Accepted: 4 March 2020; Published: 19 March 2020

Abstract: Soil organic carbon (SOC) and total nitrogen (TN) are key ecological indicators of soil quality in a given landscape. Their status, especially in drought-prone landscapes, is associated mainly with the land-use type and topographic position. This study aimed to clarify the effect of land use and topographic position on SOC and TN stocks to further clarify the ecological processes occurring in the landscape. To analyze the status of SOC and TN, we collected 352 composite soil samples from three depths in the uppermost soil (0–50 cm) in four major land-use types (bushland, cropland, grazing land, and plantation) and three topographic positions (upper, middle, and lower) at three sites: Dibatie (lowland), Aba Gerima (midland), and Guder (highland). Both SOC and TN stocks varied significantly across the land uses, topographic positions, and agro-ecosystems. SOC and TN stocks were significantly higher in bushland (166.22 Mg ha^{-1}) and grazing lands (13.11 Mg ha^{-1}) at Guder. The lowest SOC and TN stocks were observed in cropland (25.97 and 2.14 Mg ha^{-1}) at Aba Gerima, which was mainly attributed to frequent and unmanaged plowing and extensive biomass removal. Compared to other land uses, plantations exhibited lower SOC and TN stocks due to poor undergrowth and overexploitation for charcoal and firewood production. Each of the three sites showed distinct characteristics in both stocks, as indicated by variations in the C/N ratios (11–13 at Guder, 10–21 at Aba Gerima, and 15–18 at Dibatie). Overall, land use was shown to be an important factor influencing the SOC and TN stocks, both within and across agro-ecosystems, whereas the effect of topographic position was more pronounced across agro-ecosystems than within them. Specifically, Aba Gerima had lower SOC and TN stocks due to prolonged cultivation and unsustainable human activities, thus revealing the need for immediate land management interventions, particularly targeting croplands. In a heterogeneous environment such as the Upper Blue Nile basin, proper understanding of the interactions between land use and topographic position and their effect on SOC and TN stock is needed to design proper soil management practices.

Keywords: *Acacia decurrens*; Eucalyptus; drought-prone; highland; midland; lowland

---

1. Introduction

Soil organic carbon (SOC) and total nitrogen (TN) provide information on the impact of land management on soil health. The SOC stock, which is a key component and the largest carbon pool in terrestrial ecosystems, is strongly linked to nitrogen availability [1] and serves as an indicator of soil quality [2]. SOC acts as a major source or sink for atmospheric CO2 [3–6]. Globally, soil is estimated to store 3150 Pg C (1 Pg C = 1015 g C), which is four times greater than carbon storage in the terrestrial plant biomass (650 Pg C) and atmospheric (750 Pg C) pools [7]. The size of the soil carbon pool, however, is significantly controlled by the balance between the input and output of carbon in an ecosystem. Therefore, any change in the size of the SOC stock potentially affects elemental cycling, land productivity, atmospheric CO2 concentration, and thus global climate [7–9].

The amount of SOC in a terrestrial ecosystem is influenced by natural and anthropogenic factors [10]. Human-induced land-use change causes a particularly substantial loss of SOC [7,11,12]. Land-use change is associated with ecosystem carbon change [13] and drives negative impacts on climate and the environment. Numerous studies have shown that deforestation and land-use change results in land degradation and poorer soil quality [8,14–16]. In Ethiopia, the conversion of natural vegetation to croplands or plantations is increasing due to population pressure and socio-economic drivers. This has implications for biodiversity decline, land productivity, desertification, and SOC dynamics [5,17–20]. According to Assefa et al. [5], conversion of natural forest to cropland in the northern highland of Ethiopia accounted for 50% to 87% of the observed SOC reduction. Likewise, Kassa et al. [21] reported that conversion of forest and agroforestry to croplands caused an annual decline of SOC stock from 3.3 to 8.0 Mg ha^{-1} in the southwestern highlands of Ethiopia. On the other hand, reports on vegetation restoration of degraded lands in the region indicated that SOC is improved by planting Eucalyptus trees [18,22,23] or establishing exclosures [24,25].

Generally, the soil of natural vegetation has higher SOC than croplands because of its higher organic residue content [26]. However, the efficiency of SOC accumulation depends on the quality and amount of organic inputs, decomposition rate in the soil [26], and topographic position [10,27]. Topography influences SOC mainly by altering the input and output of carbon via hydrological processes, and it affects soil erosion and sediment deposition [28]. The topographic position also affects water availability, temperature regime, vegetation distribution, and soil processes [15,29].

Recently, owing to their strong influence on the sustainability of natural and agricultural ecosystems, the effects of factors such as land use, topography, and their interaction on SOC and TN stocks have attracted scientific attention at the small watershed scale [10,12,13,15,27,30]. At the regional scale, climate is the dominant factor that controls SOC and TN stocks by inducing changes in soil moisture, vegetation patterns, decomposition rate [31], microbial activity [32], and soil respiration [33]. Therefore, SOC and TN dynamics in the soil vary in response to environmental factors (both biotic and abiotic), and are sensitive to changes in climate and the local environment [1]. Thus, understanding soil carbon and nitrogen stock dynamics in different agro-ecosystems as a function of topographic position, land use, and their interaction is important for designing sustainable land management options [22,27,34] that also contribute to food security [35].

The direct and interactive effects of topography and land use on SOC and TN stocks are not well studied in the landscape of Ethiopia's Upper Blue Nile basin, which is also known as the Abay River basin and covers an area of 173,000 km^2 [36]. The climate of the region is tropical highland monsoonal [37]. The region is characterized by fragile and drought-prone areas, with diverse agro-ecosystems and severe land degradation. Although soil and water conservation practices have been used since the 1980s [38], a reduction in vegetation cover [39,40] and soil erosion induced by poor land-use management have become major challenges for ensuring food security [36].

The aim of this study was to assess the effects of major controlling factors on SOC and TN stocks in three agro-ecosystems of the Upper Blue Nile basin. The specific objectives were to (1) determine how stocks of SOC and TN vary with topographic position, land-use type, and soil depth across agro-ecosystems; (2) assess the interactive effect of land use and topographic position on SOC and TN stocks within and across agro-ecosystems; and (3) assess the current spatial distribution of SOC and TN stocks in the three agro-ecosystems of the Upper Blue Nile basin.

2. Materials and Methods

2.1. Study Sites

The study was conducted in three different agro-ecosystems of the Upper Blue Nile basin, Ethiopia (Figure 1), namely Guder, Aba Gerima, and Dibatie, representing the highland, midland, and lowland agro-ecosystems, respectively (Table 1). According to Mekonnen [41], the four dominant soil types (in the FAO classification system) in the study area are Acrisols, Leptosols, Luvisols, and Vertisols (Table 1).

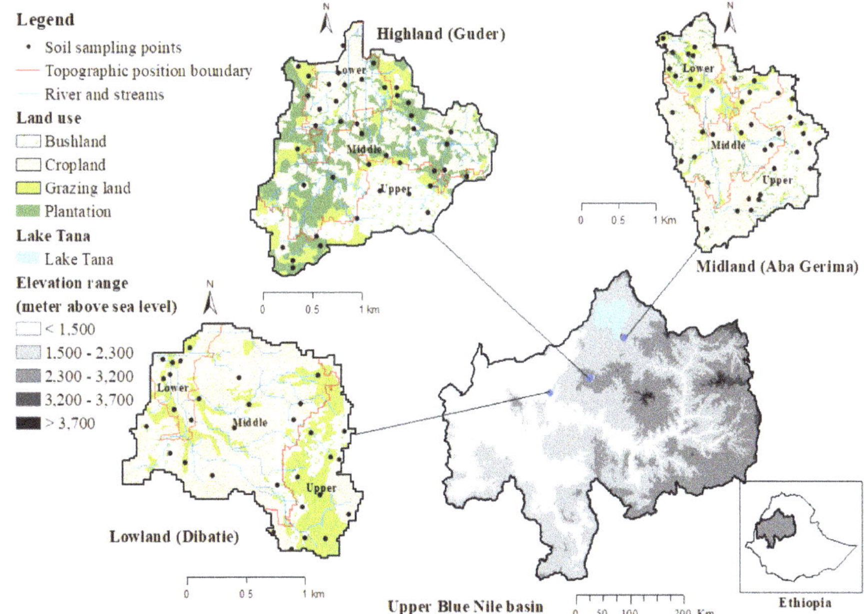

Figure 1. Location of the three study sites in the Upper Blue Nile basin, with respective land-use and drainage maps shown. The points in each watershed illustrate the distribution of sampling points with respect to land use and three topographic positions.

In the Koppen–Geiger classification [42], the climate is characterized as subtropical oceanic highland at Guder, humid subtropical at Aba Gerima, and tropical wet-dry at Dibatie. The rainfall pattern is unimodal and mostly occurs from June to September at all sites (Figure 2). The mean annual rainfall was 1022, 1343, and 2454 mm at Dibatie, Aba Gerima, and Guder, respectively. Mean annual temperature varies from 25 to 32 °C at Dibatie, from 13 to 27 °C at Aba Gerima, and from 9.4 to 25 °C at Guder [43,44].

Table 1. Site characteristics of study watersheds in the Upper Blue Nile basin.

Site Characteristics	Site (Watershed)		
	Guder (Highland)	**Aba Gerima (Midland)**	**Dibatie (Lowland)**
Longitude, latitude	11°0′35.13″ N, 36°56′7.97″ E	10°45′53.09″ N, 36°16′19.11″ E	11°39′27.26″ N, 37°30′14.21″ E
Area (ha)	343	426	246
Elevation (m a.s.l.)	2500–2800	1900–2200	1400–1700
Slope gradients (°) [a]	0–32	0–36	0–21
Topographic positions (elevation range and mode of slope (%))			
Upper	(2500–2600, 30–50)	(2200–2100, 10–20)	(1700–1600, 10–20)
Middle	(2600–2700, 10–20)	(2100–2000, 10–20)	(2100–2000, 0–10)
Lower	(2700–2800, 0–10)	(2000–1900, 0–10)	(2000–1900, 0–10)
Annual mean temperature (°C) [b]	9.4–25	13–27	25–32
Rainfall (mm yr^{-1}) [b]	1951–3424	895–2037	850–1200
Agro-ecology [c]	oceanic subtropical	humid subtropical	tropical wet-dry
Soil parent material [d]	Basalt (Quaternary)	Basalt (Oilgo pilocene)	-
Major soil types [e]	Acrisols and Leptosols	Leptosols and Luvisols	Luvisols and Vertisols
Primary soil texture [a]	clay loam	clay	clay
Sand, silt, and clay (%) [e]	30, 40, and 30	15, 30, and 55	25, 19, and 56
Selected soil properties			
pH (water)	4.2–6.5	4.7–6.8	5.8–7.4
Electric conductivity (dS m^{-1})	0.01–0.11	0.01–0.12	0.02–0.19
Cation exchange capacity (cmol kg^{-1}) [e]	21.4–65.7	23.8–26.8	23.2–48.8
Land-use types (Area, ha)	bushland (58.8), cropland(106), grazing land (47.1), plantation forest (116.5)	bushland (46.5), cropland (220), grazing land (14.6), plantation forest (38)	bushland (37.6, cropland (151), grazing land (55.3)
Dry biomass (tones ha^{-1} yr^{-1}) [f]			
Cropland (teff)	7.14	6.17	6.94
Grazing land	3.9	3.08	7.9

[a] Slope and soil texture data taken from [45]. [b] Weather data (1999–2015) was obtained from [43]. [c] Koppen–Geiger classification [42]. [d] Soil geology data taken from [46]; [e] Soil characteristics taken from [41]. [f] Dry biomass data was obtained from KAKENHI project (average from 150 plots (1 m × 3 m), 2016–2017).

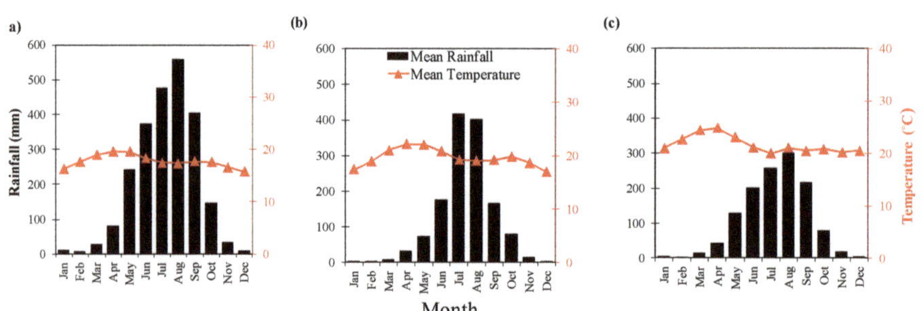

Figure 2. Climograph of Guder (**a**), Aba Gerima (**b**), and Dibatie (**c**) from 1999 to 2017.

The native tree and shrub species common at Guder are *Acacia abyssinica*, *Albizia gummifera*, *Croton macrostaches*, *Combretum molle*, *Cordia africana*, *Schefflera abyssinica*, *Dovyalis abyssinica*, and *Entada abyssinica*. Those at Aba Gerima are *A. gummifera*, *Bersama abyssinica*, *Calpurnia aurea*, *Croton macrostaches*, *Olea europaea*, *Ficus thonningii*, and *E. abyssinica*. At Dibatie, *Acacia negrii*, *Acacia sieberiana*, *Ficus sycomorus*, *Terminalia brownii*, *Terminalia schimperiana*, and *Oxytenanthra abyssinica* are common in the bushlands. *A. decurrens* at Guder and *Eucalyptus camadulensis* at Aba Gerima are the dominant exotic tree species planted as woodlots for fuelwood, charcoal, and construction wood production. Clear felling (Guder) and coppice management (Aba Gerima) are the common plantation management practices. The rotation period of the plantations at Guder and Aba Gerima is 3–5 and 5–7 years, respectively.

Rainfed, subsistence-based and mixed farming (crop cultivation and livestock rearing) is the main agricultural practice at the study sites [44]. At Guder, teff (*Eragrostis tef*), barley (*Hordeum vulgare*), wheat (*Triticum aestivum*), and potato (*Solanum tuberosum*) are grown. At Aba Gerima, teff, finger millet (*Eleusine coracana*), wheat, and maize (*Zea mays*) are cultivated. At Dibatie, maize, teff, sorghum (*Sorghum bicolor*), and groundnut (*Arachis hypogaea*) are the major food crops [43,45].

2.2. Soil Sampling

Based on the available land-use types and elevation range of the watersheds (Table 1), three topographic positions (i.e., upper, middle, and lower) were selected. Cropland and grazing land are common in all topographic positions at the three study sites, whereas plantation (*A. decurrens* or *E. camaldulensis*) at Dibatie and bushland in the lower position at all sites are not part of the current land-use systems (Figure 1). In each topographic position, four replicated land uses were measured. A total of 352 soil samples were collected from the three agro-ecosystems. The top 50 cm of soil was sampled, divided into three soil layers of 0–15, 15–30, and 30–50 cm. Soil samples were collected from five points, at the four corners and in the center of a plot (10 m × 10 m) using a hand-driven soil auger. Soil samples collected from each plot from similar layers were thoroughly mixed to obtain a composite sample (1 kg). Soil bulk density was determined separately by using a metal core cylinder (100 cm^3), which was inserted at the midpoint of the 0–15, 15–30, and 30–50 cm layers. All composite soil samples were first air-dried and then passed through a 2-mm sieve, packed, labeled, and transported to Japan for chemical analysis at the Arid Land Research Center of Tottori University.

2.3. Soil Analysis

Soil pH and electrical conductivity were measured at a 1:5 soil-to-water ratio using a pH meter (D-51, Horiba, Kyoto, Japan) and conductivity meter (ES-51, Horiba), respectively. Bulk soil density (Mg m^{-3}) was determined for core soil samples after oven-drying at 105 °C for 24 h.

2.4. Determination of SOC and TN Stocks

Five-gram subsamples of homogenized soil from each soil depth were dried at 60 °C for 48 h. From each subsample, 1 g of soil was taken, and total organic carbon and nitrogen were determined using a CN corder (Macro Corder JM1000CN, J-Science Lab, Kyoto Japan). Total carbon and nitrogen stocks (Mg ha^{-1}) down to the 50 cm soil horizon were calculated using the model of [47]:

$$\text{SOC (or TN) stock} = \text{content} \times \rho b \times d \times 10{,}000 \text{ m}^2 \text{ ha}^{-1} \times 0.001 \text{ Mg kg}^{-1}, \tag{1}$$

where SOC (or TN) stock is the soil organic carbon or total nitrogen stock (Mg ha^{-1}), content is the soil organic carbon or total nitrogen concentration (kg Mg^{-1}), ρb is the soil bulk density (Mg m^{-3}), and d is the thickness of the soil layer (m).

2.5. Data Analysis

Data with a non-normal distribution were transformed using square-root and log transformation techniques. Two-way (within agro-ecosystem) and nested three-way (between agro-ecosystems) analysis of variance were used to test the significance of mean differences in SOC and TN content and stock as dependent variables, while topographic position, land use, soil depth, and their interactions (between two or three factors) were considered as driving factors. Differences in means between groups were analyzed using Tukey's HSD (honestly significant difference) test within the Agricolae package (version 1.2-8). Statistical analyses were carried out in RStudio [48], an interface for the R software program (version 3.4.4). The significance level was set at alpha = 0.05.

3. Results

3.1. Effect of Topographic Position on SOC and TN Contents and Stocks

At Guder, SOC content in croplands increased significantly ($p < 0.05$) from the upper (10.96 mg g^{-1}) to the lower topographic position (16.68 mg g^{-1}; Figure 3a). In the case of grazing land, SOC content decreased from 22.59 mg g^{-1} in the upper position to 14.57 mg g^{-1} in the middle position and then increased to 17.40 mg g^{-1} in the lower position. For bushland and *A. decurrens* plantations, SOC content did not vary among topographic positions. TN content for bushland decreased significantly from 2.87 mg g^{-1} in the upper position to 2.39 mg g^{-1} in the middle position (Figure 3d). However, TN in cropland and grazing lands were not significantly different across topographic position.

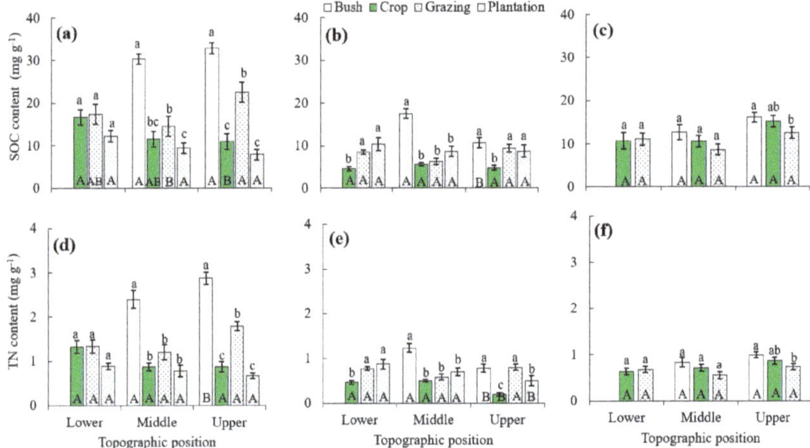

Figure 3. Soil organic carbon (SOC) and total nitrogen (TN) contents under different land uses and topographic positions at Guder (**a,d**), Aba Gerima (**b,e**), and Dibatie (**c,f**). Different lowercase letters above the bars indicate significant differences in SOC and TN among land uses in the same topographic position ($p < 0.05$); different capital letters indicate significant differences in SOC and TN among topographic positions within the same land use ($p < 0.05$). Error bars represent the standard error of the mean at alpha = 0.05.

At Aba Gerima, SOC and TN contents differed among topographic positions ($p < 0.05$; Figure 3b,e). The TN contents of bushland (0.78 mg g^{-1}), cropland (0.19 mg g^{-1}), and Eucalyptus plantations (0.49 mg g^{-1}) were significantly lower at the upper position than that at the middle and lower positions. The highest contents of SOC (17.52 mg g^{-1}) and TN (1.23 mg g^{-1}) were in bushland at the middle position, whereas croplands in the upper position showed the lowest SOC (4.78 mg g^{-1}) and TN contents (0.19 mg g^{-1}).

At Dibatie, SOC and TN contents in the upper position in bushland and cropland were 16.12 and 0.99, and 15.22 and 0.86 mg g^{-1} higher, respectively, than those in grazing land (12.58 and 0.74 mg g^{-1}; Figure 3c,f). In contrast, both SOC and TN contents were similar among land-use types in the middle and lower topographic positions.

At Guder, the SOC stock decreased from the upper to lower topographic positions in grazing land (Figure 4a). SOC stock under *A. decurrens* plantations increased significantly ($p < 0.05$), from 42.73 Mg ha^{-1} in the upper position to 44.63 Mg ha^{-1} in the middle position and 58.94 Mg ha^{-1} in the lower position. The SOC stock in bushland was highest (166.22 Mg ha^{-1}) in the upper position. The TN stock was significantly higher (13.11 Mg ha^{-1}) in grazing lands in the upper position and decreased toward the lower position (Figure 4d).

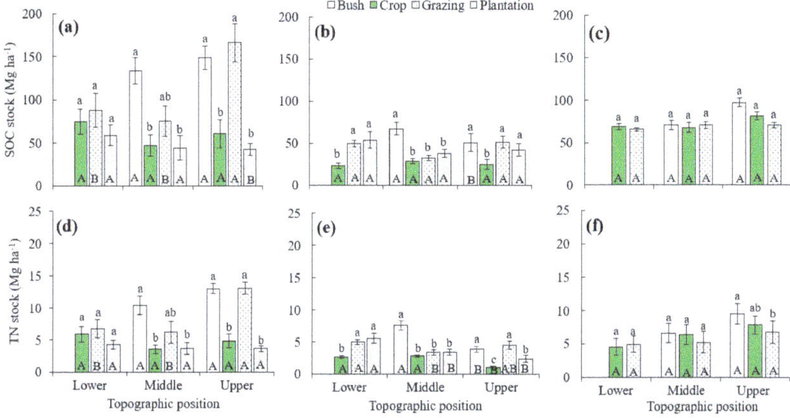

Figure 4. Soil organic carbon (SOC) and total nitrogen (TN) stocks under different land-use types and topographic positions at Guder (**a,d**), Aba Gerima (**b,e**), and Dibatie (**c,f**). Different lowercase letters above the bars indicate significant differences in SOC and TN among land-use types in the same topographic position (*p < 0.05*); different capital letters indicate significant differences in SOC and TN among topographic positions within the same land-use type ($p < 0.05$). Error bars represent the standard error of the mean at alpha = 0.05.

At Aba Gerima, TN stocks in cropland and Eucalyptus plantations increased significantly from the upper to lower positions (Figure 4e).

At Dibatie, TN stocks in bushland, cropland, and grazing lands varied significantly between the upper and the middle and lower positions (Figure 4f). The highest (9.53 Mg ha⁻¹) and lowest (4.56 Mg ha⁻¹) TN stocks were recorded in bushlands and croplands in the upper and middle positions, respectively.

In general, SOC and TN stocks at Guder, TN content and stock at Aba Gerima, and SOC and TN contents at Dibatie were influenced by topographic position ($p < 0.05$; Table 2). With the exception of SOC stock at Dibatie, both SOC and TN contents and stocks at the three study sites were strongly affected by land use ($p < 0.05$). Likewise, the interaction between topographic position and land use had a significant effect on both SOC and TN contents and stocks at Guder and Aba Gerima, whereas no significant effect was detected at Dibatie (Table 2).

Table 2. Results of two-way analysis of variance (ANOVA) for SOC and TN contents and stocks as a function of topographic position and land use in different agro-ecosystems of the Upper Blue Nile basin.

Agro-Ecosystem	Source	df	*p*-Value			
			SOC Content	TN Content	SOC Stock	TN Stock
	topographic position	2	0.278	0.093	<0.001	<0.001
Guder	land use	3	<0.001	<0.001	<0.001	<0.001
	topographic position × land use	5	0.034	0.019	<0.001	<0.001
	topographic position	2	0.562	<0.001	0.307	<0.001
Aba Gerima	land use	3	<0.001	<0.001	<0.001	<0.001
	topographic position × land use	5	0.012	0.002	<0.001	<0.001
	topographic position	2	0.010	0.003	0.046	0.005
Dibatie	land use	2	0.024	0.003	0.492	0.004
	topographic position × land use	3	0.924	0.681	0.766	0.793

Notes: Topographic positions: upper, middle, and lower; land uses: bushland, cropland, grazing land, and plantation.

3.2. SOC and TN Contents and Stocks for Different Land Uses across Soil Depths

Across soil profiles, both SOC and TN contents were slightly decreased from top to the lower soil profile at Aba Gerima (Figure 5b,e) compared with Guder (Figure 5a,d) and Dibatie (Figure 5c,f).

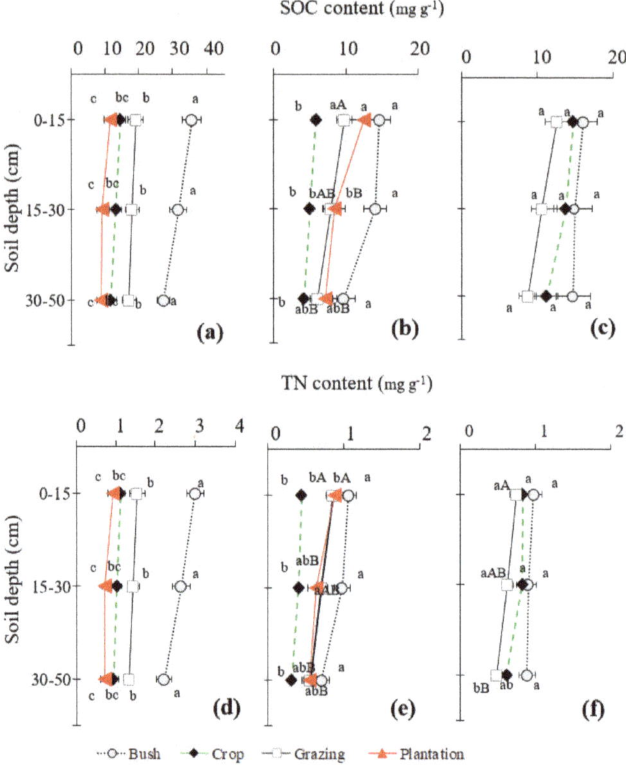

Figure 5. Soil organic carbon and total nitrogen contents in relation to land-use type at Guder (**a,d**), Aba Gerima (**b,e**), and Dibatie (**c,f**). Different lowercase letters indicate significant differences in SOC; n = 12).

The SOC and TN contents at Guder (Figure 5a,d) and Aba Gerima (Figure 5b,e) varied significantly among land uses at all soil depths ($p < 0.05$). At Dibatie, except for TN contents in the lower soil depth, there were no significant differences in SOC and TN contents among land-use types at all soil depths (Figure 5c,f).

At Guder, SOC and TN contents of bushland were significantly higher than the other land uses in all soil profiles (Figure 5a,d). At Aba Gerima, SOC contents in the 0–15 cm layer were 15, 11.3, and 6.7 times higher in bushland, plantation, and grazing land, respectively, than in cropland (Figure 5b). However, in the 15–30 and 30–50 cm layers, the SOC content in bushland was significantly greater than that in the other land uses. Similarly, TN content in the 0–15 cm layer at Aba Gerima was significantly higher ($p < 0.05$; 1.06 mg g^{-1}) in bushland than in the other land uses (Figure 5e). In the 15–30 and 30–50 cm soil layers, the TN content was 1.29, 0.81, and 0.77 times higher in bushland, grazing land, and plantation, respectively, than in cropland ($p < 0.05$). At Dibatie, there were no significant differences in SOC content among land-use types at all soil depths (Figure 5c). However, TN contents in the 30–50 cm soil layer were significantly higher in bushland and cropland than in grazing land (0.49 mg g^{-1}; $p < 0.05$; Figure 5f).

On the other hand, the SOC and TN stocks at Guder and Aba Gerima varied significantly among land uses within each soil profile ($p < 0.05$; Table S1). Significant differences in SOC stocks across soil depths within each land-use type were observed at Aba Gerima and Dibatie, whereas TN stock only varied significantly at Dibatie (Table S1). SOC stocks of grazing land and plantations showed a 0.43- and 0.44-fold decrease from the top layer to the lower layer, respectively, at Aba Gerima. The SOC stock in cropland decreased significantly across soil depths at Dibatie. TN stock in the 0–15 cm layer was higher than at 15–30 and 30–50 cm soil depths at Dibatie, whereas no significant difference was found in the TN stock between the 15–30 and 30–50 cm layers.

3.3. Effect of Agro-ecosystem on SOC and TN Contents and Stocks

SOC and TN contents and stocks in bushland and grazing land at Guder were significantly larger than those at Dibatie and Aba Gerima (Table 3). SOC content in croplands at Aba Gerima (5.01 mg g^{-1}) was significantly lower than that at Guder (13.07 mg g^{-1}) and Dibatie (13.28 mg g^{-1}). Plantation at Guder (*A. decurrens*) and Aba Gerima (Eucalyptus) had similar SOC and TN stocks (Table 3). In contrast, the TN stock in cropland was significantly higher at Dibatie (6.26 Mg ha^{-1}) than at Guder (4.77 Mg ha^{-1}) and Aba Gerima (2.14 Mg ha^{-1}).

Both SOC and TN contents in the lower and middle topographic positions were significantly higher at Guder than at Aba Gerima and Dibatie (Table S2), whereas the SOC and TN contents in the upper position were significantly lower at Aba Gerima than at Guder and Dibatie. The SOC stocks in the upper position were in the following order: Guder (104.67 Mg ha^{-1}), Dibatie (88.90 Mg ha^{-1}), Aba Gerima (41.91 Mg ha^{-1}). The SOC stock in the middle and lower positions and TN stock in all topographic positions were significantly lower at Aba Gerima, but the values for Guder and Dibatie were similar (Table S2).

There was significant variation in the C/N ratios of cropland, grazing land, and plantation among sites (Table 3). The highest (17.52) and lowest (11.03) C/N ratios were those of grazing lands at Dibatie and Aba Gerima, respectively. Eucalyptus plantations (12.15) showed a significantly higher C/N ratio than *A. decurrens* plantations (12.16). The C/N ratios in the upper position were significantly higher at Aba Gerima followed by Dibatie and Guder, whereas the C/N ratios in the middle and lower positions were significantly higher at Dibatie than at the other two sites (Table S2).

Bulk densities in bushland, cropland, and grazing land differed significantly among sites ($p < 0.05$; Table 3), whereas those of plantations at Guder (*A. decurrens*) and Aba Gerima (*Eucalyptus*) were not significantly different. Bulk density ranged from 0.90 to 1.18 Mg m^{-3}, from 1.09 to 1.19 Mg m^{-3}, and from 1.11 to 1.32 Mg m^{-3} at Guder, Aba Gerima, and Dibatie, respectively (Table 3). Bulk densities in bushland, cropland, and grazing land were significantly lower at Guder than at Aba Gerima and Dibatie (Table 3). The bulk density also varied significantly among topographic positions ($p < 0.05$; Table S2). Soils at Guder showed significantly lower soil bulk density in the lower and middle positions as compared to the upper position. Soils at Dibatie had significantly higher soil bulk density in the middle position (Table S2).

Overall, SOC and TN contents and stocks were strongly dependent on agro-ecosystem ($p < 0.05$), land use ($p < 0.05$), and the interaction between agro-ecosystem and land use ($p < 0.05$; Table 4). Topographic position ($p < 0.05$) also influenced SOC content, SOC stock, and TN stock, but not TN content. In addition, the interaction between agro-ecosystem and topographic position affected TN content and SOC and TN stocks. TN content, SOC stock, and TN stock were also strongly dependent on the interaction of agro-ecosystem, topographic position, and land use.

Table 3. SOC and TN contents and stocks in different land-use types at the three study sites.

Land Use	Site	SOC		TN		C/N Ratio	Bulk Density
		mg g^{-1}	Mg ha^{-1}	mg g^{-1}	Mg ha^{-1}		Mg m^{-3}
Bushland	Guder	31.63 (1.45) [a]	141.19 (6.74) [a]	2.63 (0.13) [a]	11.73 (0.50) [a]	12.16 (0.30) [a]	0.90 (0.01) [b]
	Aba Gerima	13.42 (1.16) [b]	59.23 (7.19) [b]	0.96 (0.09) [b]	4.13 (1.91) [b]	14.01 (0.31) [a]	1.09 (0.03) [a]
	Dibatie	15.31 (1.19) [b]	85.70 (9.81) [a,b]	0.92 (0.06) [b]	8.08 (1.51) [a]	16.16 (0.57) [a]	1.11 (0.04) [a]
Cropland	Guder	13.07 (1.04) [a]	61.00 (2.33) [a]	1.02 (0.07) [a]	4.77 (0.28) [b]	12.52 (0.32) [b]	0.96 (0.03) [b]
	Aba Gerima	5.01 (0.29) [b]	25.97 (7.04) [b]	0.39 (0.03) [b]	2.14 (1.44) [c]	17.63 (2.33) [a]	1.12 (0.01) [a]
	Dibatie	13.28 (0.96) [a]	72.62 (8.33) [a]	0.75 (0.04) [a]	6.26 (1.34) [a]	17.38 (0.82) [a]	1.12 (0.03) [a]
Grazing	Guder	18.19 (1.20) [a]	109.94 (3.69) [a]	1.44 (0.09) [a]	8.68 (0.31) [a]	12.68 (0.20) [b]	1.18 (0.04) [b]
	Aba Gerima	7.97 (0.49) [b]	44.14 (7.55) [b]	0.71 (0.04) [b]	4.17 (1.19) [b]	11.03 (0.17) [b]	1.19 (0.03) [b]
	Dibatie	10.84 (0.81) [b]	66.42 (4.69) [b]	0.63 (0.05) [b]	5.65 (1.52) [a,b]	17.52 (0.72) [a]	1.32 (0.02) [a]
Plantation	Guder	9.85 (0.87) [a]	48.79 (7.15) [a]	0.78 (0.06) [a]	3.86 (0.65) [a]	12.16 (0.41) [b]	1.02 (0.02) [a]
	Aba Gerima	9.24 (0.79) [a]	44.77 (6.33) [a]	0.68 (0.06) [a]	3.56 (0.38) [a]	17.15 (2.62) [a]	1.06 (0.03) [a]

Mean (standard error) values were calculated across the whole soil depth from 0 to 50 cm. Within a column, different letters for each land use indicate a significant difference between sites (Tukey's HSD at $p < 0.05$).

Table 4. Nested three-way ANOVA for SOC and TN contents and stocks as functions of topographic position and land use across agro-ecosystems.

Source	df	*p*-Value			
		SOC Content	**TN Content**	**SOC Stock**	**TN Stock**
Agro-ecosystem	2	<0.001	<0.001	<0.001	<0.001
Topographic position	2	0.043	0.862	0.019	<0.001
Land use	3	<0.001	<0.001	<0.001	<0.001
Agro-ecosystem × topographic position	4	0.082	<0.001	0.003	<0.001
Agro-ecosystem × land use	5	<0.001	<0.001	<0.001	<0.001
Agro-ecosystem × topographic position × land use	13	0.058	<0.001	<0.001	<0.001

Agro-ecosystems: Guder, Aba Gerima, and Dibatie; topographic positions: upper, middle, and lower; land uses: bushland, cropland, grazing land, and plantation.

4. Discussion

4.1. Effects of Land-Use Type on SOC and TN Contents and Stocks Across Topographic Positions

Climate, soil type, land use, and topography are the principal factors that control SOC and TN distributions at a regional scale [49,50]. In a small watershed, however, soil type and climate variability are commonly low [51]. Our findings confirmed that land use and topography influenced the SOC and TN storage in the three agro-ecosystems.

At Guder, the SOC content in cropland was significantly increased (Figure 3a) from the upper to lower topographic positions. In fact, the upper position of a watershed is often exposed to soil erosion, serving as a source of run-off and sediment for the lower positions [10]. In cropland, particularly, this situation has amplified the variation of SOC content in association with geomorphologic processes. Cropland in the highlands is poor in vegetation cover and experiences soil disturbance due to tillage and high biomass removal [5,22]. In contrast, SOC content in grazing land was significantly decreased from the upper to lower positions (Figure 4a). Less soil disturbance, greater vegetation cover, and organic input from grazing animals would improve the SOC in the upper position. Similarly, Mekuria et al. [25] reported better vegetation cover and biomass in communal grazing lands in the upper position than in the lower position, which is more easily accessed by livestock that induce changes in SOC. Zhu et al. [30] also found an increasing trend in SOC content for cropland and a decreasing trend in grassland from the summit to the lower part of a watershed in China. A review by Deng et al. [52] of studies conducted worldwide revealed that conversion of native vegetation to grassland significantly increased the SOC stock. In contrast, the SOC content in bushland and plantation at Guder were not affected by topographic position. This distribution pattern may be due to the generally good vegetation cover in bushland and plantations, which may reduce soil erosion in the upper position, resulting in similar SOC contents in the middle and lower positions. Likewise, Fu et al. [13] reported uniform SOC contents under different vegetative types along a hillslope on the Loess Plateau of China.

The TN content in bushland at Guder was higher in the upper than middle position, which was probably due to the presence of a large number of native leguminous shrubs (e.g., *E. abyssinica*) and trees (e.g., *A. abyssinica* and *A. gummifera*) in the bushland. These results correspond with the finding of [21], who reported high TN content in native vegetation consisting of leguminous tree species. TN stock of the grazing land was significantly higher (13.11 Mg ha^{-1}) in the upper position than in the middle and lower positions (Figure 4d), likely because grazing land in the upper position was recently converted from bushland [40], which may have stored relatively high SOC and TN stocks.

At Aba Gerima, both SOC and TN contents were significantly different among land uses (Figure 3b,e). SOC and TN contents in the middle position of bushland (17.52 and 1.23 mg g^{-1}, respectively) were higher than those in the upper position. Similarly, a study conducted in northern Ethiopia [53] reported higher SOC and TN contents in the middle position of natural vegetation. Our results could be associated with soil erosion, which is a common problem in the study area and elsewhere in Ethiopia [36,43]. Soil erosion often causes translocation of soil from the upper slope to lower area and contributes to the loss of soil organic matter [15]. Many studies elsewhere in the world [30,53,54] have reported that soil in sites of deposition has higher SOC and TN stocks.

At Dibatie, both SOC and TN contents in the upper position were significantly affected by land-use type (Figure 3c,f), which could be due to greater anthropogenic pressures in the upper than in the middle and lower topographic positions. Many members of the farming community live around the lower part of the watershed and their livelihoods depend on the bushland. This results in continuous removal of wood and bushland clearing for cropland and grazing land toward the upper position. SOC contents in bushland and cropland were 16.12 and 15.22 mg g^{-1} higher and TN contents were 0.99 and 0.86 mg g^{-1} higher than those of grazing lands. Natural vegetation at Dibatie is dominated by deciduous tree and shrub species that commonly contribute large amounts of organic matter to the soil. However, grassland is regularly burned, which substantially reduces the grass cover and induces loss of SOC and TN contents [55]. In a study in Ethiopia, [56] reported that the natural vegetation in Dibatie (Combretum–Terminalia) decreased as a result of fire.

With regard to land-use effects, the SOC stock of grazing land soil decreased significantly from the upper (162.22 Mg ha^{-1}) to middle positions (75.50 Mg ha^{-1}) at Guder. Soil bulk density in grazing lands is relatively higher as a result of livestock trampling [22,24,57]. At this site, a bulk density of 1.18 Mg/m^3 was recorded in the grazing land (Table 3). At Aba Gerima, a high SOC stock was stored in the middle position of bushland. Similarly, at Dibatie bushlands showed higher SOC and TN contents than those of the other land-use types.

4.2. Effect of Soil Depth on SOC and TN Contents and Stocks

At Guder, SOC and TN contents at 0–15 cm soil depth in bushland were higher than those of other land-use types (Figure 5a,d). Bushland comprises a sizable proportion of native vegetation, and the bushes, shrubs, and trees contain a substantial amount of wood biomass with a lower decomposition rate, which could improve the organic input and contribute more to soil SOC and TN. These results are similar to those of previous studies [10,21] that reported higher SOC and TN contents in the surface soil under native vegetation as compared to that of other land uses. Therefore, conversion of bushland to another land-use type may cause a substantial amount of SOC loss from the surface soil, as reported by studies conducted elsewhere [3,4,21].

Plantation (*A. decurrens* woodlot) contained lower surface SOC and TN contents than we expected (Figure 5a,d). Tesfaye et al. [58] reported lower SOC and TN contents in *A. decurrens* plantations in the central highlands of Ethiopia, which reflects the complete removal of plant residues from the woodlots. The plantations were established on previous cropland areas, but due to prolonged soil disturbance and soil erosion, this land was no longer able to support crop production. Thus, farmers had to change the cropland to plantations as a result of poor soil fertility and degradation [59,60]. Because plantations are commonly used for charcoal production, both the above- and belowground biomass is completely removed at the end of a rotation cycle (~3–5 years). According to Sultan et al. [61], plantations have high stand density (<1 m spacing), no understory vegetation cover, poor infiltration, and high runoff, all of which could contribute to their lower SOC and TN contents.

Similarly, in the 0–15 cm layer of Aba Gerima, cropland has significantly less surface SOC and TN contents than bushland, plantation, and grazing land (Figure 5b,e). This difference may be due to croplands having less organic input than areas with more vegetation. However, plantations at Aba Gerima had SOC contents comparable to those of cropland and grazing lands. These differences in SOC content from the plantations at Guder are likely induced by the differences in species and woodlot management. Unlike the Acacia plantations at Guder, the plantations at Aba Gerima consist of *Eucalyptus camaldulensis*, and tree harvesting operations do not include the belowground biomass. A study in northern Ethiopia revealed that Eucalyptus plantations had a better potential to restore SOC content than did cropland and grazing land [5,22], and [62] reported that the conversion of cropland to Eucalyptus plantations ameliorates soil degradation in central Ethiopia. Moreover, Assefa et al. [63] reported that the amount of fine root biomass in Eucalyptus plantations was higher than that of cropland and grazing land.

In the lower soil depths (15–30 and 30–50 cm) at Guder, SOC and TN contents were similar to those of the surface layer, probably largely due to plant roots and exudates, dissolved organic matter, bioturbation, and translocation of particulate organic matter [64]. This result is in line with the finding of [62], who reported a similar trend across soil depths. At Dibatie, soil depth generally had no effect on SOC content, however in the lower depths TN content was higher in bushland than in grazing land. This may be because of regular burning of the surface cover in woodland, which is the most common soil fertility problem in the lowlands of northwestern Ethiopia [65], as well as leaching and lower temperature in the subsurface layer [55].

Land use had a significant effect on both SOC and TN stocks across the entire 50-cm soil profile at Guder (Table S1). The topsoil layer of bushland stored significantly greater SOC and TN stocks than that of plantation, which may be largely due to less carbon input from litter biomass, roots, and residues, including understory biomass in plantations [10,12,66,67]. At the lower two depths, however, bushland and grazing land had the highest SOC and TN stocks. Similarly, at Aba Gerima, SOC and TN stocks of cropland were lower than those of the tree- and grass-based systems of bushland, plantation, and grazing land. Many studies have reported that cropland stores the lowest SOC and TN stocks [27,52]. In the 30–50 cm soil layer, bushland also showed higher SOC and TN accumulation than cropland (Table S1). At Dibatie, however, SOC and TN stocks were similar at all soil depths, except for the TN stock in the lower soil depth. This could be due to the practice of burning woodland (as discussed above).

4.3. Effect of Agro-Ecosystem on SOC and TN Contents and Stocks

Agro-ecology had a significant effect on SOC and TN contents and stocks (Table 4). The soil under bushland and grazing land had lower SOC and TN contents at Aba Gerima and Dibatie than at Guder. In different ecosystems, climate strongly affects the soil carbon and nitrogen by controlling vegetation productivity and organic matter decomposition [68]. Guder had higher mean annual precipitation and was cooler than the other two sites (Figure 2). Similar studies also reported that areas with high mean annual precipitation and lower mean annual temperature tend to accumulate large amounts of SOC and TN [5,49,68]. The SOC and TN contents in cropland were lower at Aba Gerima than at Guder and Dibatie. This result clearly indicated that cropland at Aba Gerima had less organic input and poor physical protection, including vegetation cover, which plays a substantial role in organic matter stabilization in cultivated land [69]. In another study of agro-ecosystems of the Upper Blue Nile basin, Ebabu et al. [43] reported greater soil loss for cropland at Aba Gerima than that at Guder and Dibatie. However, plantations had similar SOC and TN contents at Guder and Aba Gerima.

The SOC stock in bushland (141.19 Mg ha^{-1}) and grazing land (109.94 Mg ha^{-1}) was greater at Guder than at Aba Gerima. These values are comparable with previous reports of SOC stocks of 69–239 and 67–109 Mg ha^{-1} in natural vegetation and grazing land to 50 cm depth in the northwest highlands of Ethiopia [5,62], but they are markedly higher than the values reported by [12], who recorded SOC stocks of 52 and 39 Mg ha^{-1} to 50 cm depth in grazing and shrub land of northern Ethiopia, respectively. These values are also lower than the estimated mean of tropical sites (216 Mg ha^{-1}; [35]) and the global average (254 Mg ha^{-1}; [70]). However, cropland had higher SOC and TN stocks at Dibatie than those at Guder and Aba Gerima, which could be related to the different farming system at Dibatie. Unlike at Guder and Aba Gerima, crop residues are not collected in the field at Dibatie, which could be contributing to the SOC accumulation. In addition, the bulk density in cropland at Dibatie was higher than that at the other sites (Table 3). On the other hand, cropland at Dibatie is a new land-use type, having been converted from woodlands (Combretum–Terminalia) recently. In southern Ethiopia, [71] reported that soil under Combretum–Terminalia vegetation stored higher carbon stock than the aboveground biomass.

4.4. Implications of SOC and TN Stocks as Indicators for Sustainable Land Management in the Upper Blue Nile Watershed

At the watershed scale, the effects of topographic position and land use on SOC and TN stocks were not consistent. At Guder, Aba Gerima, and Dibatie, topographic position and land use, land use, and topographic position, respectively, were the dominant factors that affected SOC stock (Table 2). However, TN stock in all agro-ecosystems was affected by topographic position and land use. Thus, by maintaining the same land uses at Guder, both stocks of SOC and TN could be enhanced by topographic position, whereas converting bushland and grazing land to A. decurrens woodlots would likely diminish the SOC and TN stored in the soil. At Aba Gerima, conversion of cropland to Eucalyptus plantation had a positive impact on SOC and TN [5,62]. Plantation had lower SOC and TN stocks due to poor undergrowth and litter removal [72]. The interaction of land use and topographic position showed a significant effect on SOC and TN stocks at Guder and Aba Gerima (Table 2), indicating that the variation in topography and land use may simultaneously affect different soil processes including soil erosion and the accumulation and decomposition of organic matter [10,30].

Across the agro-ecosystems, topographic position and land use were the main factors influencing SOC and TN stocks, but agro-ecosystem also showed a significant interactive effect with topographic position and land use on the SOC and TN stocks (Table 4). Among agro-ecosystems, SOC and TN stocks were higher at Guder, followed by Dibatie and Aba Gerima (Figure 4). In addition to vegetation composition, the hydrological regime, soil formation processes, and climate (temperature and precipitation) are important factors that affect the SOC [68], which in turn influences soil respiration [33]. In this study, Guder has higher mean annual precipitation and lower mean annual temperature (Figure 2a) than Aba Gerima (Figure 2b) and Dibatie (Figure 2c). Agro-ecosystems in cooler and moister climates accumulate high SOC and have a low rate of soil respiration [33] and limited microbial activity [50]. A warm and moist agro-ecosystem such as Dibatie, however, tends to store moderate SOC stocks due to high biomass production (Table 1) and greater soil respiration. Aba Gerima has low SOC and TN stocks, likely as a result of severe soil erosion, prolonged crop cultivation, and poor land management. The C/N ratio varied from 11–13 at Guder to 10–21 at Aba Gerima and 15–18 at Dibatie. The C/N ratio is commonly considered as an indicator of microbial activity and quality of soil organic matter [73]. Similar to the SOC and TN stocks, the C/N ratio also varied among land-use types, agro-ecosystems, and topographic positions.

5. Conclusions

This study clearly demonstrated that SOC and TN stocks varied significantly across land-use types and topographic positions of different agro-ecosystems. Poor and environmentally damaging land management practices tended to reduce SOC and TN in soil. Interactive effects of topographic position and land-use types on SOC and TN stocks were significant at Guder and Aba Gerima. Bushland at Guder accumulated a substantial amount of SOC and TN stocks. Cropland at Aba Gerima had poor SOC and TN stocks. Compared to other land-use types, the soil of *A. decurrens* plantation was the lowest in SOC and TN, due to high biomass removal and improper silvicultural management. However, *E. camaldulensis* plantations at Aba Gerima had a positive impact on SOC and TN stocks. Across agro-ecosystems, Guder and Dibatie accumulated larger SOC and TN stocks than those of Aba Gerima.

Overall, land use was a crucial factor influencing SOC and TN, both within and across the sites. However, the effect of topographic position was more pronounced across watersheds than within them. Aba Gerima showed lower SOC and TN stocks due to prolonged crop cultivation and mismanagement of the landscape. This calls for immediate land management interventions, particularly targeting croplands. Our findings highlight the importance of assessing SOC and TN stocks when designing evidence-based land management options in the Upper Blue Nile basin.

Supplementary Materials: The following are available online at http://www.mdpi.com/2071-1050/12/6/2425/s1. Table S1: SOC and TN stocks at three soil depths under different land-use types at Guder, Aba Gerima, and Dibatie. Table S2: SOC and TN contents and stocks at the three topographic positions of the study sites.

Author Contributions: Conceptualization, G.A. and N.H.; data curation, G.A. and K.E.; validation, A.T. (Atsushi Tsunekawa) and N.H.; methodology, G.A., M.W., T.M., and T.T.; formal analysis, G.A.; investigation, G.A.; resources A.T. (Atsushi Tsunekawa) and N.H.; writing—original draft preparation, G.A.; writing—review and editing, A.T. (Atsushi Tsunekawa), E.A., N.H., T.T., M.W., K.E., and M.L.B.; supervision, A.T. (Atsushi Tsunekawa), N.H., T.T., and M.T.; project administration, A.T. (Atsushi Tsunekawa), E.A., N.H., and A.T. (Asaminew Tassew); funding acquisition, A.T. (Atsushi Tsunekawa), E.A., and N.H. All authors have read and agreed to the published version of the manuscript.

Funding: This research was funded by the Science and Technology Research Partnership for Sustainable Development (grant no. JPMJSA1601), Japan Science and Technology Agency/Japan International Cooperation Agency.

Acknowledgments: We are grateful to Anteneh Wubet and Agerselam Gualie for the facilitation of our field and laboratory work. We also thank the Arid Land Research Center of Tottori University for providing a convenient research environment and facilities throughout our work.

Conflicts of Interest: The authors declare no conflict of interest.

References

1. Chen, L.-F.; He, Z.-B.; Du, J.; Yang, J.-J.; Zhu, X. Patterns and environmental controls of soil organic carbon and total nitrogen in alpine ecosystems of northwestern China. *Catena* **2016**, *137*, 37–43. [CrossRef]

2. Bünemann, E.K.; Bongiorno, G.; Bai, Z.; Creamer, R.E.; De Deyn, G.; de Goede, R.; Fleskens, L.; Geissen, V.; Kuyper, T.W.; Mäder, P.; et al. Soil quality—A critical review. *Soil Bio. Biochem.* **2018**, *120*, 105–125. [CrossRef]

3. Zhang, J.; Wang, X.; Wang, J. Impact of land use change on profile distributions of soil organic carbon fractions in the Yanqi Basin. *Catena* **2014**, *115*, 79–84. [CrossRef]

4. Martin, D.; Lal, T.; Sachdev, C.B.; Sharma, J.P. Soil organic carbon storage changes with climate change, landform and land use conditions in Garhwal hills of the Indian Himalayan mountains. *Agri. Ecosyst. Environ.* **2010**, *138*, 64–73. [CrossRef]

5. Assefa, D.; Rewald, B.; Sandén, H.; Rosinger, C.; Abiyu, A.; Yitaferu, B.; Godbold, D.L. Deforestation and land use strongly effect soil organic carbon and nitrogen stock in Northwest Ethiopia. *Catena* **2017**, *153*, 89–99. [CrossRef]

6. Lal, R. Managing Soils and Ecosystems for Mitigating Anthropogenic Carbon Emissions and Advancing Global Food Security. *Biosci.* **2010**, *60*, 708–721. [CrossRef]

7. Fan, S.; Guan, F.; Xu, X.; Forrester, D.; Ma, W.; Tang, X. Ecosystem Carbon Stock Loss after Land Use Change in Subtropical Forests in China. *Forests* **2016**, *7*, 142. [CrossRef]

8. Poeplau, C.; Don, A.; Vesterdal, L.; Leifeld, J.; Wesemael, B.V.; Schumacher, J.; Gensior, A. Temporal dynamics of soil organic carbon after land-use change in the temperate zone—Carbon response functions as a model approach. *Glob. Change Bio.* **2011**, *17*, 2415–2427. [CrossRef]

9. Lützow, M.V.; Kögel-Knabner, I.; Ekschmitt, K.; Matzner, E.; Guggenberger, G.; Marschner, B.; Flessa, H. Stabilization of organic matter in temperate soils: Mechanisms and their relevance under different soil conditions—A review. *Euro. J. Soil Sci.* **2006**, *57*, 426–445. [CrossRef]

10. Sun, W.; Zhu, H.; Guo, S. Soil organic carbon as a function of land use and topography on the Loess Plateau of China. *Ecol. Eng.* **2015**, *83*, 249–257. [CrossRef]

11. IPCC. *Climate Change: The Physical Science Basis. Contribution of Working Group I to the Fifth Assessment Report of the Intergovernmental Panel on Climate Change*; Cambridge Univ. Press: Cambridge, UK, 2013.

12. Gelaw, A.M.; Singh, B.R.; Lal, R. Soil organic carbon and total nitrogen stocks under different land uses in a semi-arid watershed in Tigray, Northern Ethiopia. *Agri. Ecosyst. Environ.* **2014**, *188*, 256–263. [CrossRef]

13. Fu, X.; Shao, M.; Wei, X.; Horton, R. Soil organic carbon and total nitrogen as affected by vegetation types in Northern Loess Plateau of China. *Geoderma* **2010**, *155*, 31–35. [CrossRef]

14. Gelaw, A.M.; Singh, B.R.; Lal, R. Organic Carbon and Nitrogen Associated with Soil Aggregates and Particle Sizes Under Different Land Uses in Tigray, Northern Ethiopia. *Land Degrad. Dev.* **2015**, *26*, 690–700. [CrossRef]

15. Yimer, F.; Ledin, S.; Abdelkadir, A. Changes in soil organic carbon and total nitrogen contents in three adjacent land use types in the Bale Mountains, south-eastern highlands of Ethiopia. *For. Ecol. Manag.* **2007**, *242*, 337–342. [CrossRef]

16. Meshesha, D.T.; Tsunekawa, A.; Tsubo, M.; Ali, S.A.; Haregeweyn, N. Land-use change and its socio-environmental impact in Eastern Ethiopia's highland. *Reg. Environ. Change* **2014**, *14*, 757–768. [CrossRef]

17. Alem, S.; Pavlis, J. Conversion of grazing land into Grevillea robusta plantation and exclosure: Impacts on soil nutrients and soil organic carbon. *Environ. Mon. Ass.* **2014**, *186*, 4331–4341. [CrossRef]

18. Lemenih, M.; Olsson, M.; Karltun, E. Comparison of soil attributes under Cupressus lusitanica and Eucalyptus saligna established on abandoned farmlands with continuously cropped farmlands and natural forest in Ethiopia. *For. Ecol. Manag.* **2004**, *195*, 57–67. [CrossRef]

19. Guteta, D.; Abegaz, A. Dynamics of selected soil properties under four land uses in Arsamma watershed, Southwestern Ethiopian Highlands. *Phys. Geog.* **2017**, *38*, 83–102. [CrossRef]

20. Chen, H.; Zhang, W.; Wang, K.; Hou, Y. Soil organic carbon and total nitrogen as affected by land use types in karst and non-karst areas of northwest Guangxi, China. *J. Sci. Food Agri.* **2012**, *92*, 1086–1093. [CrossRef]

21. Kassa, H.; Dondeyne, S.; Poesen, J.; Frankl, A.; Nyssen, J. Impact of deforestation on soil fertility, soil carbon and nitrogen stocks: The case of the Gacheb catchment in the White Nile Basin, Ethiopia. *Agri. Ecosyst. Environ.* **2017**, *247*, 273–282. [CrossRef]

22. Teferi, E.; Bewket, W.; Simane, B. Effects of land use and land cover on selected soil quality indicators in the headwater area of the Blue Nile basin of Ethiopia. *Environ. Mon. Ass.* **2016**, *188*, 83. [CrossRef] [PubMed]

23. Feyisa, K.; Beyene, S.; Angassa, A.; Said, M.Y.; de Leeuw, J.; Abebe, A.; Megersa, B. Effects of enclosure management on carbon sequestration, soil properties and vegetation attributes in East African rangelands. *Catena* **2017**, *159*, 9–19. [CrossRef]

24. Mekuria, W.; Langan, S.; Noble, A.; Johnston, R. Soil Restoration after seven Years of Exclosure Management in Northwestern Ethiopia. *Land Degrad. Dev.* **2017**, *28*, 1287–1297. [CrossRef]

25. Mekuria, W.; Wondie, M.; Amare, T.; Wubet, A.; Feyisa, T.; Yitaferu, B. Restoration of degraded landscapes for ecosystem services in North-Western Ethiopia. *Heliyon* **2018**, *4*, e00764. [CrossRef] [PubMed]

26. Solomon, D.; Lehmann, J.; Zech, W. Land use effects on soil organic matter properties of chromic luvisols in semi-arid northern Tanzania: Carbon, nitrogen, lignin and carbohydrates. *Agri. Ecosyst. Environ.* **2000**, *78*, 203–213. [CrossRef]

27. Dessalegn, D.; Beyene, S.; Ram, N.; Walley, F.; Gala, T.S. Effects of topography and land use on soil characteristics along the toposequence of Ele watershed in southern Ethiopia. *Catnea* **2014**, *115*, 47–54. [CrossRef]

28. Dialynas, Y.G.; Bastola, S.; Bras, R.L.; Billings, S.A.; Markewitz, D.; Richter, D.d. Topographic variability and the influence of soil erosion on the carbon cycle. *Glob. Biogeochem. Cycles* **2016**, *30*, 644–660. [CrossRef]

29. Wang, Y.; Fu, B.; Lü, Y.; Chen, L. Effects of vegetation restoration on soil organic carbon sequestration at multiple scales in semi-arid Loess Plateau, China. *Catena* **2011**, *85*, 58–66. [CrossRef]

30. Zhu, H.; Wu, J.; Guo, S.; Huang, D.; Zhu, Q.; Ge, T.; Lei, T. Land use and topographic position control soil organic C and N accumulation in eroded hilly watershed of the Loess Plateau. *Catena* **2014**, *120*, 64–72. [CrossRef]

31. Conant, R.T.; Ryan, M.G.; Ågren, G.I.; Birge, H.E.; Davidson, E.A.; Eliasson, P.E.; Evans, S.E.; Frey, S.D.; Giardina, C.P.; Hopkins, F.M.; et al. Temperature and soil organic matter decomposition rates—Synthesis of current knowledge and a way forward. *Glob. Change Bio.* **2011**, *17*, 3392–3404. [CrossRef]

32. Crowther, T.W.; Thomas, S.M.; Maynard, D.S.; Baldrian, P.; Covey, K.; Frey, S.D.; van Diepen, L.T.; Bradford, M.A. Biotic interactions mediate soil microbial feedbacks to climate change. *Proc. Natl. Acad. Sci. USA* **2015**, *112*, 7033–7038. [CrossRef] [PubMed]

33. Wang, W.; Fang, J. Soil respiration and human effects on global grasslands. *Glob. Planet. Change* **2009**, *67*, 20–28. [CrossRef]

34. Takoutsing, B.; Weber, J.C.; Tchoundjeu, Z.; Shepherd, K. Soil chemical properties dynamics as affected by land use change in the humid forest zone of Cameroon. *Agrofor. Sys.* **2015**, *90*, 1089–1102. [CrossRef]

35. Lal, R. Soil Carbon Sequestration Impacts on Global Climate Change and Food Security. *Science* **2004**, *304*, 1623–1627. [CrossRef] [PubMed]

36. Haregeweyn, N.; Tsunekawa, A.; Poesen, J.; Tsubo, M.; Meshesha, D.T.; Fenta, A.A.; Nyssen, J.; Adgo, E. Comprehensive assessment of soil erosion risk for better land use planning in river basins: Case study of the Upper Blue Nile River. *Sci Total Environ.* **2017**, *574*, 95. [CrossRef]

37. Gebremicael, T.G.; Mohamed, Y.A.; Betrie, G.D.; van der Zaag, P.; Teferi, E. Trend analysis of runoff and sediment fluxes in the Upper Blue Nile basin: A combined analysis of statistical tests, physically-based models and landuse maps. *J. Hydrol.* **2013**, *482*, 57–68. [CrossRef]

38. Sultan, D.; Tsunekawa, A.; Haregeweyn, N.; Adgo, E.; Tsubo, M.; Meshesha, D.T.; Masunaga, T.; Aklog, D.; Ebabu, K. Analyzing the runoff response to soil and water conservation measures in a tropical humid Ethiopian highland. *Phys. Geog.* **2017**, *38*, 423–447. [CrossRef]

39. Sisay, K.; Thurnher, C.; Belay, B.; Lindner, G.; Hasenauer, H. Volume and Carbon Estimates for the Forest Area of the Amhara Region in Northwestern Ethiopia. *Forests* **2017**, *8*, 122. [CrossRef]

40. Berihun, M.L.; Tsunekawa, A.; Haregeweyn, N.; Meshesha, D.T.; Adgo, E.; Tsubo, M.; Masunaga, T.; Fenta, A.A.; Sultan, D.; Yibeltal, M. Exploring land use/land cover changes, drivers and their implications in contrasting agro-ecological environments of Ethiopia. *Land Use Policy* **2019**, *87*, 104052. [CrossRef]

41. Mekonnen, G. *Soil Characterization Classification and Mapping of Three Twin Watersheds in the Upper Blue Nile basin, Ethiopia*; Amhara Design and Supervision Works Enterprise: Bahir Dar, Ethiopia, 2016.

42. Peel, M.C.; Finlayson, B.L.; McMahon, T.A. Updated world map of the Köppen-Geiger climate classification. *Hydrol. Earth Sys. Sci. Discuss.* **2007**, *4*, 439–473. [CrossRef]

43. Ebabu, K.; Tsunekawa, A.; Haregeweyn, N.; Adgo, E.; Meshesha, D.T.; Aklog, D.; Masunaga, T.; Tsubo, M.; Sultan, D.; Fenta, A.A.; et al. Effects of land use and sustainable land management practices on runoff and soil loss in the Upper Blue Nile basin, Ethiopia. *Sci. Total Environ.* **2019**, *648*, 1462–1475. [CrossRef]

44. Nigussie, Z.; Tsunekawa, A.; Haregeweyn, N.; Adgo, E.; Nohmi, M.; Tsubo, M.; Aklog, D.; Meshesha, D.T.; Abele, S. Factors influencing small-scale farmers' adoption of sustainable land management technologies in north-western Ethiopia. *Land Use Policy* **2017**, *67*, 57–64. [CrossRef]

45. Yibeltal, M.; Tsunekawa, A.; Haregeweyn, N.; Adgo, E.; Meshesha, D.T.; Aklog, D.; Masunaga, T.; Tsubo, M.; Billi, P.; Vanmaercke, M.; et al. Analysis of long-term gully dynamics in different agro-ecology settings. *Catena* **2019**, *179*, 160–174. [CrossRef]

46. Poppe, L.; Frankl, A.; Poesen, J.; Admasu, T.; Dessie, M.; Adgo, E.; Deckers, J.; Nyssen, J. Geomorphology of the Lake Tana basin, Ethiopia. *J. Maps* **2013**, *9*, 431–437. [CrossRef]

47. Ellert, B.H.; Bettany, J.R. Calculation of organic matter and nutrients stored in soils under contrasting management regimes. *Can J. Soil Sci.* **1995**, *75*, 529–538. [CrossRef]

48. R Core Team. *R: A Language and Environment for Statistical Computing*; R Foundation for Statistical Computing: Vienna, Austria, 2015; ISBN 3-900051-07-0.

49. Li, D.; Niu, S.; Luo, Y. Global patterns of the dynamics of soil carbon and nitrogen stocks following afforestation: A meta-analysis. *New Phytol.* **2012**, *195*, 172–181. [CrossRef] [PubMed]

50. Wiesmeier, M.; Urbanski, L.; Hobley, E.; Lang, B.; von Lützow, M.; Marin-Spiotta, E.; van Wesemael, B.; Rabot, E.; Ließ, M.; Garcia-Franco, N.; et al. Soil organic carbon storage as a key function of soils—A review of drivers and indicators at various scales. *Geoderma* **2019**, *333*, 149–162. [CrossRef]

51. Chai, H.; Yu, G.R.; He, N.P.; Wen, D.; Li, J.; Fang, J.P. Vertical distribution of soil carbon, nitrogen, and phosphorus in typical Chinese terrestrial ecosystems. *Chi. Geog. Sci.* **2015**, *25*, 549–560. [CrossRef]

52. Deng, L.; Zhu, G.-Y.; Tang, Z.-S.; Shangguan, Z.-P. Global patterns of the effects of land-use changes on soil carbon stocks. *Glob. Ecol. Conserv.* **2016**, *5*, 127–138. [CrossRef]

53. Berihu, T.; Girmay, G.; Sebhatleab, M.; Berhane, E.; Zenebe, A.; Sigua, G.C. Soil carbon and nitrogen losses following deforestation in Ethiopia. *Agron. Sustain. Dev.* **2016**, 37. [CrossRef]

54. Ma, W.; Li, Z.; Ding, K.; Huang, B.; Nie, X.; Lu, Y.; Xiao, H. Soil erosion, organic carbon and nitrogen dynamics in planted forests: A case study in a hilly catchment of Hunan Province, China. *Soil Tillage Res.* **2016**, *155*, 69–77. [CrossRef]

55. Knicker, H. How does fire affect the nature and stability of soil organic nitrogen and carbon? A review. *Biogeochem* **2007**, *85*, 91–118. [CrossRef]

56. van Breugel, P.; Friis, I.; Demissew, S.; Lillesø, J.-P.B.; Kindt, R. Current and Future Fire Regimes and Their Influence on Natural Vegetation in Ethiopia. *Ecosystems* **2015**, *19*, 369–386. [CrossRef]

57. Don, A.; Schumacher, J.; Freibauer, A. Impact of tropical land-use change on soil organic carbon stocks—A meta-analysis. *Glob. Change Bio.* **2011**, *17*, 1658–1670. [CrossRef]

58. Tesfaye, M.A.; Bravo-Oviedo, A.; Bravo, F.; Kidane, B.; Bekele, K.; Sertse, D. Selection of Tree Species and Soil Management for Simultaneous Fuelwood Production and Soil Rehabilitation in the Ethiopian Central Highlands. *Land Degrad. Dev.* **2015**, *26*, 665–679. [CrossRef]
59. Nigussie, Z.; Tsunekawa, A.; Haregeweyn, N.; Adgo, E.; Nohmi, M.; Tsubo, M.; Aklog, D.; Meshesha, D.T.; Abele, S. Factors Affecting Small-Scale Farmers' Land Allocation and Tree Density Decisions in an Acacia decurrens-Based taungya System in Fagita Lekoma District, North-Western Ethiopia. *Small-Scale For.* **2016**, *16*, 219–233. [CrossRef]
60. Wondie, M.; Mekuria, W. Planting of Acacia decurrens and Dynamics of Land Cover Change in Fagita Lekoma District in the Northwestern Highlands of Ethiopia. *Mt. Res. Dev.* **2018**, *38*, 230–239. [CrossRef]
61. Sultan, D.; Tsunekawa, A.; Haregeweyn, N.; Adgo, E.; Tsubo, M.; Meshesha, D.T.; Masunaga, T.; Aklog, D.; Fenta, A.A.; Ebabu, K. Impact of Soil and Water Conservation Interventions on Watershed Runoff Response in a Tropical Humid Highland of Ethiopia. *Environ. Manag.* **2018**, *61*, 860–874. [CrossRef]
62. Tesfaye, M.A.; Bravo, F.; Ruiz-Peinado, R.; Pando, V.; Bravo-Oviedo, A. Impact of changes in land use, species and elevation on soil organic carbon and total nitrogen in Ethiopian Central Highlands. *Geoderma* **2016**, *261*, 70–79. [CrossRef]
63. Assefa, D.; Rewald, B.; Sandén, H.; Godbold, D. Fine Root Dynamics in Afromontane Forest and Adjacent Land Uses in the Northwest Ethiopian Highlands. *Forests* **2017**, *8*, 249. [CrossRef]
64. Twongyirwe, R.; Sheil, D.; Majaliwa, J.G.M.; Ebanyat, P.; Tenywa, M.M.; van Heist, M.; Kumar, L. Variability of Soil Organic Carbon stocks under different land uses: A study in an afro-montane landscape in southwestern Uganda. *Geoderma* **2013**, *193–194*, 282–289. [CrossRef]
65. Lemenih, M.; Feleke, S.; Tadesse, W. Constraints to smallholders production of frankincense in Metema district, North-western Ethiopia. *J. Arid Environ.* **2007**, *71*, 393–403. [CrossRef]
66. Stockmann, U.; Adams, M.A.; Crawford, J.W.; Field, D.J.; Henakaarchchi, N.; Jenkins, M.; Minasny, B.; McBratney, A.B.; Courcelles, V.d.R.d.; Singh, K.; et al. The knowns, known unknowns and unknowns of sequestration of soil organic carbon. *Agric. Ecosyst. Environ.* **2013**, *164*, 80–99. [CrossRef]
67. Wang, T.; Kang, F.; Cheng, X.; Han, H.; Ji, W. Soil organic carbon and total nitrogen stocks under different land uses in a hilly ecological restoration area of North China. *Soil Tillage Res.* **2016**, *163*, 176–184. [CrossRef]
68. Deng, L.; Liu, G.B.; Shangguan, Z.P. Land-use conversion and changing soil carbon stocks in China's 'Grain-for-Green' Program: A synthesis. *Glob Change Bio.* **2014**, *20*, 3544–3556. [CrossRef]
69. Liu, S.; Dong, Y.; Cheng, F.; Yin, Y.; Zhang, Y. Variation of soil organic carbon and land use in a dry valley in Sichuan province, Southwestern China. *Ecol. Eng.* **2016**, *95*, 501–504. [CrossRef]
70. Batjes, N.H. Total carbon and nitrogen in the soils of the world. *Euro. J. Soil Sci.* **1996**, *47*, 151–163. [CrossRef]
71. Tesfaye, M.; Negash, M. Combretum—Terminalia vegetation accumulates more carbon stocks in the soil than the biomass along the elevation ranges of dryland ecosystem in Southern Ethiopia. *J. Arid Environ.* **2018**, *155*, 59–64. [CrossRef]
72. Temesgen, D.; Gonzálo, J.; Turrión, M.B. Effects of short-rotation Eucalyptus plantations on soil quality attributes in highly acidic soils of the central highlands of Ethiopia. *Soil Use Manag.* **2016**, *32*, 210–219. [CrossRef]
73. Ostrowska, A.; Porębska, G. Assessment of the C/N ratio as an indicator of the decomposability of organic matter in forest soils. *Ecol. Ind.* **2015**, *49*, 104–109. [CrossRef]

Article

Effects of Landscape Changes on Soil Erosion in the Built Environment: Application of Geospatial-Based RUSLE Technique

Bilal Aslam [1], Ahsen Maqsoom [2], Shahzaib [2], Zaheer Abbas Kazmi [3,*], Mahmoud Sodangi [3], Fahad Anwar [3], Muhammad Hassan Bakri [3], Rana Faisal Tufail [2] and Danish Farooq [4]

[1] Department of Earth Sciences, Quaid-i-Azam University, Islamabad 45320, Pakistan; bilalaslam45@gmail.com

[2] Department of Civil Engineering, COMSATS University Islamabad, Wah Campus 47040, Pakistan; ahsen.maqsoom@ciitwah.edu.pk (A.M.); FA16-CVE-077@cuiwah.edu.pk (S.); faisal.tufail@ciitwah.edu.pk (R.F.T.);

[3] Department of Civil & Construction Engineering, Imam Abdulrahman Bin Faisal University, Dammam 31441, Saudi Arabia; misodangi@iau.edu.sa (M.S.); faafarooqi@iau.edu.sa (F.A.); mhbakri@iau.edu.sa (M.H.B.)

[4] Department of Transport Technology and Economics, Budapest University of Technology and Economics, Stoczek u. 2, H-1111 Budapest, Hungary; farooq.danish@mail.bme.hu

* Correspondence: zakazmi@iau.edu.sa

Received: 24 June 2020; Accepted: 20 July 2020; Published: 22 July 2020

Abstract: The world's ecosystem is severely affected by the increase in the rate of soil erosion and sediment transport in the built environment and agricultural lands. Land use land cover changes (LULCC) are considered as the most significant cause of sediment transport. This study aims to estimate the effect of LULCC on soil erosion potential in the past 20 years (2000–2020) by using Revised Universal Soil Loss Equation (RUSLE) model based on Geographic Information System (GIS). Different factors were analyzed to study the effect of each factor including R factor, K factor, LS factor, and land cover factor on the erosion process. Maps generated in the study show the changes in the severity of soil loss in the Chitral district of Pakistan. It was found out that 4% of the area was under very high erosion risk in the year 2000 which increased to 8% in the year 2020. An increase in agricultural land (4%) was observed in the last 20 years which shows that human activities largely affected the study area. The outcomes of this study will help the stakeholders and regulatory decision makers to control deforestation and take other necessary actions to minimize the rate of soil erosion. Such an efficient planning will also be helpful to reduce the sedimentation in the reservoir of hydraulic dam(s) constructed on Chitral river, which drains through this watershed.

Keywords: sediment transport; soil erosion; RUSLE (Revised Universal Soil Loss Equation model); human activities

---

1. Introduction

World land resources are declining day by day due to soil erosion, so much that it has become the main focus of researchers and engineers [1]. Different studies have been carried out for the sustainability of natural habitat [2–5]. It is a loss of soil rich with nutrients that affects the productivity and sustainability of the original soil [6]. Approximately 80% of agricultural areas are facing higher rates of soil loss, and transported sediments severely affect the natural and built environment. Erosion of soil depletes the storage capacity of reservoirs and dams which ultimately decreases the power generation capacity. It also disturbs the agricultural output and aquatic life by polluting water in the rivers [7].

According to Chuenchum et al. [7], about 2.5 to 4 billion tons of soil is annually eroded worldwide. There are many influencing factors of soil erosion including slope, elevation, rainfall, plane curvature, drainage density, lithology, and lineaments. However, land use and climatic changes are the two most significant factors which affect the sediment transport to river tributaries. It is predicted that human actions will disturb the climate and land use land cover (LULC) significantly. Therefore, it is very important to evaluate the effect of these changes on soil erosion potential [8]. According to the past research, it has been observed that landscape characteristics are responsible for about 65% to 74% of changes in sediment yield and soil erosion. Changes in streamflow discharge due to land use also increase the intensity of soil erosion. Usually, it is observed that areas with more grassland are less vulnerable to soil erosion, whereas arable lands are more susceptible to soil loss [9].

Soil degradation is considered a serious issue due to human actions. Human beings have largely contributed to each influencing factor of land use [10]. It may be due to natural factors including different geomorphological and climatic conditions [11]. Variations in natural vegetation are observed as the first effect of this impact, irrespective of human contribution in the natural environment [12]. When LULC is changed to agricultural land from natural vegetation especially in mountainous regions, it fallouts with an increase in soil erosion rate due to crop production [13]. Land-use changes usually occur during agricultural development, which lead to the hydrological responses that further increases the rate of soil loss. Increased deforestation and growing agricultural and urban land in the tropical areas influence the intensity of soil erosion. It also plays an important role in the performance of hydraulic structures. An increased soil erosion and sedimentation fills the reservoirs of hydraulic dams, which ultimately decreases the power generation capacity.

Land use land cover change (LULCC) is the major influencing aspect that contributes to the increasing rate of soil erosion. It harms the environment by disturbing the supply of water, the capacity of storage basin, agricultural yield, and availability of freshwater in the area [8]. Land cover such as plantation directly affects soil erosion [14,15]. A generalized analysis of land use may be difficult if there is a lack of consistent time-series data for land use or land cover. Recently, scholars are interested in highlighting the effect of the environment on the erosion process. Their sole focus is on the effects of erosion including soil production and reducing the production of crops and affecting water quality by carrying nutrients, heavy metal impurities, and pesticides to surface water bodies. Sediment transport is also responsible for affecting channel, floodplain morphology, and sedimentation of reservoirs [16]. The above-mentioned impacts on soil erosion due to LULC changes can be minimized by estimating the loss of soil. For the estimation of average yearly soil loss, the RUSLE model is most commonly used by researchers, engineers, and planners. Due to its simplicity, ease of use and integration of different factors affecting soil erosion, RUSLE model is preferred over many other methods used for the estimation of soil loss [17,18]. It consists of six parameters including slope length factor (LS), erosion control practices factor (P), soil erodibility factor (K), cover management factor (C), and rainfall erosivity factor (R), which make it possible to calculate average yearly soil loss. The value of each factor is adapted for statistical and empirical data with the integration of GIS or is taken from the previous literature. The soil erosion can also be calculated by the integration of ArcGIS, MATLAB and SolidWorks software [19]. RUSLE in conjunction with GIS has become an effective tool for assessing soil erosion. Many researchers used the integration of RUSLE model with GIS in their study to estimate the rate of soil erosion [20–22]. Soil and water conservation practices (SWCPs) have a positive impact on soil properties. A relevant study has been conducted in Ethiopia [23].

A study was conducted by Sharma et al. [8] on the effect of LULC on erosion process in the reservoir from 1989–2004. This study illustrated that a slight increase in mean soil loss was observed. In 1989, it was 12.11 t/ha/y and in 2004 it became 13.2 t/ha/y. Results indicated that deforestation and increased wasteland in higher slopes increased the mean soil loss rate in 15 years. Another research in the Western Polish Carpathians shows that due to the increase in plantation and a decrease in the rate of cultivation, the rate of eroded soil decreased in the last 160 years (1846–2009). The rate of soil erosion in the year of 1846 was 18.13 tons/ha/y and it decreased to 4 tons/ha/y in the year 2009 [14].

Ouyang et al. [9] demonstrated the behavior of land-use changes, which shows that it has a more severe effect on erosion rate than changes in the properties of soil. For the estimation of soil loss in the Kelantan River basin, Abdulkareem et al. [24] used the USLE (universal soil loss equation) model based on GIS. Arable areas have a greater susceptibility to erosion process; research [25] illustrated that arable lands are approximately ten times more vulnerable to soil loss than orchards. In recent years, many studies have been conducted to analyze the impact of LULCC on soil erosion and sediment transport rate [14,15,20,26,27].

Soil loss is estimated by the nature of erosion process and data availability of the factors that contribute towards erosion. In the past, soil erosion estimation was done by using different methods ranging from factor-based approaches to process-based models. Now, RUSLE model is the most widely used considering its simplicity, and its input parameters are easily available. Investigation of impact of LULCC on soil erosion can be done by analyzing the effect of land use land cover changes (LULCC) on soil erosion potential and discharge of sediments through historic satellite images [8]. In this study, the main purpose is to figure out the potential of soil loss in the mountainous region, which changes due to an increase in agriculture production and decrease in the forest cover. For this purpose, a GIS-based RUSLE model is used to estimate the long-term effect of LULC on soil erosion. The results of the study are demonstrated in the form of temporal severity maps, highlighting the areas with high land use changes and their impact on soil erosion intensity. The quantitative estimation of soil loss due to LULCC is important for researchers and planners to take necessary actions to minimize the rate of soil erosion. Such an efficient planning will also be helpful to reduce the sedimentation in the reservoir of hydraulic dam(s) constructed on Chitral river, which drains through this watershed.

2. Methodology

2.1. Study Area

The northernmost part of Pakistan is the Chitral district which is situated in the world's largest mountains (Figure 1). Its coordinates are 35°53′15″ N and 71°48′01″ E. Its neighbors include Afghanistan, China, Central Asian States, and Northern Areas of Gilgit. Chitral district is 322 km away from the provincial capital of Khyber Pakhtunkhwa. The DEM (digital elevation model) (Figure 2) clearly shows that the study area has rugged mountainous terrain with variation in altitude from 1053 m to 7695 m. In the north-east, it is bounded by the Karakoram, in the south by Hindu Raj Range and north-west by Hindu Kush mountains. Moreover, there are 40 peaks within an area of 14,850 km and altitude ranges from 1094 m at Arandu to 7726 m at Tirchmir. The Chitral district is composed of several valleys. The Chitral Mastuj valley is the largest and most important valley, expanding from the Afghan border to Arandu on the southern tip. Other important valleys include Terich, Shishi, Owir, Mulkhow, Lotkoh, Laspur, Torkhow, and Ashrat. The main Chitral valley is 354 km long and its width varies from 180 m to 4800 m. The whole study area is drained into Chitral river through several tributary channels. From the origin of the glacier Chianter, the river enters Afghanistan at Arandu. The study area is a mountainous tract that receives about 10–25 mm rain per month. The minimum and maximum temperature remain between 21 °C and 37 °C, respectively. Since high mountains prevent much of the monsoonal wind and moisture from reaching the study area, the summer season usually receives less amount of precipitation. Therefore, the region is not considered favorable for vegetation growth. On 16 July 1973, a maximum discharge of 1586 m³/s was recorded while March 10, 1964 witnessed a minimum value of 46 m³/s. The river flows undeviatingly the whole year because of the melting of glaciers and snow. During the season of monsoon, i.e., July–September, some extra runoff can be recorded. The river siphons a wide area of a steep slope and widespread area covered with snow, which contributes half of the discharge to the Kabul River. For irrigation and hydropower generation, a dam was constructed on the Chitral River having a maximum output of 250,000 kW. The location of the dam is about 30 km west of Peshawar. Due to the extensive silting, the reservoir is almost full. Throughout summers and winters, the power generation capacity of the dam has dropped to 64,000–20,000 kW, correspondingly.

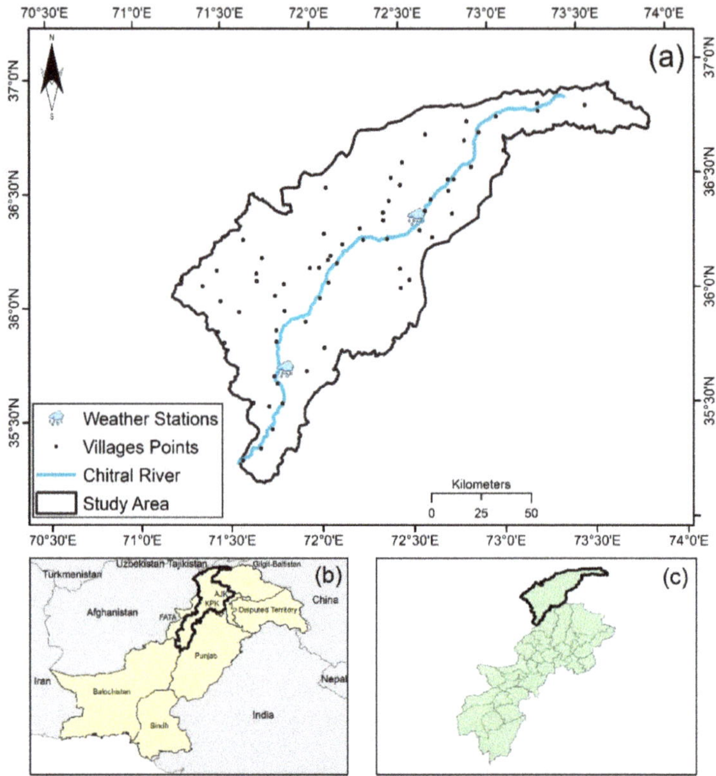

Figure 1. (a) Study area (Chitral District) with locations of weather stations and villages; (b) KPK Province of Pakistan in which the study area lies; (c) Location of study area in KPK Province.

2.2. LULC

For the assessment of LULC, Landsat satellite data were obtained. Landsat is the largest program run by NASA/USGS for earth's imagery. Landsat 7 was used to develop the land cover maps of 2000 and 2010 while Landsat 8 was used for 2020. All three land covers were developed from satellite imageries having zero cloud cover, taken during the period of least snow cover in the area. Along with this data, practical samples were also collected from the field (study area). Based on Landsat satellite data and samples, supervised classification for the years 2000, 2010, and 2020 is made. Accuracy assessment is based on the following two parameters.

2.3. Precipitation

Precipitation data is required for the calculation of rain erosivity factor (R). This factor justifies the extent and intensity of every single rainfall throughout a year. This study utilized Satellite Rainfall Data of GPM (Global Precipitation Measurement). This satellite data is the product of NASA (National Aeronautics and Space Administration). It is widely used because of its enhanced accuracy. The annual rainfall data for the two meteorological stations situated in the study area (mentioned in Figure 1) was obtained from the Pakistan Meteorological Department. It was integrated with the satellite data to validate the results.

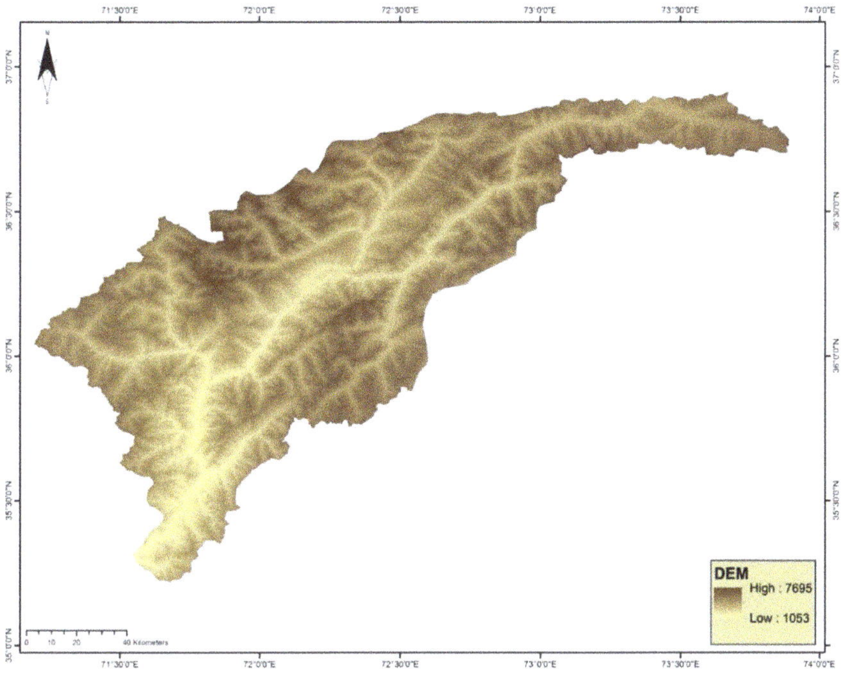

Figure 2. Digital Elevation Model (DEM) of the study area.

2.4. RUSLE Model

The soil loss due to water is most commonly estimated by using the RUSLE model because it is simple, with easy availability of input data, and wide-ranging of its applicability. In this study, RUSLE model is used in combination with GIS to calculate the LULCC. Modeling of soil erosion is an important part of current techniques used in the study of geomorphology along with other practices like photointerpretation, rainfall simulation, and GIS [28,29]. This is described by the following equation,

$$E = R.K.LS.C.P \tag{1}$$

where E is the amount of average annual soil loss (tons/ha/y), R is rainfall erosivity factor (MJ mm/ha/h/y), K is soil erodibility factor (tons/ha)/(MJ mm/ha/h), LS is a topographic factor (dimensionless), C is vegetation cover and management factor (dimensionless), and P is soil and water conservation factor (dimensionless).

Rainfall erosivity factor is the most important constraint of the erosion process [30]. Analytical calculation of the R factor is impossible because detailed long-term precipitation data is not available. So, R factor is estimated with the help of other methods by using rainfall data available for it [31]. Modified Fournier Index (MFI), the most commonly used empirical method, was used in this research

$$MFI = \sum_{j=1}^{12} \frac{p_i^2}{P} \tag{2}$$

where, p_i is monthly i-month precipitation (mm) and P is yearly precipitation (mm).

Modified Fournier index and an R factor of RUSLE model are strongly linearly correlated [20]. Licznar [32] determined the correlation between the R factor of RUSLE and MFI for Poland and the power-law equation is defined as

$$R = 0.2265.MFI^{1.2876} \tag{3}$$

This equation has been verified, validated [8] and applied in many studies targeting different parts of Pakistan [4,33,34]. Therefore, it can be assumed that this equation will provide accurate results for the area selected in this study.

There is a minor change in monthly and annual rainfall of two different periods, so they have the same R factor. The R factors of both from equation 3 and erosivity index are compared. Latocha et al. [28] in the Polish Mountains and Sudety and Drzewiecki et al. [35] in Polish Carpathians calculated the quantity of soil loss by using RUSLE model. The factor K denotes the erodibility of soil associated with vulnerability to runoff rate and soil loss. The following equation is used to find K factor,

$$K = 2.77.10^{-6}.M^{1.14}.(12 - OM) + 0.043.(S - 2) + 0.033(P - 3) \tag{4}$$

where M is particle size parameter (0.002–0.1 mm and 0.002–2.0 mm particle size), OM is organic matter percentage content, S is class of soil structure and P is class of soil permeability. Validation of Equation (4) was not possible due to the unavailability of data. Therefore, soil texture map of the area was used and the values were assigned according to [8,33,36].

LS is the long slope which shows the surface topography. In this study, the equation proposed by Mitasova et al. [37] is used to find out the LS factor. Mancino et al. [38,39] have already used this equation in their research. DEM is used for calculation of flow accumulation and steepness of the slope.

$$LS = (M + 1)(As|22.13)^{m}(sin\beta|0.0896)^{n} \tag{5}$$

where As is an area of un-slope drainage per width of contour, β is slope gradient (radians), m (0.4–0.6), and n (1.0–1.4) are exponential factors and they depend on the type of dominant erosion. ArcMap GIS software 10.2 is used to calculate the LS factor.

$$LS = POWER([flow\ accumulation].cell\ size|22.13, 0.6).POWER(sin([slope].0.01745)|0.0896, 1.3) \tag{6}$$

C represents the vegetation cover and management factor and it is used for the determination of the effect of management actions on soil erosion intensity. The NDVI indices have been used for the extraction of vegetative area, NDWI indices for water body extraction and NDBI for the built-up area. Accuracy of maps was validated from grouped samples. The erosion control practice factor (P) is generally calculated for areas with erosion control practices. Since no such practices were in place in the study area, a constant raster value is used for soil and water conservation factor P [31,40]. This value ranges from 0 to 1.

3. Results

3.1. Dynamics of Land Use Land Cover

General classification accuracy of 79%, 81% and 84% for the LULC maps of 2000, 2010 and 2020 respectively were obtained by using different image processing techniques in collaboration with hybrid classification techniques (Table 1). The kappa coefficient of accuracy also improved throughout these years. The results of the year 2020 are more accurate as compared to the previous years. The classes of agriculture in the valley and water bodies have the highest accuracy.

The study area is classified into eight categories, that is, agriculture in sloping valley, agriculture in the valley, bare areas, natural herbaceous shrubs, natural high shrubs, natural trees, snow and ice, and water bodies. The study area is 1,449,120.7 ha and LULC was estimated for years of 2000, 2010, and 2020. Figure 3 clearly shows that the studied watershed is largely covered by natural herbaceous shrubs. The highly thick natural herbaceous shrubs and snow and ice are the two most plentiful land-use types, while a small percentage of the entire watershed is covered by bare areas, natural high shrubs, and natural trees. A very small area was occupied by agriculture in sloping valley, agriculture in valley and water bodies. Satellite examination helped in deriving the thematic maps of the last two decades (20 years). The composition or total extent of the individual LULC category/class is listed in Table 2.

Table 1. Accuracy of land use classes.

Land Use Classes	2020		2010		2000	
	Class Accuracy (%)	Overall Accuracy (%) and Kappa Coefficient	Class Accuracy (%)	Overall Accuracy (%) and Kappa Coefficient	Class Accuracy (%)	Overall Accuracy (%) and Kappa Coefficient
Agriculture in Valley	88		86		85	
Agriculture in Sloping Valley	78		76		76	
Bare Areas	89	84% and 0.82	87	81% and 0.8	84	79% and 0.77
Natural Herbaceous Shrubs	82		80		79	
Natural High Shrubs	80		78		79	
Snow and Ice	82		77		73	
Natural Trees	79		78		77	
Water Bodies	91		87		84	

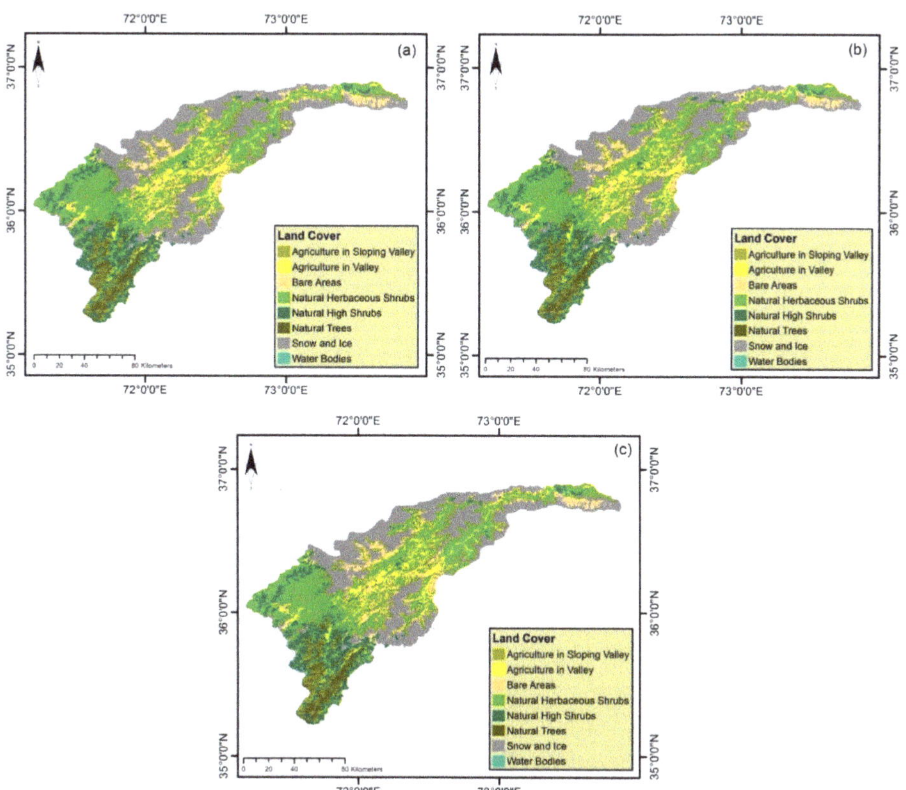

Figure 3. LULC (land use land cover) map of study area (**a**) 2020, (**b**) 2010, (**c**) 2000.

Table 2. Land use land cover changes from 2000 to 2020.

Land Use Classes	2020		2010		2000	
	Area (Ha)	Area (%)	Area (Ha)	Area (%)	Area (Ha)	Area (%)
Agriculture in Valley	39,234.5	3%	28,982.4	2%	14,491.2	1%
Agriculture in Sloping Valley	51,107.9	4%	14,491.2	1%	14,491.2	1%
Bare Areas	161,053.9	11%	130,420.9	9%	115,929.7	8%
Natural Herbaceous Shrubs	515,637.5	36%	550,665.9	38%	565,157.2	39%
Natural High Shrubs	135,184.6	9%	86,947.2	6%	72,456.1	5%
Snow and Ice	465,076.1	32%	507,192.2	35%	521,683.6	36%
Natural Trees	80,471.7	5%	101,438.4	7%	115,929.7	8%
Water Bodies	1354.4	1%	28,982.4	2%	28,982.4	2%
Total	1,449,120.7	100%	1,449,120.7	100%	1,449,121	100%

It is obvious from Table 2 and Figure 4 that natural herbaceous shrubs and snow and ice remained the two most dominant land use classes throughout the study period. Together, these two classes occupy 75% of the area. Table 3 shows the changes in soil erosion of every individual class for all the studied years.

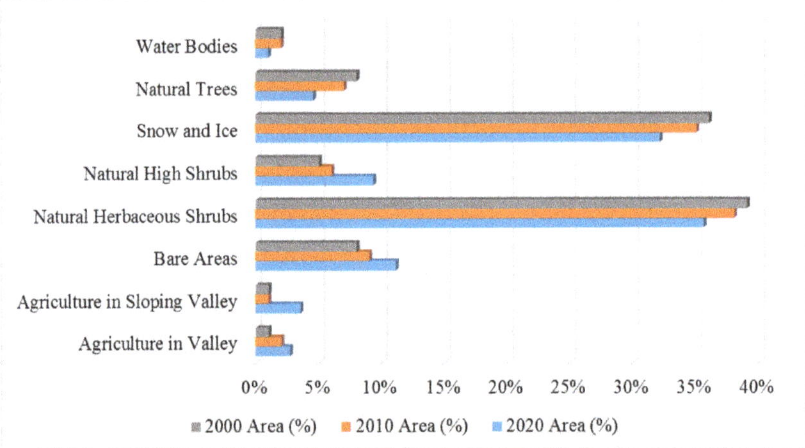

Figure 4. The trend of LULC from 2000 to 2020.

Table 3. Statistical analysis of land use classes.

Land Use Classes	Soil Erosion Year	Statistics (All Units in tons/ha/year)			
		Mean	Min	Max	STD
Agriculture in Valley	2020	144.09	5.97	1578.242	203.87
	2010	122.66	4.21	1464.702	173.43
	2000	111.14	3.57	1370.052	143.54
Agriculture in Sloping Valley	2020	198.9	6.45	5681.93	562.02
	2010	155.25	4.91	5242.39	513.59
	2000	123.57	4.48	4865.9	468.21
Bare Areas	2020	177.04	5.03	5467.25	461.49
	2010	164.5	4.57	4963.82	414.54
	2000	152.15	4.12	4531.28	398.43

Table 3. *Cont.*

Land Use Classes	Soil Erosion Year	Statistics (All Units in tons/ha/year)			
		Mean	Min	Max	STD
Natural Herbaceous Shrubs	2020	593.53	5.44	158,647.45	7067.94
	2010	605.96	5.87	159,302.21	6746.44
	2000	611.4	6.12	159,623.75	6783.98
Natural High Shrubs	2020	116.32	5.64	8421.65	403.38
	2010	104.89	4.32	7988.97	367.45
	2000	103.57	4.22	7701.54	313.54
Snow and Ice	2020	1276.08	5.4	335,719.74	15,947.62
	2010	1152.43	4.99	334,365.09	15,562.19
	2000	1033.03	4.53	333,126.55	14,907.43
Natural Trees	2020	124.84	6.173	4767.86	286.48
	2010	115.44	5.54	4511.21	234.65
	2000	112.01	5.12	4186.56	212.47
Water Bodies	2020	70.23	5.64	377.412	85.31
	2010	63.8	4.65	361.982	73.54
	2000	59.17	4.13	340.552	63.56

The area occupied by natural herbaceous shrubs is reduced to 515,637.5 ha (36%) in 2020 as compared to 550,665.9 ha (38%) in 2010 and 565,157.2 ha (39%) in 2000. This reduction is approximately 8.76% in comparison to the original area in 2000. The snow and ice is the second major LULC class during the period of study. It also shows decrement. The area covered by snow and ice is reduced to 465,076.1 ha (32%) in 2020 as compared to 507,192.2 ha (35%) in 2010 and 521,683.6 ha (36%) in 2000.

There is a gain in the bare area during the study period. Content of minerals and organic matter in soil is suitable for soil quality because it affects the soil functioning positively [8]. The bare areas are mostly rich in organic matter and mineral content as compared to the agricultural land. This gain in the area is 28% in 2020 as compared to the total bare area in 2000. Natural high shrubs increased to 135,184.5 ha (9%) as compared to 86,947.2 ha (6%) in 2010 and 72,456.1 ha (5%) in 2000. It shows an increment of 33.33% as compared to the original area in 2000. Water bodies of this region are reduced in these last 20 years. The water bodies decreased to 1354.4 ha (1%) in 2020 as compared to 28,982.4 ha (2%) in 2010 and 2000. Natural trees of the study area also showed a decrement of 30.5% as compared to the original area in 2000. These natural trees decreased to 80,471.7 ha (5%) in 2020 as compared to 101,438.4 ha (7%) in 2010 and 115,929.7 ha (8%) in 2000.

In contrast to all these land use classes, agriculture shows an increment in both valley and sloping valley. Agriculture in the valley increased to 39,234.5 ha (3%) in 2020 as compared to 29,892.4 ha (2%) in 2010 and 14,491.2 ha (1%) in 2000. The increase in agriculture in the valley is gradual. The increase in agriculture in the sloping valley is not gradual. In the first decade, there was no change in agriculture in the sloping valley, but it rapidly increased in the second decade. It increased to 51,107.9 ha (4%) in 2020 as compared to 1491.2 ha (1%) in 2010 and 2000. Table 4 and Figure 5 show the land use class-wise soil erosion of every individual class in detail. It provides the ease to see and understand the changes throughout the study period because of its sub-divisions.

3.2. Estimation of Sediment Yield and Soil Erosion

The mean soil erosion of the entire watershed was 9.21 tons/ha/y in 2000, 12.43 tons/ha/y in 2010, and 15.63 tons/ha/y in 2020. Figures 6 and 7 show the spatial variation and distribution of different RUSLE factors and also the resultant map of soil erosion for all three years. The Warsak Dam which is the reservoir part of the watershed does not contribute to the net soil erosion, but it behaves as a sink for eroded particles of soil which was excluded from further analysis of soil erosion. The range of R factor was found to be 349.769 MJ mm ha^{-1} h^{-1} year^{-1} as highest and 197.347 MJ mm ha^{-1} h^{-1}

year^{-1} as the lowest value, and variation in its spatial view is shown in the Figure 6a. The range of K factor value lies between 0.1 and 0.4. The spatial distribution of the K factor is shown in Figure 6b. The calculated values of the LS factor using SRTM ranges from a minimum value of 0.2 for flat terrain to a maximum value of 6.3 for high-elevation areas, predominantly for hill slope region. The spatial distribution of the LS factor is shown in Figure 6c.

Table 4. Land use class-wise soil erosion severity.

Land Use Classes	Soil Erosion Year	Soil Loss Class (%)				
		Very Low	Low	Moderate	High	Very High
Agriculture in Valley	2020	41	16	17	19	7
	2010	38	23	29	6	4
	2000	32	26	31	7	4
Agriculture in Sloping Valley	2020	51	21	17	9	2
	2010	47	20	20	11	2
	2000	41	19	24	13	3
Bare Areas	2020	7	18	26	35	14
	2010	18	22	24	27	9
	2000	21	23	26	23	7
Natural Herbaceous Shrubs	2020	39	18	21	11	11
	2010	41	23	30	4	2
	2000	40	31	20	5	4
Natural High Shrubs	2020	50	22	17	9	2
	2010	53	20	15	10	2
	2000	51	19	17	10	3
Snow and Ice	2020	61	26	7	5	1
	2010	57	21	12	6	4
	2000	60	25	8	5	2
Natural Trees	2020	39	18	21	11	11
	2010	41	23	30	4	2
	2000	40	31	20	5	4
Water Bodies	2020	61	26	7	5	1
	2010	57	31	7	4	1
	2000	60	25	8	5	2

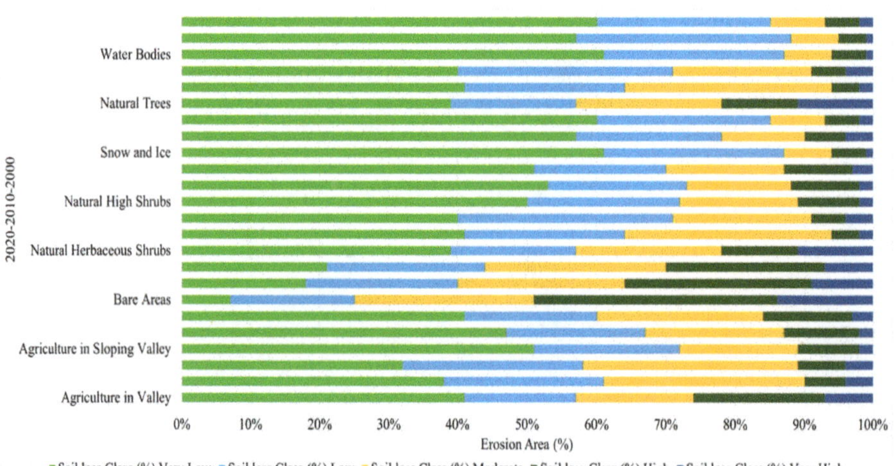

Figure 5. Land use class-wise severity classes with a 10-year gap according to the area affected.

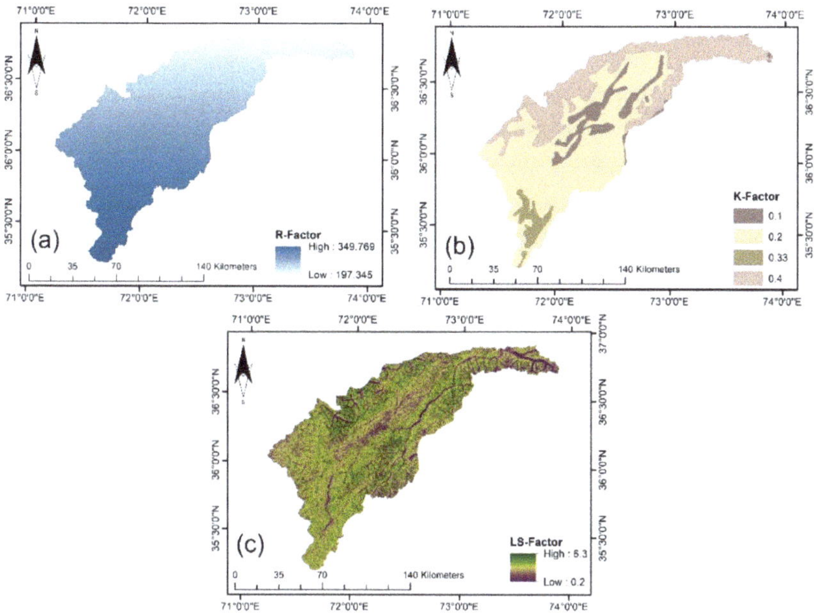

Figure 6. The RUSLE (Revised Universal Soil Loss Equation) factors for the study area (**a**) R factor, (**b**) K factor, (**c**) LS factor.

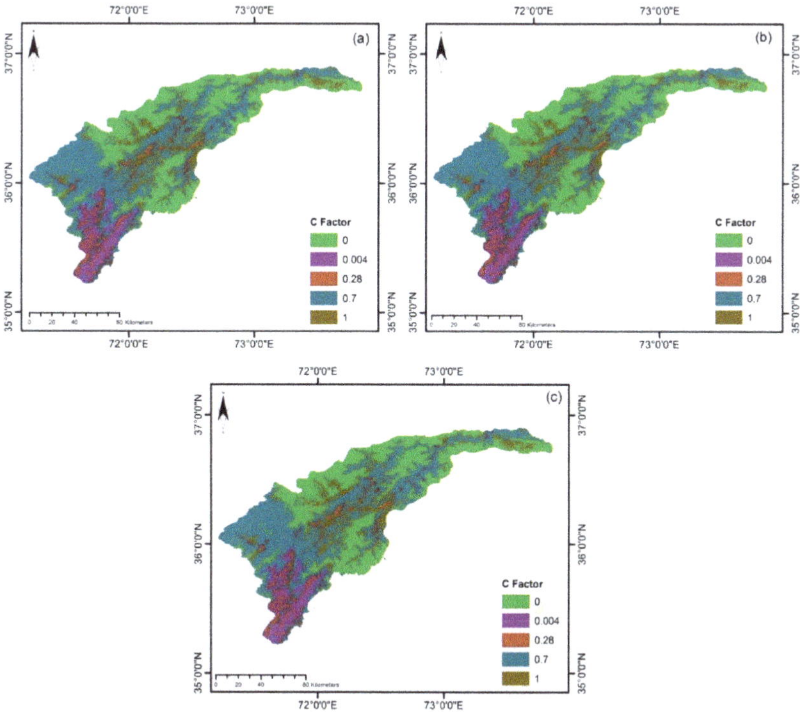

Figure 7. C factor map of the study area (**a**) 2020, (**b**) 2010, (**c**) 2000.

The spatial distribution of p factor is not generated because there are no precautionary or conservation measures being taken against the soil erosion in this region. The values of four parameters of RUSLE out of five remain constant but the value of management (C factor) shows variation both spatially and temporally. The spatial and temporal distributions of the C factor are presented in Figure 7a–c. The values of the C factor range from 0 to 1 for all three years of study, but changes in LULC make their spatial distributions different. Sediment yield of the watershed was obtained by putting all the five parameters into the RUSLE, and the obtained results are illustrated in Figure 8a–c. The figures assist us to extract the information that the central part of the study area is predominantly affected by the sediment yield. The percentage of the affected area kept changing throughout 20 years. In the year 2000, it was 20%, 50% in 2010, and then changed to 80% in 2020 (Table 5). Due to anthropogenic activities, all those areas which fall in the high class in 2000 and 2010 have been transformed into a very high soil erosion hazard zone. Hence, it can be said that man-made activities further increase the potential of soil erosion in areas which are already vulnerable to soil erosion. This increased percentage of the affected area is alarming and needs to be addressed properly.

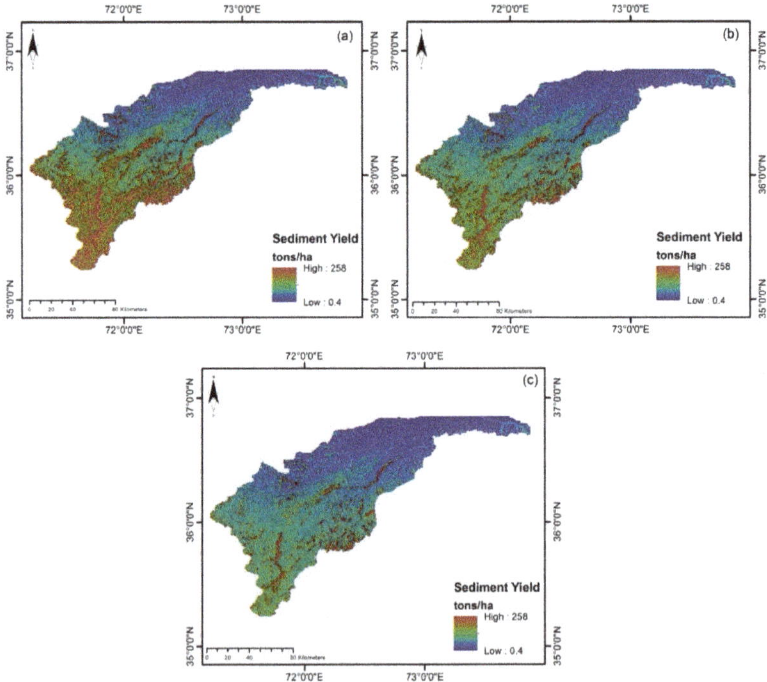

Figure 8. Sediment yield map of the study area (**a**) 2020, (**b**) 2010, (**c**) 2000.

Table 5. Soil erosion severity classes showing the area affected from 2000 to 2020.

Soil Erosion Severity Class	Soil Loss (tons/ha/year)	2020		2010		2000	
		Total Area (ha)	Area (%)	Total Area (ha)	Area (%)	Total Area (ha)	Area (%)
Very Low	<5	623,122	43%	536,175	37%	521,684	36%
Low	5–10	173,895	12%	275,333	19%	333,298	23%
Moderate	10–20	304,315	21%	362,280	25%	362,280	25%
High	20–50	231,859	16%	202,877	14%	173,895	12%
Very High	>50	115,930	8%	724,56	5%	579,65	4%
Total		1,449,121	100%	1,449,121	100%	1,449,121	100%

The soil erosion of the studied watershed showed a variation on the spatial scale to some extent in all three years of this study. Singh et al. [41] suggested that the soil erosion maps of the studied years are to be classified into four soil erosion classes from a management perception such as low, medium, high, and very high. The area affected by the soil erosion was divided into five categories (Figure 9) to have a better and clear understanding. Quantification of the soil hazard is very important to identify the hazard level and hotspot regions. Usually, hazard is defined in terms of very low, low, moderate, high and very high [42]. Ranges for the soil loss tolerance level for the study area are defined by OECD [43]. Multiple researches, carried out in several parts of the region, same as study area, followed the similar soil erosion classification method [44–46]. Therefore, similar quantification technique is followed in this study while keeping ranges of this region in mind. It was categorized as very low (less than 5 tons/ha/y), low (5–10 tons/ha/y), moderate (10–20 tons/ha/y), high (20–50 tons/ha/y), and very high (more than 50 tons/ha/y). The results of the difference between the three maps are presented in Table 5. The area affected by soil erosion showed increment. It is obvious from the increment that the land cover use kept changing throughout these years. There is a net change in the total area under the very low, low, moderate, and high categories by 7%, −11%, −4%, and 4%, respectively. The percentage of affected areas under a very high category was 4% in the year 2000, and it changed to 5% in 2010 and 8% in 2020. The area under very high category is increased by 4% in the comparison between 2000 and 2020, which is a major issue of concern. The change in soil erosion and its spatial distribution is depicted in Figure 10.

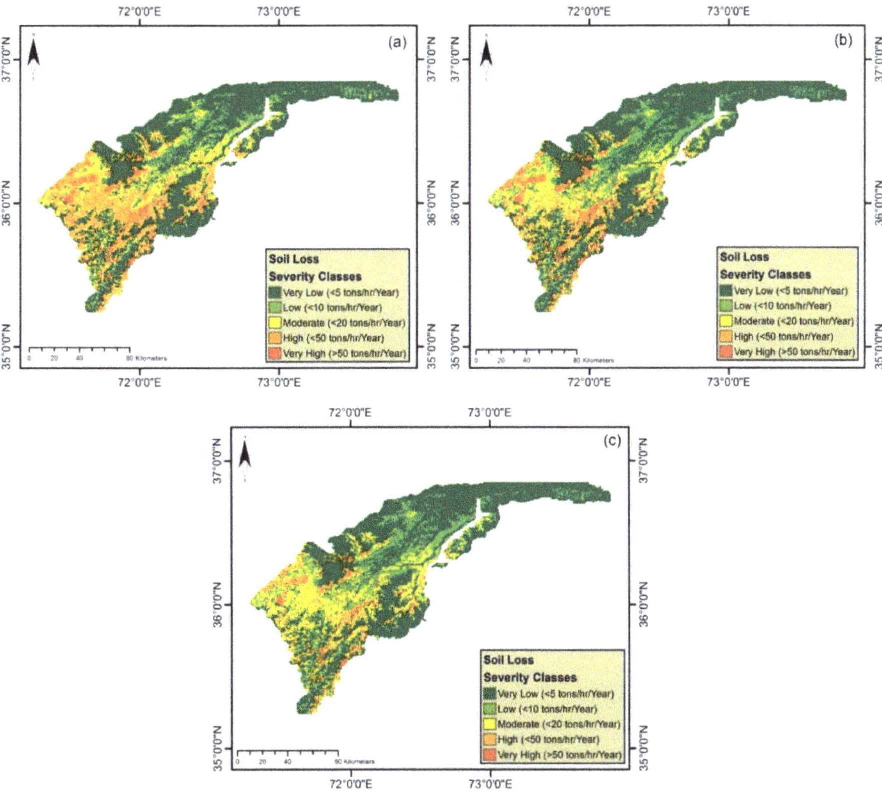

Figure 9. Soil loss severity map of the study area (**a**) 2020, (**b**) 2010, (**c**) 2000.

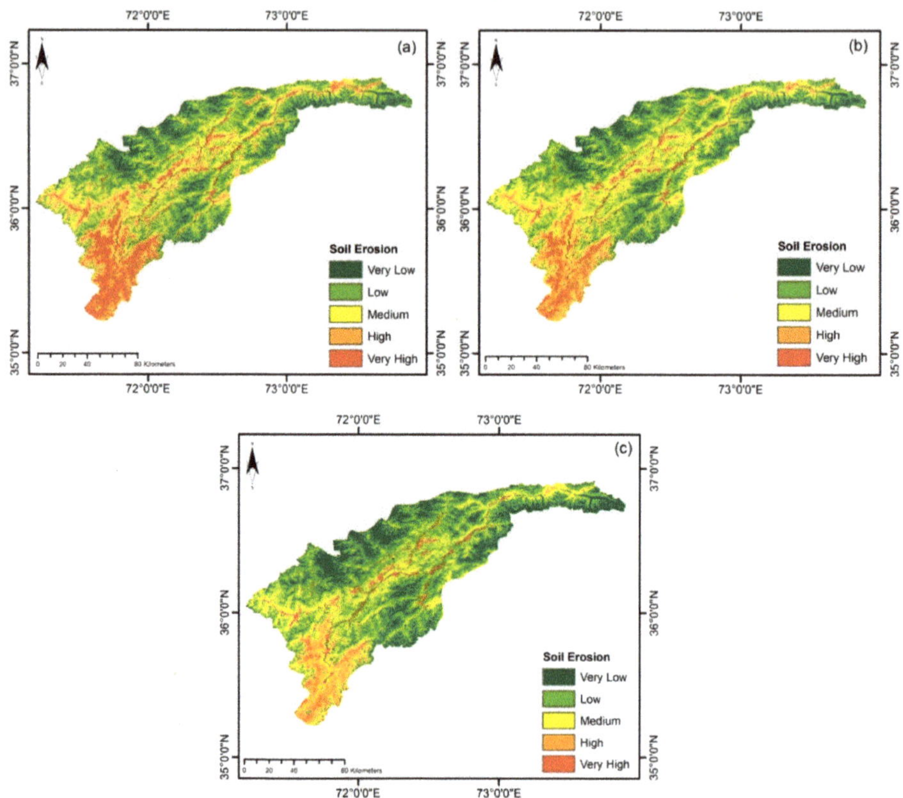

Figure 10. Soil erosion map of Chitral district (**a**) 2020, (**b**) 2010, (**c**) 2000.

The study area is a mountainous region. The figure shows that changes were severe in areas with high topographic potential for soil erosion during the study period. Degraded forestland, predominantly in foothill regions due to increasing human activities, is one of the main reasons behind this severe soil erosion. Invulnerable areas, land use changes in the intensification of agricultural activities and deforestation make the land more prone to soil erosion. From these results, it can be inferred that any land transition to cropland would be harmful, as it was the major source of sedimentation. On the other hand, the forest was the most effective barrier to soil loss.

With the passage of time and advancement in the knowledge of land, the cover has become essential to overcome the issues related to biogeochemical cycles, biodiversity, deterioration of environmental quality, loss of productive ecosystems, and loss of agricultural lands. The main reason behind the LULC changes includes rapid growth in population, rural-to-urban migration, reclassification of rural areas as urban areas, lack of valuation of ecological services, poverty, ignorance of biophysical limitations, and use of ecologically incompatible technologies. The present study area of Chitral is a mountainous and developing town. During the past few decades, the study area had witnessed substantial increase in population, economic growth, industrialization, and transportation activities that harmed environmental health of the region.

4. Discussion

The accuracy of the generated maps has improved throughout the study period. One of the main reasons is that the advancement in technology has enabled us to produce improved GIS, which can encounter more complex problems and issues. The results of the year 2020 are more accurate as compared to the previous years. The study conducted by [8] shows the same trend of improvement in accuracy but another study [47] shows that the accuracy for the maps in 2002 was 81.6% and it decreased to 80.5 percent for the year 2009. Thus, accuracy may depend upon one's personal skills and availability of data.

Due to the involvement of multiple data sets, we used the latest technologies like remote sensing and GIS to quantify LULC. Interpretation of remote sensing imagery, GIS and existing study area conditions enabled us to classify the study area into eight categories, that is, agriculture in sloping valley, agriculture in the valley, bare areas, natural herbaceous shrubs, natural high shrubs, natural trees, snow and ice, and water bodies. To determine the hazard potential and vulnerable hotspots, such zonation has been made in several studies. Those studies are taken as base line. Four concepts (regenerative economics, nature-based solutions, connectivity, and systems thinking) were introduced by Keesstra et al. [48] to explain the neutrality of land degradation. A study in the past includes strategies to reach UN Sustainable Development Goals more effectively [49]. Visser et al. [50] discussed the process to achieve Sustainable Development Goals (SDGs) by using transitions for soil-water system. Another study was conducted in the past to examine the geography of soil [51]. The classification of same nature has already been made in previous researches [8,47]; some other classes can also be found in the literature [52].

Table 2 shows the area covered by every land use class for the studied duration. It shows reduction in area covered by agriculture in valley, agriculture in sloping valley, bare areas, and natural high shrubs. The area occupied by natural herbaceous shrubs was reduced. The possible reason for the reduction in natural herbaceous shrubs might be the degradation of land because of nutrient depletion by intense soil erosion due to the lack of required soil conservation measures. The major class of snow and ice also shows decrement. This reduction is the consequence of global warming, the greenhouse effect, and the changing climates due to the mentioned reasons [53]. The class of bare area shows an increased percentage. It might be due to the settlement of more people in urban areas which leaves the rural areas bare [54]. Degraded forest, river sand, bare soil, heavily eroded land, and rock outcrop are included in the bare areas described in this study [8]. Natural high shrubs increased because the study area is governed by high altitude, receiving comparatively more rainfall, which excels the growth of natural high shrubs. The area covered by the water bodies is reduced due to lack of conservation practices. The agriculture in the area is increased, which is not very unusual considering the increasing demand of population and food. The highest soil erosion has a strong connection with cultivated land [8]. The strata of the cultivated soil are disturbed when it is ploughed. This disturbance of soil strata makes it more vulnerable to soil erosion. The increase in agricultural production is possible by the use of fertilizers turning uncultivable land to cultivable lands. Our study shows contrast to many other studies in which agriculture area or cultivated land is decreased due to many reasons such as conversion of agricultural land to settlements and to other land use classes such as forest [14].

Different land use types have great impact on soil erosion. The rate of soil erosion may change from area to area. It is not fixed that it will increase every time. It depends on conservation practices and management against soil erosion. Calculation of soil erosion can be divided into two categories, i.e., one for short period and the other for long period. Studies for short-term soil erosion rates are conducted in the Polish Carpathians, [35,55–60]. The Szymbark IG&SO PAS Research Station (49°38′04″ N, 21°07′08″ E), located in the Polish Carpathians, where soil erosion measurements covered the last 30 years is an exception [61–64]. A similar analysis was conducted by Latocha et al. [30] for soil erosion for a long-term period (in the last 150 years). However, their study was in the Sudetes Mountains, where after World War II, immigration actions led to depopulation and then forest expansion. This consequently decreased the soil erosion rates [14]. Our study is for the short period of 20 years and it shows that the rate of soil erosion increased from 2000 to 2020. The increment in the mean soil erosion may be due to slight transitions in land and land use. The main reason for increase in the mean soil erosion of the watershed over the studied period could be allocated to the drop in percentage of forest area, the increment of bare areas, and increased cultivation practice in areas that are more prone and vulnerable to soil erosion. It is important to remember that the change in the mean soil erosion may significantly affect the sedimentation process of the reservoir. LULC dynamics in the watershed result in more sedimentation in the reservoir, which would affect its life negatively. LULC dynamics obtained in this study indicate that the studied watershed is affected by moderate management and some other local human activities. RUSLE model was used to evaluate the subsequent effects of LULC dynamics on soil erosion potential.

This study shows increment in soil erosion rates. Changes in land use and land cover contributed to this increment. The LULC change was rapid in the second decade of the 21st century in the study area. It shows that the area is under many human activities such as a reduction in natural trees, snow, and ice, and increment in agriculture in plain and sloping areas as well. Increase in soil erosion can also be attributed to changes in the land cover, which are reflected by NDVI, and in turn changing the C factor values which strongly affect the soil erosion. Global climate system may be altered due to LULC changes. Nearly two-thirds of the precipitation that falls on the earth's surface is returned to the atmosphere via evaporation. Precipitation amount and intensity and potential evapo-transpiration is directly affected by climate change, and it indirectly affects plant water-use efficiency through altering plant growth rate and species composition, which result in change of land cover. The figures show that the areas with high soil erosion potential were mostly located along the foothill regions in the southern part and some portions of the central watershed during studied years. This study shows contrast to many previously conducted studies such as [14,47]. In these studies, the rates of soil erosion decreased with changes in land use. It is obvious from all these references that soil erosion rates may change and differ for different land, land uses and land covers depending upon the topography, settlement trends and meteorology of the area. The outcomes of this study will help the stakeholders and regulatory decision makers to control deforesting, controlled agricultural farming and to take other necessary actions to minimize the rate of soil erosion. Such an efficient planning will also be helpful to reduce the sedimentation in the reservoir of hydraulic dam(s) (Warsak Dam) constructed on Chitral river, which drains through this watershed. Apart from its local implications, this study may also provide useful guidelines for the other parts of the globe having similar climate and geographical settings.

5. Conclusions

The study presented in this paper emphasizes on rapid assessment of LULC changes and their consequences on regional soil erosion potential in a mountainous watershed of Chitral, Pakistan. For this purpose, RUSLE model is used in combination with GIS, which is proved to be an effective tool to extract land use land cover changes. The results of the study demonstrate that LULC changes were significant during the period from 2000 to 2020. The risks of soil erosion have been increased by 4%. The mean erosion potential for the whole watershed was increased from 9.2 tons/ha/y in 2000 to 15.63 tons/ha/y in 2020. A significant expansion in bare areas, agriculture in the valley, and agriculture in sloping valley has been noticed. On the other hand, there is a decrease in areas covered by water bodies and natural trees. These results indicate the significant impact of different land cover factors and human activities on LULC change, and consequently, on soil erosion.

Massive deforestation and increased agricultural land could be treated as most important reasons for increase in soil loss during the study period. This demands an immediate attention of the stakeholders, regulatory decision makers, and environmental management groups to propose comprehensive guidelines for possible LULCC. The farmers need to be provided with the capacity development programs to raise their level of awareness. An efficient system to control soil erosion will be equally helpful to reduce the sedimentation and increase the life and efficiency of hydraulic dam(s) constructed on Chitral river, which drains through this watershed. The soil loss hazard maps generated in this study could also be helpful to develop strategies to overcome food and water problems in the coming decades.

6. Limitations and Future Studies

In the future, the effect of LULC changes can be estimated in other hilly and plain areas. Effects of all other influencing factors like slope, elevation, and rainfall should also be estimated by integrating RUSLE and GIS. Global climate system may be altered due to LULC changes. Nearly two-thirds of the precipitation that falls on the surface of earth is returned to the atmosphere via evaporation. The prediction about the changes in the precipitations can be made depending upon LULC changes in the study area. Additionally, LULC alters the amount of CO_2 in the atmosphere. A study can be conducted to estimate the change of CO_2 in the atmosphere due to LULCC. The storage capacity of the Warsak Dam is decreasing day by day due to changes in LULC which directly contributes to soil erosion. Thus, there is a need to propose some control structures in Chitral river by using BIM (building information modeling) to control soil erosion.

Author Contributions: Conceptualization, B.A. and A.M.; methodology, B.A.; software, S. and R.F.T.; validation, Z.A.K., A.M. and M.S.; formal analysis, Z.A.K.; investigation, B.A.; resources, F.A. and M.H.B.; data curation, D.F.; writing—original draft preparation, B.A. and A.M.; writing—review and editing, Z.A.K. and M.S.; visualization, S. and R.T.F.; supervision, A.M. and Z.A.K.; project administration, B.A. All authors have read and agreed to the published version of the manuscript.

Funding: This research received no external funding.

Conflicts of Interest: The authors declare no conflict of interest.

References

1. Chen, P.; Feng, Z.; Mannan, A.; Chen, S.; Ullah, T. Assessment of Soil Loss from Land Use/Land Cover Change and Disasters in the Longmen Shan Mountains, China. *Appl. Ecol. Environ. Res.* **2019**, *17*, 11233–11247. [CrossRef]
2. Hussnain, M.Q.; Waheed, A.; Wakil, K.; Pettit, C.J.; Hussain, E.; Naeem, M.A.; Anjum, G.A. Shaping up the Future Spatial Plans for Urban Areas in Pakistan. *Sustainability* **2020**, *12*, 4216. [CrossRef]
3. Stock, J.L.; Chusid, J.M. Urbanizing India's Frontier: Sriganganagar and Canal-Town Planning on the Indus Plains. *Plan. Perspect.* **2020**, *35*, 253–276. [CrossRef]
4. Maqsoom, A.; Aslam, B.; Khalil, U.; Ghorbanzadeh, O.; Ashraf, H.; Tufail, R.F.; Farooq, D.; Blaschke, T. A Gis-Based Drastic Model and an Adjusted Drastic Model (Drastica) for Groundwater Susceptibility Assessment Along the China–Pakistan Economic Corridor (Cpec) Route. *ISPRS Int. J. Geo-Inf.* **2020**, *9*, 332. [CrossRef]
5. Aslam, B.; Ismail, S.; Ali, I. A Gis-Based Drastic Model for Assessing Aquifer Susceptibility of Safdarabad Tehsil, Sheikhupura District, Punjab Province, Pakistan. *Model. Earth Syst. Environ.* **2020**, 1–11. [CrossRef]
6. Martínez-Mena, M.; Carrillo-López, E.; Boix-Fayos, C.; Almagro, M.; García Franco, N.; Díaz-Pereira, E.; Montoya, I.; de Vente, J. Long-Term Effectiveness of Sustainable Land Management Practices to Control Runoff, Soil Erosion, and Nutrient Loss and the Role of Rainfall Intensity in Mediterranean Rainfed Agroecosystems. *Catena* **2020**, *187*, 104352. [CrossRef]
7. Chuenchum, P.; Xu, M.; Tang, W. Estimation of Soil Erosion and Sediment Yield in the Lancang–Mekong River Using the Modified Revised Universal Soil Loss Equation and GIS Techniques. *Water* **2020**, *12*, 135. [CrossRef]
8. Sharma, A.; Tiwari, K.N.; Bhadoria, P.B.S. Effect of Land Use Land Cover Change on Soil Erosion Potential in an Agricultural Watershed. *Environ. Monit. Assess.* **2011**, *173*, 789–801. [CrossRef]
9. Ouyang, W.; Wu, Y.; Hao, Z.; Zhang, Q.; Bu, Q.; Gao, X. Combined Impacts of Land Use and Soil Property Changes on Soil Erosion in a Mollisol Area under Long-Term Agricultural Development. *Sci. Total Environ.* **2018**, *613*, 798–809. [CrossRef]
10. Wiejaczka, Ł.; Olędzki, J.R.; Bucała-Hrabia, A.; Kijowska-Strugała, M. A Spatial and Temporal Analysis of Land Use Changes in Two Mountain Valleys: With and without Dam Reservoir (Polish Carpathians). *Quaest. Geogr.* **2017**, *36*, 129–137. [CrossRef]
11. Wilbanks, T.J. Socio-Economic Scenario Development for Climate Change Analysis. In *CIRED Working Paper DT/WP*; CIRED Working Paper DT/WP No. 2010-23; Centre International de Recherche sur l'Environnement et le Développement: Nogent-sur-Marne, France, 2010.

12. Bucała, A.; Budek, A.; Kozak, M. The Impact of Land Use and Land Cover Changes on Soil Properties and Plant Communities in the Gorce Mountains (Western Polish Carpathians), During the Past 50 Years. *Z. Geomorphol. Suppl. Issues* **2015**, *59*, 41–74. [CrossRef]

13. Nearing, M.; Polyakov, V.O.; Nichols, M.H.; Hernandez, M.; Li, L.; Zhao, Y.; Armendariz, G. Slope–Velocity Equilibrium and Evolution of Surface Roughness on a Stony Hillslope. *Hydrol. Earth Syst. Sci. Discuss.* **2017**, *21*, 3221–3229. [CrossRef]

14. Kijowska-Strugała, M.; Bucała-Hrabia, A.; Demczuk, P. Long-Term Impact of Land Use Changes on Soil Erosion in an Agricultural Catchment (in the Western Polish Carpathians). *Land Degrad. Dev.* **2018**, *29*, 1871–1884. [CrossRef]

15. Buttafuoco, G.; Conforti, M.; Aucelli, P.P.C.; Robustelli, G.; Scarciglia, F. Assessing Spatial Uncertainty in Mapping Soil Erodibility Factor Using Geostatistical Stochastic Simulation. *Environ. Earth Sci.* **2012**, *66*, 1111–1125. [CrossRef]

16. De Moor, J.J.W.; Verstraeten, G. Alluvial and Colluvial Sediment Storage in the Geul River Catchment (the Netherlands)—Combining Field and Modelling Data to Construct a Late Holocene Sediment Budget. *Geomorphology* **2008**, *95*, 487–503. [CrossRef]

17. Adhikary, P.P.; Tiwari, S.P.; Mandal, D.; Lakaria, B.L.; Madhu, M. Geospatial Comparison of Four Models to Predict Soil Erodibility in a Semi-Arid Region of Central India. *Environ. Earth Sci.* **2014**, *72*, 5049–5062. [CrossRef]

18. Botterweg, P.; Leek, R.; Romstad, E.; Vatn, A. The Eurosem-Gridsem Modeling System for Erosion Analyses under Different Natural and Economic Conditions. *Ecol. Model.* **1998**, *108*, 115–129. [CrossRef]

19. Bayat, F.; Monfared, A.B.; Jahansooz, M.R.; Esparza, E.T.; Keshavarzi, A.; Morera, A.G.; Fernández, M.P.; Cerdà, A. Analyzing Long-Term Soil Erosion in a Ridge-Shaped Persimmon Plantation in Eastern Spain by Means of Isum Measurements. *Catena* **2020**, *183*, 104176. [CrossRef]

20. Ganasri, B.P.; Ramesh, H. Assessment of Soil Erosion by Rusle Model Using Remote Sensing and Gis— A Case Study of Nethravathi Basin. *Geosci. Front.* **2016**, *7*, 953–961. [CrossRef]

21. Conforti, M.; Buttafuoco, G. Assessing Space–Time Variations of Denudation Processes and Related Soil Loss from 1955 to 2016 in Southern Italy (Calabria Region). *Environ. Earth Sci.* **2017**, *76*, 457. [CrossRef]

22. Özşahin, E.; Eroğlu, İ. Soil Erosion Risk Assessment Due to Land Use/Land Cover Changes (Lulcc) in Bulgaria from 1990 to 2015. *Alinteri J. Agric. Sci.* **2019**, *34*, 1–8.

23. Guadie, M.; Molla, E.; Mekonnen, M.; Cerdà, A. Effects of Soil Bund and Stone-Faced Soil Bund on Soil Physicochemical Properties and Crop Yield under Rain-Fed Conditions of Northwest Ethiopia. *Land* **2020**, *9*, 13. [CrossRef]

24. Abdulkareem, J.H.; Pradhan, B.; Sulaiman, W.N.A.; Jamil, N.R. Prediction of Spatial Soil Loss Impacted by Long-Term Land-Use/Land-Cover Change in a Tropical Watershed. *Geosci. Front.* **2019**, *10*, 389–403. [CrossRef]

25. Chen, Z.; Wang, L.; Wei, A.; Gao, J.; Lu, Y.; Zhou, J. Land-Use Change from Arable Lands to Orchards Reduced Soil Erosion and Increased Nutrient Loss in a Small Catchment. *Sci. Total Environ.* **2019**, *648*, 1097–1104. [CrossRef] [PubMed]

26. Dissanayake, D.; Morimoto, T.; Ranagalage, M. Accessing the Soil Erosion Rate Based on Rusle Model for Sustainable Land Use Management: A Case Study of the Kotmale Watershed, Sri Lanka. *Model. Earth Syst. Environ.* **2019**, *5*, 291–306. [CrossRef]

27. Yang, Y.; Li, Z.; Li, P.; Ren, Z.; Gao, H.; Wang, T.; Xu, G.; Yu, K.; Shi, P.; Tang, S. Variations in Runoff and Sediment in Watersheds in Loess Regions with Different Geomorphologies and Their Response to Landscape Patterns. *Environ. Earth Sci.* **2017**, *76*, 517. [CrossRef]

28. Latocha, A.; Szymanowski, M.; Jeziorska, J.; Stec, M.; Roszczewska, M. Effects of Land Abandonment and Climate Change on Soil Erosion—An Example from Depopulated Agricultural Lands in the Sudetes Mts., Sw Poland. *Catena* **2016**, *145*, 128–141. [CrossRef]

29. Leh, M.; Bajwa, S.; Chaubey, I. Impact of Land Use Change on Erosion Risk: An Integrated Remote Sensing, Geographic Information System and Modeling Methodology. *Land Degrad. Dev.* **2013**, *24*, 409–421. [CrossRef]

30. Prasannakumar, V.; Vijith, H.; Abinod, S.; Geetha, N. Estimation of Soil Erosion Risk within a Small Mountainous Sub-Watershed in Kerala, India, Using Revised Universal Soil Loss Equation (Rusle) and Geo-Information Technology. *Geosci. Front.* **2012**, *3*, 209–215. [CrossRef]

31. Efthimiou, N.; Lykoudi, E.; Karavitis, C. Soil Erosion Assessment Using the Rusle Model and Gis. *Eur. Water* **2014**, *47*, 15–30.

32. Licznar, P. Prognozowanie Erozyjności Deszczy W Polsce Na Podstawie Miesięcznych Sum Opadów. *Arch. Ochr. Środowiska* **2004**, *30*, 29–39.

33. Ullah, S.; Ali AIqbal, M.; Javid, M.; Imran, M. Geospatial Assessment of Soil Erosion Intensity and Sediment Yield: A Case Study of Potohar Region, Pakistan. *Environ. Earth Sci.* **2018**, *77*, 705. [CrossRef]

34. Samie, A.; Abbas, A.; Azeem, M.M.; Hamid, S.; Iqbal, M.A.; Hasan, S.S.; Deng, X. Examining the Impacts of Future Land Use/Land Cover Changes on Climate in Punjab Province, Pakistan: Implications for Environmental Sustainability and Economic Growth. *Environ. Sci. Pollut. Res.* **2020**, *27*, 25415–25433. [CrossRef] [PubMed]

35. Drzewiecki, W.; Wężyk, P.; Pierzchalski, M.; Szafrańska, B. Quantitative and Qualitative Assessment of Soil Erosion Risk in Małopolska (Poland), Supported by an Object-Based Analysis of High-Resolution Satellite Images. *Pure Appl. Geophys.* **2014**, *171*, 867–895. [CrossRef]

36. Maqsoom, A.; Aslam, B.; Hassan, U.; Kazmi, Z.A.; Sodangi, M.; Tufail, R.F.; Farooq, D. Geospatial Assessment of Soil Erosion Intensity and Sediment Yield Using the Revised Universal Soil Loss Equation (Rusle) Model. *ISPRS Int. J. Geo-Inf.* **2020**, *9*, 356. [CrossRef]

37. Mitasova, H.; Mitas, L.; Brown, W.M.; Johnston, D.M. *Terrain Modeling and Soil Erosion Simulations for Fort Hood and Fort Polk Test Areas*; Report for US Army Construction Engineering Research Laboratory, Geographic Modeling and Systems Laboratory; University of Illinois at Urbana-Champaign: Urbana, IL, USA, 1999.

38. Mancino, G.; Nolè, A.; Salvati, L.; Ferrara, A. In-between Forest Expansion and Cropland Decline: A Revised Usle Model for Soil Erosion Risk under Land-Use Change in a Mediterranean Region. *Ecol. Indic.* **2016**, *71*, 544–550. [CrossRef]

39. Vickers, N.J. Animal Communication: When I'm Calling You, Will You Answer Too? *Curr. Biol.* **2017**, *27*, 713–715. [CrossRef]

40. Panagos, P.; Borrelli, P.; Meusburger, K.; van der Zanden, E.H.; Poesen, J.; Alewell, C. Modelling the Effect of Support Practices (P-Factor) on the Reduction of Soil Erosion by Water at European Scale. *Environ. Sci. Policy* **2015**, *51*, 23–34. [CrossRef]

41. Singh, G.; Babu, R.; Narain, P.; Bhushan, L.S.; Abrol, I.P. Soil Erosion Rates in India. *J. Soil Water Conserv.* **1992**, *47*, 97–99.

42. Mahmood, A.; Oweis, T.; Ashraf, M.; Aftab, M.; Khan Aadal, N.; Ahmad, I.; Sajjad, M.R.; Majid, A. Improving Land and Water Produc Vi Es in the Dhrabi Watershed. In *Assessments and Options for Improved Productivity and Sustainability of Natural Resources in Dhrabi Watershed, Pakistan*; ICARDA: Aleppo, Syria, 2012; p. 27.

43. Directorate for Food, Agriculture, Fisheries Development; Policies and Environment Division. *Environmental Indicators for Agriculture: Methods and Results: Executive Summary: [Agriculture and Food]*; OECD: Paris, France, 2000; Volume 3.

44. Farhan, Y.; Nawaiseh, S. Spatial Assessment of Soil Erosion Risk Using Rusle and Gis Techniques. *Environ. Earth Sci.* **2015**, *74*, 4649–4669. [CrossRef]

45. Nasir, A.; Uchida, K.; Ashraf, M. Estimation of Soil Erosion by Using Rusle and Gis for Small Mountainous Watersheds in Pakistan. *Pak. J. Water Resour.* **2006**, *10*, 11–21.

46. Bashir, S.; Baig, M.A.; Ashraf, M.; Anwar, M.M.; Bhalli, M.N.; Munawar, S. Risk Assessment of Soil Erosion in Rawal Watershed Using Geoinformatics Techniques. *Sci. Int.* **2013**, *25*, 583–588.

47. Alkharabsheh, M.M.; Alexandridis, T.K.; Bilas, G.; Misopolinos, N.; Silleos, N. Impact of Land Cover Change on Soil Erosion Hazard in Northern Jordan Using Remote Sensing and Gis. *Procedia Environ. Sci.* **2013**, *19*, 912–921. [CrossRef]

48. Keesstra, S.; Mol, G.; Leeuw, J.D.; Okx, J.; Cleen, M.D.; Visser, S. Soil-Related Sustainable Development Goals: Four Concepts to Make Land Degradation Neutrality and Restoration Work. *Land* **2018**, *7*, 133. [CrossRef]

49. Keesstra, S.D.; Bouma, J.; Wallinga, J.; Tittonell, P.; Smith, P.; Cerdà, A.; Montanarella, L.; Quinton, J.N.; Pachepsky, Y.; Putten, W.H.V.D. The Significance of Soils and Soil Science towards Realization of the United Nations Sustainable Development Goals. *Soil* **2016**, *2*, 111–128. [CrossRef]

50. Visser, S.; Keesstra, S.; Maas, G.; Cleen, M.D. Soil as a Basis to Create Enabling Conditions for Transitions Towards Sustainable Land Management as a Key to Achieve the Sdgs by 2030. *Sustainability* **2019**, *11*, 6792. [CrossRef]

Sustainability **2020**, *12*, 5898

51. Rodrigo-Comino, J.; Senciales, J.M.; Bolinches, A.C.; Brevik, E.C. The Multidisciplinary Origin of Soil Geography: A Review. *Earth-Sci. Rev.* **2018**, *177*, 114–123. [CrossRef]

52. Li, L.; Wang, V.; Liu, C. Effects of Land Use Changes on Soil Erosion in a Fast Developing Area. *Int. J. Environ. Sci. Technol.* **2014**, *11*, 1549–1562. [CrossRef]

53. Schneider, S.H. The Greenhouse Effect: Science and Policy. *Science* **1989**, *243*, 771–781. [CrossRef]

54. Schneider, A. Monitoring Land Cover Change in Urban and Peri-Urban Areas Using Dense Time Stacks of Landsat Satellite Data and a Data Mining Approach. *Remote Sens. Environ.* **2012**, *124*, 689–704. [CrossRef]

55. Święchowicz, J. Spłukiwanie Gleby Na Użytkowanych Rolniczo Stokach Pogórskich W Latach Hydrologicznych 2007–2008 W Łazach (Pogórze Wiśnickie). *Prace i Studia Geogr.* **2010**, *45*, 243–263.

56. Swięchowicz, J. Water Erosion on Agricultural Foothill Slopes (Carpathian Foothills, Poland). *Z. Geomorphol. Suppl. Issues* **2012**, *56*, 21–35. [CrossRef]

57. Nowak, A.; Tokarczyk, N. Evaluation of Soil Resilience to Anthropopressure in Łosie Village (Lower Beskids Mts)—Preliminary Results. *Ekológia (Bratislava)* **2013**, *32*, 138–147. [CrossRef]

58. Froehlich, W. Natezenie Erozji Gleb Na Stokach Beskidzkich W Swietle Badan Metodami Klasycznymi I Radioizotypowymi. *Zeszyty Naukowe Akademii Rolniczej w Krakowie. Sesja Naukowa* **1997**, *48*, 35–46.

59. Froehlich, W. Importance of Splash in Erosion Process within a Small Flysch Catchment Basin. *Stud. Geomorphol. Carpatho-Balcanica* **1980**, *14*, 77–112.

60. Higgitt, D.L.; Froehlich, W.; Walling, D.E. Applications and Limitations of Chernobyl Radiocaesium Measurements in a Carpathian Erosion Investigation, Poland. *Land Degrad. Dev.* **1992**, *3*, 15–26. [CrossRef]

61. Bochenek, W.; Gil, E. Zróżnicowanie Spływu Powierzchniowego I Spłukiwania Gleby Na Poletkach Doświadczalnych O Różnej Długości (Szymbark, Beskid Niski). *Prace i Studia Geograficzne* **2010**, *45*, 265–278.

62. Demczuk, P.; Gil, E. The Application of the Usle Model in the Automatic Mapping of Soil Erosion in the Bystrzanka Catchment (Flysh Carpathian). *Quaest. Geogr.* **2009**, *28*, 2.

63. Kijowska-Strugała, M. *Transport Zawiesiny W Warunkach Zmieniającej Się Antropopresji W Zlewni Bystrzanki (Karpaty Fliszowe)*; IGiPZ PAN: Warszawa, Poland, 2015; Volume 247.

64. Kijowska-Strugala, M.; Demczuk, V. Impact of Land Use Changes on Soil Erosion and Deposition in a Small Polish Carpathians Catchment in Last 40 Years. *Carpathian J. Earth Environ. Sci.* **2015**, *10*, 261–270.

 *sustainability*

Article

Farming Influence on Physical-Mechanical Properties and Microstructural Characteristics of Backfilled Loess Farmland in Yan'an, China

Lina Ma [1,2,3], Shengwen Qi [1,2,3,*], Bowen Zheng [1,2,3], Songfeng Guo [1,2,3,*], Qiangbing Huang [4] and Xinbao Yu [5]

[1]   Key Laboratory of Shale Gas and Geoengineering, Institute of Geology and Geophysics, Chinese Academy of Sciences, Beituchengxilu 19, Chaoyang District, Beijing 100029, China; malina@mail.iggcas.ac.cn (L.M.); zhengbowen@mail.iggcas.ac.cn (B.Z.)
[2]   College of Earth and Planetary Sciences, University of Chinese Academy of Sciences, Yuquanlu 19, Shijingshan District, Beijing 100049, China
[3]   Innovation Academy for Earth Science, Chinese Academy of Sciences, Beituchengxilu 19, Chaoyang District, Beijing 100029, China
[4]   Department of Geological Engineering, Chang'an University, Xi'an 710054, China; dcdgx24@chd.edu.cn
[5]   Department of Civil Engineering, University of Texas at Arlington, 416 Yates St., Arlington, TX 76019, USA; Xinbao@uta.edu
*   Correspondence: qishengwen@mail.iggcas.ac.cn (S.Q.); guosongfeng@mail.iggcas.ac.cn (S.G.); Tel.: +86-010-82998055 (S.Q.); +86-13439078159 (S.G.)

Received: 27 May 2020; Accepted: 30 June 2020; Published: 8 July 2020

Abstract: A gigantic project named Gully Land Consolidation (GLC) was launched in the hill-gully region of the Chinese Loess Plateau in 2011 to cope with land degradation and create new farmlands for cultivation. However, as a particular kind of remolded loess, the newly created and backfilled farmland may bring new engineering and environmental problems because the soil structure was disturbed and destroyed. In this study, current situations and characteristics of GLC are introduced. Test results show that physical-mechanical properties and microstructural characteristics of backfilled loess of one-year and five-year farmland are significantly affected by the Gully Land Consolidation project. Compared to natural loess, the moisture content, density, and internal friction angle of backfilled loess increase. On the contrary, the porosity, plasticity index, particle size index, and cohesion index decrease. Through SEM tests, it is observed that the particles of backfilled loess are rounded, with large pores filled with crushed fine particles, which results in skeleton strength weakness among particles and pores. The pore size distribution (PSD) of the four types of loess (Q_3 loess, Q_2 loess, one-year farmland, and five-year farmland) was measured using mercury intrusion porosimetry (MIP) tests, showing that the pore size of Q_3 loess is mainly mesopores 4000–20,000 nm in size, accounting for 67.5%. The Q_2, five-year, and one-year farmland loess have mainly small pores 100–4000 nm in size, accounting for 52.5%, 51.7%, and 71.7%, respectively. The microscopic analysis shows that backfill action degrades the macropores and mesopores into small pores and micropores, leading to weak connection strength among soil particles, which further affects the physical-mechanical properties of loess. The disturbance of backfilled loess leads to an obvious decrease in cohesion and a slight increase in internal friction compared to natural loess. The farming effect becomes prominent with increased backfill time, while the loess soil moisture content increases gradually. Both the cohesion and internal friction of the backfilled loess soil decrease to different degrees. This study is helpful to investigate sustainable land use in the Chinese Loess Plateau and similar areas.

Keywords: Gully Land Consolidation; backfilled loess; physical-mechanical property; microstructural characteristic; pore size distribution

1. Introduction

The Chinese Loess Plateau (CLP) covers an area of 640,000 km^2 in the upper and middle reaches of the Yellow River. It also has multiple crisscrossing hills and gullies with fractured geological structures [1]. Meanwhile, the semiarid climate, with only 464.1 mm of average annual precipitation, contributes to the fragility of the ecosystem in the CLP [2]. Owing to aridity, sparse vegetation, and concentrated rainstorms, the CLP is facing serious problems of soil erosion and water loss, becoming the most vulnerable ecological environment in China [3–5].

Since 1949, the ecological management of the CLP has gone through several important stages, such as "slope management," "integration management of gully and slope," "small watershed management," and the "Grain for Green (GFG) project," which increased vegetation coverage and improved ecological quality [6,7]. In particular, since the Chinese government implemented the GFG project in 1999 to convert farmland to forests, shrubland, and grassland, the vegetation coverage rate of the CLP increased significantly, from 31.6% in 1999 to about 65% in 2017 [8], thereby reducing soil erosion and water loss to a certain extent [9,10]. Most notably, the land consolidation measures dramatically transformed the ecology and landscape of the CLP, turning it from yellow to green [2,11]. However, in recent years, Chinese scientists and government officials have noticed problems: There have been significant reductions in arable land induced by the large-scale implementation of the GFP on the Loess Plateau. Especially in Yan'an, the total area of arable land has reduced by more than half, creating an urgent need for new farmland and a shortage in grain production, which are the main obstacles in sustainable agricultural development [12,13].

On the other hand, Yan'an city has 44,000 gullies that are 500 m or more in length and 20,900 gullies that are 1 km or more in length. These numerous gullies have a land creation capacity in Yan'an of about 1.5 million acres [11]. To offset the loss of cropland and cope with land degradation, Shaanxi Province, which has 13% of the area of the CLP, launched a megaproject called Gully Land Consolidation (GLC) in 2011 [2,14–16], aimed at creating 26.67 × 10^4 ha of farmland in creek valleys from 2011 to 2020, where the total investment has reached ¥30 billion [17]. The GLC project was officially listed as a major national land improvement project in 2013 [18]. Yan'an City created 3.33 × 104 ha of farmland from 2013 to 2017 [19]. The GLC project has become an important measure in increasing areas of cultivated land and to expand land resources. Meanwhile, it also effectively reduces the loss of soil and water in trenches, which can largely alleviate the problem of land shortage and support agricultural modernization [20].

As new comprehensive management, the main approaches of GLC include reshaping valleys by incising foot slopes, filling gullies and stream channels, constructing or rebuilding drainage canals, dams, and reservoirs, and creating flat farmlands [2,11,19]. The backfilled loess from farmland in the GLC is a special type of remolded soil. After field investigations, although layered excavation and mechanical compaction were used in the construction process, these operations and management still lack unified and standard guides. Some essential indicators that affect the arability of farmland soils, such as the layered thickness of backfilled soils, tamping methods, and soil compaction, were not adequately demonstrated and implemented. This results in uneven compaction of newly constructed farmland and vast differences of microstructure among backfilled loess. The microstructure is the fundamental structural unit of soil, including grain morphology, grain contact patterns, pore characteristics, and pore size distribution (PSD) [21–24]. Different types of loess have significantly different microstructural characteristics, even if the same type also shows distinctly different microstructure under different pressure, moisture content, dry density, and other conditions. Furthermore, the macroscopic and mechanical properties of soils, such as shear behavior, compressibility, collapsibility, and permeability, are controlled by its microstructural characteristics [25–27]. Under the influence of rainfall and irrigation, collapse subsidence is likely to occur, causing some new engineering and environmental problems.

However, there are few studies on the engineering characteristics of backfilled loess in the GLC in China and also around the world. Based on this, this paper selected two gullies of Chunshuyaozi (CSYZ) and Shijiagou (SJG) as study areas, which were typical GLC sites in the "National Soil and

Water Conservation Demonstration Park of South Gully." Four types of loess samples were collected with minimal disturbance, including natural samples of Middle Pleistocene Lishi loess (Q_2) and Upper Pleistocene Malan loess (Q_3) from the excavated slope and backfilled samples from two farmlands located in the reshaped CSYZ and SJG gullies. Farmland in the CSYZ gully has been backfilled for five years, and that in the SJG gully for one year. Moreover, a series of routine soil tests were performed on the four types of loess samples, including basic physical property tests, direct shear tests, and compression tests. Meanwhile, this study also investigated the microstructural characteristics of the loess by conducting scanning electron microscope (SEM) and mercury intrusion porosimetry (MIP) tests. By comparing the physical-mechanical properties and microstructural characteristics of four types of loess samples, this paper shows the evolution of backfilled loess and provides a basis for sustainable land use in the CLP.

2. Materials and Methods

2.1. Study Site

Chunshuyaozi (CSYZ) gully is located in Chunshuyaozi village, Louping town, Ansai district, Yan'an city, Shaanxi Province, and is where the GLC project was implemented in 2013. After the farmland construction was completed in one year, the farmland was immediately planted. So far, it has been five years since CSYZ farmland was backfilled. Figure 1a,b shows satellite images of CSYZ before and after the GLC, taken on 21 July 2010 and 12 March 2019, respectively. Through comparison, it was found that the morphology of these slopes and trenches changed significantly after excavation and backfill.

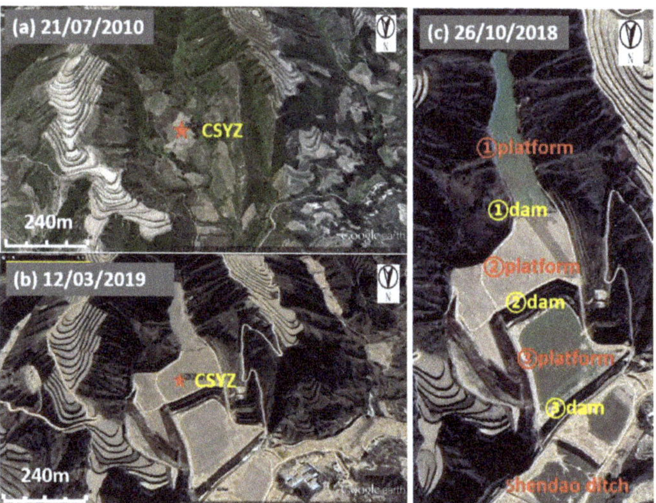

Figure 1. Satellite images of Chunshuyaozi (CSYZ) gully (**a**) before and (**b**) after the Gully Land Consolidation (GLC) project. (**c**) Morphology of CSYZ gully.

The main direction of the CSYZ gully is north–south, with the head on the south side (Figure 1c). The farmlands are composed of three platforms and three sedimentary dams. The first-level platform is elongated and irregular, located at the south of the gully, and is 40 m wide in the south and 100 m wide in the north. A field survey showed that the major plant was alfalfa. Close to the north side of the first-level platform is the first-level dam, which is about 2.5 m high and nearly 50 m wide. The second-level platform is irregular and polygonal, with a length of 300 m in the east–west direction and about 130 m in the north–south direction. The main crop in this platform is black wolfberry.

The secondary dam on the north side of the secondary platform has a height of about 15 m and a length of about 230 m in the east–west direction and consists of four steps. The third-level platform is an irregular square and resembles a dustpan. It is about 200 m long from north to south and about 170 m from east to west, and is mainly planted with canola. The gully mouth area is the third dam with a length of about 300 m from east to west and a height of about 5 m. Intersecting with the CSYZ gully is the Shendao gully in a nearly NW direction.

The Chang'an University team revealed the main stratigraphy of the CSYZ gully from new to old: Q_3 loess, Q_2 loess, Hipparion red soil of the Pliocene Series, and underlying sandy mudstone of Zaoyuan section, Yan'an Formation in the lower Jurassic (J_1). The thickness of Q_3 loess was mainly between 15 and 25 m, and some parts can reach 30 m. The thickness of Q2 loess ranged from 30 to 60 m. Loess in gullies mainly included backfilled loess of Q_3 and Q_2, between 10 and 25 m depth. The groundwater level of the original slope in this study area was 72 m deep. The thickness of backfilled farmland was between 12 and 22 m due to the undulations of the original channel [28].

Shijiagou (SJG) gully, located in the east of South Gully village, Louping township, Ansai district, Yan'an city, Shaanxi Province, began to implement the GLC project in 2019. Figure 2a,b shows satellite images of SJG before and after the GLC, taken on 13 February 2010 and 12 March 2019, respectively. The direction trend of SJG gully is NE 45°, with a length of 600 m. The backfilled farmlands are composed of five platforms and five sedimentary dams. A field survey showed that the major plant in SJG was alfalfa. The SJG gully is just 1.5 km away from the CSYZ gully in the southeast direction, and thus the strata are similar.

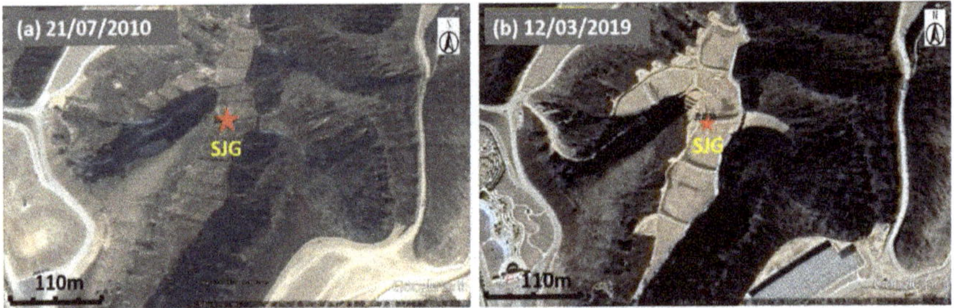

Figure 2. Satellite images of Shijiagou (SJG) gully (**a**) before and (**b**) after GLC.

2.2. Soil Sampling

Loess soil samples were taken from 14 sampling sites of the excavated slopes and the backfilled farmlands in the study area, eight sites in the CSYZ gully, and six sites in the SJG gully. Two parallel samples were taken from each site. For the CSYZ gully, the sampling site labeled Q_3 is Q_3 loess located in the middle-upper part of the southern slope, which is about 15 m from the crest of the slope in the vertical direction and 1 m from the slope surface in the horizontal direction. The sampling site labeled Q_2 is Q_2 loess located in the lower part of the western slope, which is about 55 m from the crest of the slope in the vertical direction and 1 m from the slope surface in the horizontal direction. The sampling sites labeled F-C-1 to F-C-6 are backfilled loess, and samples were taken from about 0.5 m depth. Among them, F-C-1 is on the northern side of the second-level farmland, F-C-2 and F-C-3 are on the middle-upper and lower part of the secondary dam, respectively, and F-C-4 to F-C-6 are on the southern side of the three-level platform, from east to west (Figure 3).

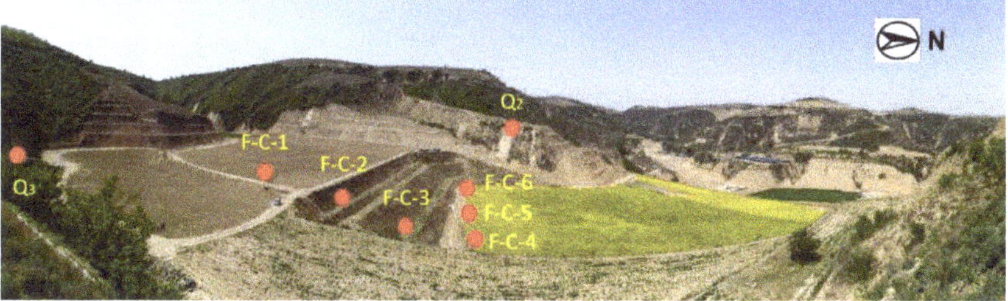

Figure 3. Plan of CSYZ gully and sampling location.

For the SJG gully, sample sites labeled F-S-1, F-S-2, and F-S-3 are located on the west side of the fifth-level platform, respectively, from south to north, and sample sites labeled F-S-4 to F-S-6 are on the south side, from west to east. The sampling depth was about 0.5 m (Figure 4).

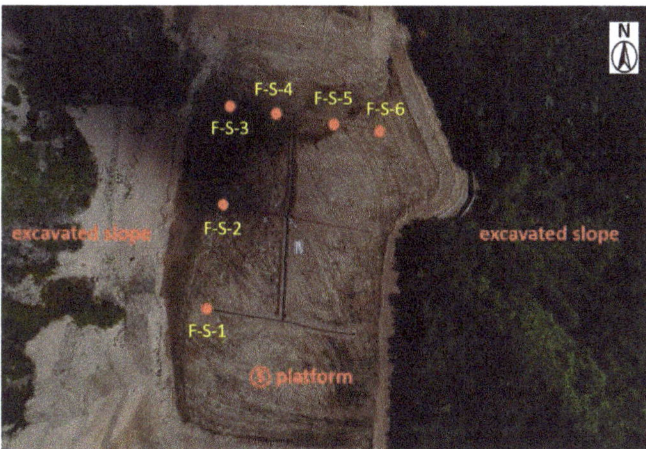

Figure 4. Plan of SJG gully and sampling location.

All soils were sampled by an undisturbed sampling method. Through experience, we found that the advantages of PVC sampling cylinders are that they are sealable, impervious, and lightweight compared to traditional iron cylinders. PVC pipes were cut into short tubes 16 cm in height and 11 cm in diameter in advance. Cardboard plates 11 cm in diameter were made from cardboard boxes to be used as caps for the PVC tubes. The sampling process and method are as follows (Figure 5): First, about the top 0.5 m surface layer was removed at the sampling site, and the soil was excavated to make an intact soil block about $25 \times 20 \times 20$ cm (Figure 5a,b). A PVC tube was placed on top of the soil block and gently pushed into the block while the surrounding soil was trimmed off (Figure 5c,d). Then, multiple layers of preservative film were used to completely seal the PVC tube in order to preserve the soil moisture from evaporation loss. Cardboard plates were taped to both ends of the tube to secure the soil sample and prevent disturbance (Figure 5e). The tube was marked with the sample number and placed in a box to be transported to the laboratory.

Figure 5. Sampling process: (**a**) Excavation of soil block from sampling site, (**b**) cutting to suitable size, (**c**) pressing the soil sample, (**d**) sampling by PVC pipe, (**e**) sealing the soil sample.

2.3. Experimental Method and Design

Tests of basic physical properties on 28 soil samples from 14 sampling sites were conducted, including moisture content, density, specific gravity, liquid limit, plastic limit, and particle analysis, and the following properties were obtained through calculation: Dry density, porosity, saturation, plasticity index, and characteristic index of granularity. All test methods were based on the Geotechnical Test Method Standard GB/T 50123-1999 [29]. Specific gravity was tested with the pycnometer method, and the liquid–plastic limit was tested using the liquid–plastic limit combined determination method. The particle analysis tests were performed by a Malvern Mastersizer 3000 laser particle size analyzer. Samples of Q_3, Q_2, F-C-4, and F-S-2 were selected for consolidated and undrained direct shear tests. A total of 14 soil samples of natural and backfilled loess were tested for consolidation and compression. The porosity ratios were obtained, and coefficients of compressibility and modulus of compressibility were calculated. According to the regulations of the test standard, each site should have two specimens, and if the difference between the two values is within the error range, the average of the two is taken as the result. According to the tests, properties of natural Q_2 and Q_3 loess are relatively stable, and the values of the two specimens are very close. For SEM and MIP tests, samples of F-C-4 and F-S-2 were chosen to represent CSYZ and SJG loess.

A compression experiment is generally used to study the compressibility of soil. The compression tests of all samples were carried out with GZQ-1 pneumatic automatic consolidometer. Coefficient of compressibility (a) and modulus of compressibility (Es) were respectively calculated by Equations (1) and (2):

$$\alpha = \frac{e_1 - e_2}{p_2 - p_1}, \tag{1}$$

where p_1 is the original pressure, defined as 100 kPa; p_2 is the summation pressure, defined as 200 kPa; and e_1 and e_2 are corresponding void ratios.

$$E_s = \frac{1 + e_1}{\alpha}, \tag{2}$$

Scanning electron microscopy (SEM) and mercury intrusion porosimetry (MIP) tests are two well-accepted and effective methods of studying the microstructure characteristics of soils. SEM can obtain microscopic images of the soil surface, which are generally used for qualitative analysis of soil structures. MIP is generally used for quantitative analysis of pore characteristics by measuring the size and distribution of micropores in the soil. In this paper, SEM and MIP tests were performed on loess of Q_3, Q_2, CSYZ, and SJG. The micrographs of particle morphology were analyzed by means of SEM, and the pore size distribution characteristics were measured by MIP. The characteristics of micromorphology, soil grain morphology, grain contact patterns, types of porosity, and pore size distribution were comprehensively analyzed, and the evolution process of microstructure characteristics of backfilled farmland loess with increased backfilled time was discussed.

The apparatus used in SEM tests was an FEI Nova Nano SEM field emission scanning electron microscope. The test steps were as follows: (1) The sample was cut into a cuboid with dimensions of 2 cm × 1 cm × 1 cm, air-dried in a light-shielded place, then split into two parts by hand along the middle position, and a flat and fresh side was selected as the observation surface. (2) After the sample was evacuated under vacuum pumping, the surfaces were treated with gold spray. Then the sample was fixed on the target plate with conductive tape and transferred to the observation room. (3) The instrument software was opened to find the right position to observe the sample. The focus was adjusted to obtain optimal images with appropriate brightness at selected degrees of magnification. This experiment was observed at four magnifications: 200×, 400×, 800×, and 1600×. In this paper, representative images of 200× and 800× were selected for analysis.

The test principle of MIP is as follows. This method is based on the assumption that the measured pores are cylindrical. As a noninvasive liquid with liquid surface tension, mercury cannot enter pores without pressure. During the mercury intrusion process, the work of moving mercury into the pores is converted to overcome the work done by the surface tension. Hence the Washburn equation for the external pressure and pore radius was deduced [30] as Equation (3):

$$P = -\frac{2\gamma \cos \theta}{r},\tag{3}$$

where P is the applied pressure (kPa); r is the pore radius (nm); γ is the surface tension of mercury, 0.485 N/m; and θ is the contact angle of the mercury and the surface of the soil grain, 140°. It can be known from the formula that pressure is inversely proportional to pore size. The greater the applied pressure, the smaller the pore radius that mercury can enter. Mercury intrusion starts from large pores, then mesopores, and finally small pores. By the volume of mercury intrusion at a certain pressure, the volume of the corresponding pore sizes can be determined [31]. The instrument used in this test was a Micromeritics Auto Pore IV 9500, which can provide a pressure range of 3–4 × 10^5 kPa, with a corresponding pore size range of 3 nm–450 μm, and the accuracy is 1 μL. The samples were cut into cuboids of 1 × 1 × 1 cm and air-dried in a light-shielded place.

3. Results

3.1. Tests of Basic Physical Properties

Basic physical test results can be summarized as follows (Table 1). The results (Figure 6a) show that the moisture content of F-C is between 16.5% and 18.4%, and F-S is between 19.1% and 24.4%. Compared with natural loess, there is a significant increase in moisture content of backfilled loess. At the same time, the moisture content of F-C is generally lower than that of F-S, and the scope of the former is less than that of the latter. The results of density tests show a similar tendency (Figure 6b). The density of F-C is about 1.77–1.95 g/cm^3, and that of F-S is about 1.82–2.08 g/cm^3. On average, the density of backfilled loess shows a significant increase compared with the natural loess, and the increase in SJG is greater than that in CSYZ.

Table 1. Basic physical indices of 14 loess groups.

Soil Sample	Number	Moisture Content (%)	Density (g/cm³)	Specific Gravity	Dry Density (/cm³)	Porosity
Natural loess	Q_3	9.7	1.50	2.70	1.37	0.97
	Q_2	15.0	1.78	2.69	1.55	0.74
Backfilled loess of CSYZ	F-C-1	17.2	1.95	2.71	1.66	0.63
	F-C-2	18.4	1.77	2.71	1.49	0.81
	F-C-3	16.8	1.85	2.71	1.58	0.71
	F-C-4	17.1	1.84	2.71	1.57	0.72
	F-C-5	17.2	1.94	2.71	1.66	0.64
	F-C-6	16.5	1.87	2.71	1.61	0.69
Backfilled loess of SJG	F-S-1	20.4	2.08	2.73	1.73	0.58
	F-S-2	20.1	1.91	2.73	1.59	0.72
	F-S-3	24.4	1.93	2.73	1.55	0.76
	F-S-4	22.2	1.89	2.73	1.55	0.77
	F-S-5	21.6	1.99	2.73	1.64	0.67
	F-S-6	19.1	1.82	2.73	1.53	0.79

Soil Sample	Number	Saturation (%)	Liquid Limit	Plastic Limit	Plasticity Index
Natural loess	Q_3	26.87	17.8	27.6	9.8
	Q_2	54.68	17.1	28.0	10.9
Backfilled loess of CSYZ	F-C-1	74.13	18.7	28.0	9.3
	F-C-2	61.35	17.3	30.3	13.0
	F-C-3	64.04	18.0	27.8	9.8
	F-C-4	63.95	18.4	28.2	9.8
	F-C-5	73.15	17.4	25.8	8.4
	F-C-6	64.96	16.9	25.3	8.4
Backfilled loess of SJG	F-S-1	95.98	17.6	26.0	8.4
	F-S-2	76.57	16.4	24.8	8.4
	F-S-3	87.69	17.5	26.0	8.5
	F-S-4	79.21	17.6	26.1	8.5
	F-S-5	88.25	19.1	25.4	6.3
	F-S-6	66.30	19.5	25.8	6.3

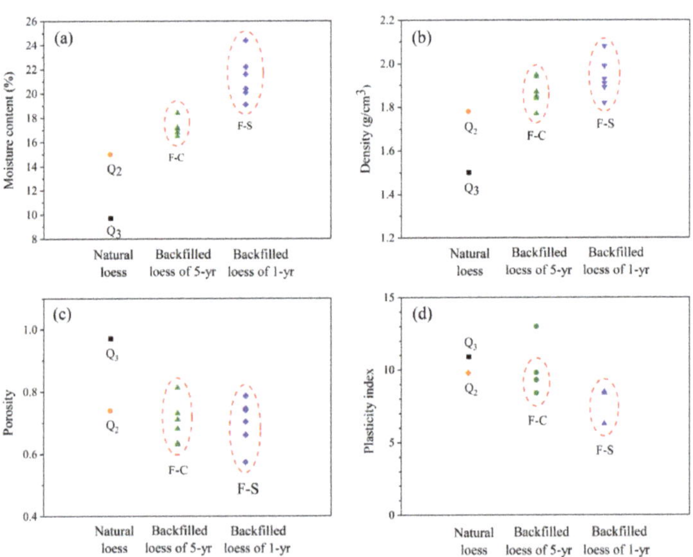

Figure 6. Comparison of basic physical properties: (**a**) moisture content, (**b**) density, (**c**) porosity, and (**d**) plasticity index.

On the contrary, the porosity of backfilled loess is significantly lower than that of natural loess (Figure 6c). The porosity ratio of Q_3 and Q_2 is 0.97 and 0.74, respectively. The porosity ratio of F-C varies from 0.63–0.81, and that of F-S ranges from 0.58 to 0.79. The mean porosity of F-C increased compared with that of F-S. Similarly, the plasticity index of backfilled loess is also lower than that of natural loess from slopes (Figure 6d). The values of F-C samples are mainly distributed in the range of 8.4–9.8, while the values of F-S samples have an obvious decrease and are mainly distributed in the range of 6.3–8.5.

3.2. Grain Size Distribution

Figure 7 shows the results of grain analysis tests. It can be seen that the grain size accumulation curves of 14 samples is smooth and continuous, with a granularity distribution ranging from 0.523 μm to 111 μm. Among all curves, Q_3 is located on the far left of the graph, Q_2 and F-C are mostly concentrated in the middle part, and F-S is mostly distributed on the right. It can be seen from Table 2 that Q_3 has four maximums of characteristic grain size indices, d_{60}, d_{50}, d_{30}, and d_{10}. Overall, although there are no significant differences, the indices of SJG are slightly higher than those of CSYZ.

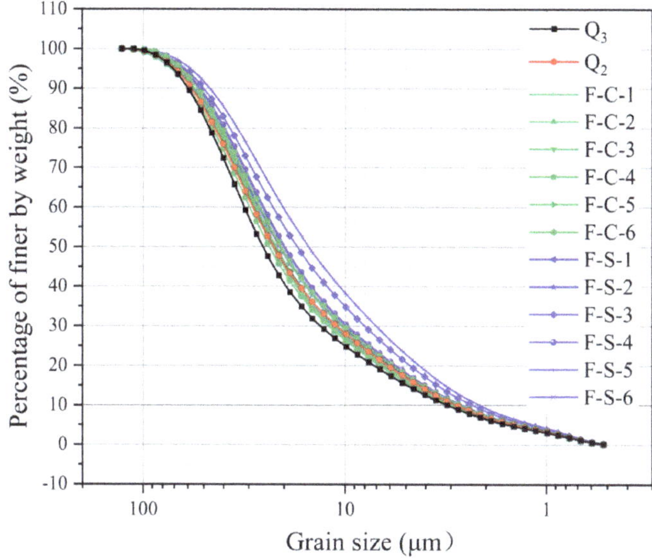

Figure 7. Grain size accumulation curves of 14 samples.

Table 2. Characteristic indices of granularity of 14 loess groups.

Soil Sample	Number	d_{60} μm	d_{50} μm	d_{30} μm	d_{10} μm	C_u	C_c
Natural loess	Q_3	31.50	26.00	14.00	3.10	10.16	2.01
	Q_2	28.00	23.00	11.00	2.90	9.66	1.49
Backfilled loess of CSYZ	F-C-1	27.50	21.50	10.00	2.70	10.19	1.35
	F-C-2	23.00	18.00	8.00	2.30	10.00	1.21
	F-C-3	25.00	20.00	9.80	2.60	9.62	1.48
	F-C-4	20.00	16.00	7.00	2.10	9.52	1.17
	F-C-5	28.00	23.00	12.50	3.00	9.33	1.86
	F-C-6	26.00	21.00	11.00	2.80	9.29	1.66
Backfilled loess of SJG	F-S-1	28.00	23.00	12.00	3.00	9.33	1.71
	F-S-2	28.00	23.00	12.00	3.00	9.33	1.71
	F-S-3	29.50	24.00	13.00	3.00	9.83	1.91
	F-S-4	26.00	20.00	10.00	2.70	9.63	1.42
	F-S-5	28.00	22.00	12.00	2.90	9.66	1.77
	F-S-6	27.00	21.00	10.00	2.70	10.00	1.37

Note: d_{10}, d_{30}, d_{50}, d_{60} are values corresponding to 10%, 30%, 50%, and 60% finer by weight, respectively; C_u is uniformity coefficient, $C_u = \frac{d_{60}}{d_{10}}$; and C_c is curvature coefficient, $C_c = \frac{d_{30}^2}{d_{10} \times d_{60}}$

In addition, the uniformity coefficient (C_u) and curvature coefficient (C_c) are two important indices for evaluating grain grading. In the tests, all soil samples satisfied $C_u \geq 5$ and $C_c = 1$–3, which indicates well-graded particle sizes of these soils. Comparing the indices of natural and backfilled loess, the latter decreased, and the decrease of CSYZ is greater (Table 2).

3.3. Direct Shear Tests

The shear strength indices of the four types of samples were obtained through direct shear tests (Table 3 and Figure 8). The results indicate that the cohesion of Q_3, Q_2, F-C, and F-S is 20.11 kPa, 34.51 kPa, 13.61 kPa, and 18.64 kPa and the internal friction angle is 23.7°, 28.8°, 31.0°, and 33.8°, respectively. Compared with natural loess, the cohesion of backfilled loess decreased, and the decrease of F-C is much larger than that of F-S, the internal friction increased, and the increase of F-C is less than that of F-S.

Table 3. Indices of shear strength for Q_3, Q_2, F-C, F-S samples.

Soil Sample	Number	Normal Stress (kPa)	Shear Strength (kPa)	Cohesion (kPa)	Internal Friction Angle (°)
Natural loess	Q_3	50 100 150 250	48.2 63.9 75.3 136.5	20.11	23.7
	Q_2	50 100 150 250	67.0 80.9 120.5 172.9	34.51	28.8
Backfilled loess of CSYZ	F-C-4	50 100 150 250	40.2 63.6 121.4 157.0	13.61	31.0
Backfilled loess of SJG	F-S-2	50 100 150 250	53.2 85.19 118.3 187.2	18.64	33.8

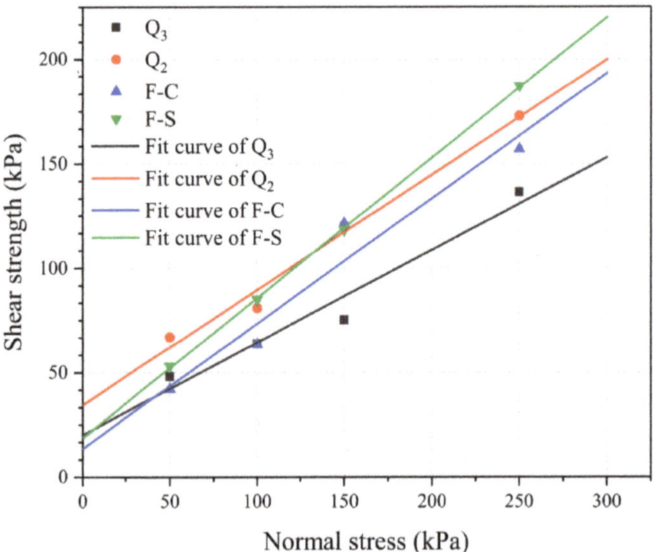

Figure 8. Normal stress–shear strength curves of Q_3, Q_2, F-C, F-S samples.

3.4. Compression Tests

Using coefficients of compressibility and modulus of compressibility as criteria to evaluate the compressibility of soils, it can be found that loess of Q_3, Q_2, and F-S has an intermediate level of compressibility, and F-C is at a low level (Table 4). The porosity ratio of F-S is lower than that of Q_3, but compressibility is greater (Figure 9), indicating that the compressibility of soil is related not only to the total pore volume, but also to the age and water content of the soil, the shape of soil particles, the strength of soil skeletons, and the pore distribution. Therefore, it is necessary to further study the microstructural characteristics of loess.

Table 4. Indices of compressibility for 14 loess groups.

Number		Q_3	Q_2	F-C-1	F-C-2	F-C-3	F-C-4	F-C-5	F-C-6
α	MPa^{-1}	0.16	0.12	0.08	0.07	0.08	0.07	0.09	0.05
E_S	MPa	12.45	14.30	19.55	27.00	18.45	23.40	18.55	33.05
Number		F-S-1	F-S-2	F-S-3	F-S-4	F-S-5	F-S-6		
α	MPa^{-1}	0.14	0.19	0.38	0.22	0.17	0.22		
E_S	MPa	11.40	9.05	9.40	7.90	9.90	7.95		

Note: α is the coefficient of compressibility and E_S is the modulus of compressibility.

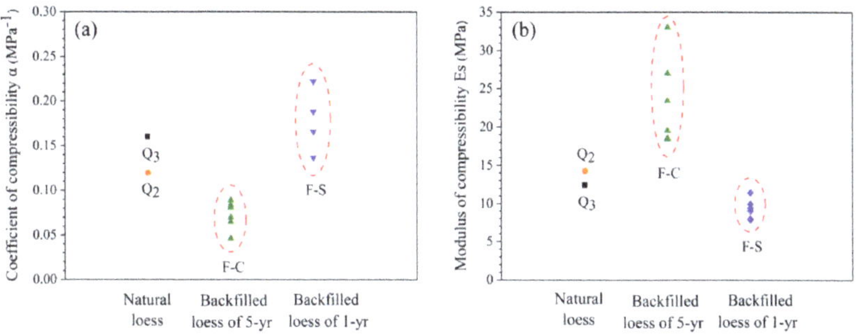

Figure 9. (**a**) Coefficient of compressibility, (**b**) modulus of compressibility.

3.5. Scanning Electron Microscope Tests

The SEM tests indicate that the microstructure of Q_3 loess is loose, with clear grains and outlines (Figure 10a,b). The shape of the grains is granular, and surfaces presented with friable minerals. The particles mainly had point-to-point contact with each other. The intergranular pores are developed, in which scaffold pores are predominant, and the connectivity among pores is pretty good (Figure 10a,b). Compared to the structure of Q_3, the structure of Q_2 loess is much denser, and the particle outlines are blurred. The particle forms mainly include aggregates and clots, and surfaces are mainly cemented face-to-face. The pores of Q_2 are reduced, and most are mosaic and intra-particle pores (Figure 10c,d). The grain distribution of the F-C backfilled loess is more uniform, and its structure is denser than that of the Q_3 loess. The aggregated clay particles are mainly attached on the surfaces, showing that soil particles tended to be rounded when softened by water. The particles have both point-to-point and face-to-face contact patterns. F-C is dominated by mosaic pores and a small number of scaffold pores. It can be seen that some small particles filled in the large gaps (Figure 10e,f). The structure of the F-S backfilled loess is also dense, while the grain distribution is not uniform. Similarly, the aggregated clay particles are mainly attached on the surfaces, and the particles have both point-to-point and face-to-face contact patterns. There are more mosaic and intra-particle pores. High magnification shows that there is structural damage among the soil particles, and some flocculent particles caused by water softening filled into the pores (Figure 10g,h).

Figure 10. SEM images of Q₃, Q₂, F-C, and F-S samples: (**a**) Q_3 under 200×, (**b**) Q_3 under 800×, (**c**) Q_2 under 200×, (**d**) Q_2 under 800×, (**e**) F-C under 200×, (**f**) F-C under 800×, (**g**) F-S under 200×, and (**h**) F-S under 800×.

3.6. Mercury Intrusion Porosimetry Tests

Figure 11 shows the curves of cumulative intrusion vs. pore size, and the volume of cumulative intrusion represents the cumulative volume of the pore. It can be seen that the largest volume of mercury intrusion is 0.3041 mL/g for the Q_3 loess, followed by 0.2420 mL/g for the Q_2 loess, and 0.1883 mL/g and 0.1785 mL/g for F-C and F-S. The corresponding porosity ratio of the four groups is calculated as 0.77, 0.61, 0.48, and 0.46. It can be found that the porosity of the backfilled loess is significantly reduced, which is consistent with the SEM test results.

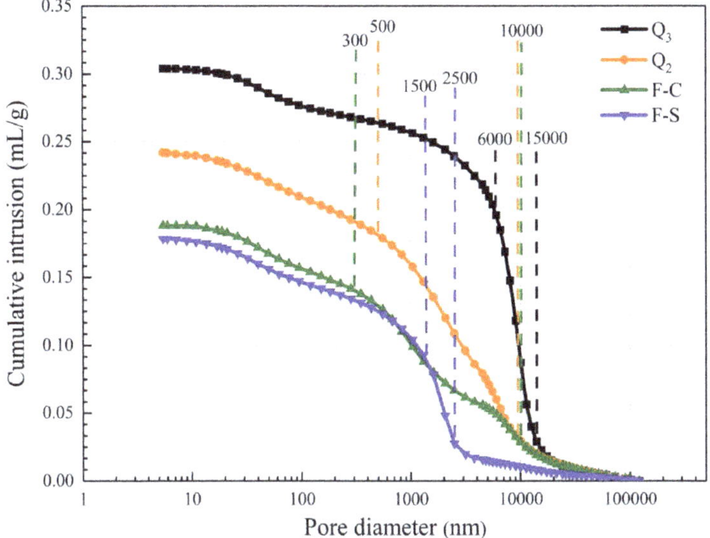

Figure 11. Cumulative intrusion vs. pore size of Q_3, Q_2, F-C, and F-S samples.

Moreover, the slope of the cumulative intrusion vs. pore size curve of Q_3 loess is abrupt in the range of pore diameters from 6000 to 15000 nm, indicating that as the pore diameter decreased, the amount of mercury intrusion changed significantly. Therefore, it can be deduced that the pore diameter of sample Q_3 is mainly distributed in the range of 6000–15,000 nm, and this range is called the dominant pore diameter. In the same way, the curve of F-S loess also has a stage of steep slope, and its dominant pore diameter ranges from 1500 to 2500 nm. However, the curves of Q_2 loess and F-C loess are relatively gentle, and the dominant pore diameter is distributed in the ranges of 500–10,000 and 300–10,000 nm, respectively. The characteristics of the dominant pore diameter are not significant, and the pore distribution is relatively uniform. At the same time, it can be found that compared with natural loess, the range of dominant pore diameters for backfilled loess moves in the direction of the smaller pore size.

Figure 12 shows curves of pore size distribution (PSD) of four types of loess. Since PSD curves for Q3 and F-S have a significant peak, they are named single-peak curves [32,33]. The peak diameter of Q_3 is 10,062 nm, and F-S is 1613 nm. However, for loess of Q_2 and F-C, there are two peaks on the PSD curves, which are named double-peak curves. The two peak values of Q_2 are 1615 and 6582 nm, and those of F-C are 832 and 8053 nm.

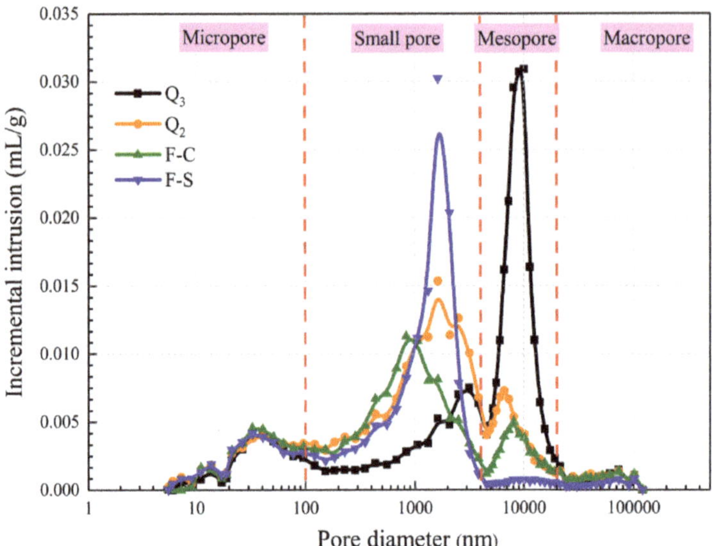

Figure 12. Pore size distribution curves of Q_3, Q_2, F-C, and F-S samples.

4. Discussion

SJG gully is about 1.5 km away from CSYZ gully, and the main strata are basically similar. There is not much difference in the elevations at the sampling sites between CSYZ and SJG, which are 1170–1185 m and 1168–1172 m. respectively. Hence, the assumption of this study is that the contribution of Q_2 and Q_3 loess in backfilled sites for CSYZ and SJG would be similar. A field survey showed that the major plant in SJG was alfalfa, and found that some crops are not growing well in the backfilled loess. Alfalfa is a plant with strong vitality in northern China that can adapt to a newly backfilled farmland environment well. Therefore, farmers always plant alfalfa on newly backfilled farmland to improve the soil. Based on the above, the effects of farming and backfilled time on soil properties were comparatively studied.

4.1. Analysis of Physical-Mechanical Properties

From Table 2 and Figure 6, the test results show that the moisture content and density of backfilled loess increased compared with natural loess. On the contrary, the porosity and plasticity indices decreased. With increased backfill time, almost all of the above-mentioned physical properties converged toward the natural loess. This reflects that backfilled soils tend to recover to their natural state when subjected to environmental processes such as wet/dry and seasonal cycles.

The analysis of grain size distribution (Figure 7) shows that there was breakage of a few large soil particles in the GLC process, and the mean grain size of backfilled farmland loess decreased (Table 2). Since the sampling depth of 0.5 m is within the influence range of plant cultivation [34–36], farming will affect the properties of soils. Some studies have shown the importance of soil functioning and soil–plant interactions [37–39]. For this study, under the influence of farm operations such as cultivation, irrigation, and digging, loess particle distribution tended to be homogenized.

Compared with natural loess, the cohesion of backfilled loess decreases, and the decrease of F-C is much larger than that of F-S, while the internal friction increases, and the increase of F-C is less than that of F-S. The reason may be that backfilled loess was disturbed in the GLC process, and the cementation and electrostatic attraction among soil particles were destroyed while the occlusal effects between particles were strengthened, so the cohesion decreased a little, and the internal friction angles increased a little. It should be noted that compared with F-C (with five years of farming), F-S (with one year of farming) has larger cohesion and internal friction as well, which indicates that with rainfall and irrigation, the water content gradually increased (Table 1), and the cohesion and internal friction angles gradually decreased under farming.

4.2. Analysis of Microstructural Characteristics

The SEM tests indicate that the grains of backfilled loess (F-C and F-S) are tightly compacted, and the distribution of grains and pores is much denser than that of natural loess (Q_2 and Q_3), thus backfilled loess has a relatively higher internal friction angle than natural loess. However, the cement between the particles has been damaged by disturbance, and the backfilled loess has relatively lower cohesion than natural loess.

There are many studies on the classification of pores in loess at home and abroad [40–43], but there is still no unified conclusion. In this study, we divided pores into four groups according to the PSD characteristics (Figure 12): Macropores (pore diameter >20 μm), mesopores (4–20 μm), small pores (0.1–4 μm), and micropores (<0.1 μm). It can be seen from Figure 12 that the four types of loess have similar volume contents of macropores and micropores, but obvious differences in the distribution of mesopores and small pores. Calculating the percentages of the four groups of pores according to the ratio of certain pore volume to total pore volume (Table 5), it can be seen that the Q_3 loess is mainly dominated by mesopores of 4000–20,000 nm, accounting for 67.5%, and the Q_2, F-C, and F-S loess have mainly small pores of 100–4000 nm, accounting for 52.5%, 51.7%, and 71.7%, respectively. This indicates that some large and medium pores in the soil were destroyed and turned into small pores and micropores as a result of the engineering of Gully Land Consolidation.

Table 5. Volume percentages of four categories of pores for Q_3, Q_2, F-C, and F-S samples.

Type of Loess	Volume Percentages (%)			
	Macropores (>20 μm)	Mesopores (4–20 μm)	Small Pores (0.1–4 μm)	Micropores (<0.1 μm)
Q_3	4.3	67.5	18.3	9.9
Q_2	6.1	26.6	52.5	14.8
F-C	6.7	22.8	51.7	19.9
F-S	3.4	5.5	71.7	19.5

4.3. Suggestion for Gully Land Consolidation

GLC changes the macroscopic properties and microstructure of loess, especially the moisture content, density, and pore size distribution. Soil moisture is a key factor influencing soil nutrient movement and soil quality in the semiarid Loess Plateau [44–46]. Meanwhile, the surface loess of backfilled farmland undergoes mechanical action, resulting in increased density and pore ratio. However, dense soil is not conducive to the growth of plant roots. Field surveys showed that most crops do not grow well in backfilled loess in the first few years. Alfalfa, as a kind of plant with strong vitality in northern China, can adapt to this new land environment well. Therefore, farmers switched to planting alfalfa on backfilled farmland in subsequent years. Under the influence of farmers' operations such as cultivation, irrigation, and digging, characteristics of backfilled loess tend to resemble those of natural loess, which will become loose and fertile. Hence, this study suggests that newly constructed farmland may not be suitable for planting crops. Farmers can make the soil fertile by planting vigorous grasses or shrubs, like alfalfa and robinia pseudoacacia. At the same time, it is important to control the compactness of soil during construction.

5. Conclusions

Test results show that physical-mechanical properties and microstructural characteristics of backfilled loess of one-year and five-year farmland are significantly affected by Gully Land Consolidation.

(1) Compared with natural loess, the moisture content, and density of backfilled loess increase. On the contrary, the pore ratio, plasticity index, and particle size index decrease. Additionally, with the development of filling time, the physical indices of backfilled loess tend to resemble those of natural loess, and the fluctuation amplitudes gradually decrease.

(2) The microstructure of soil samples observed by SEM tests indicates that natural loess has a certain skeleton strength with a relatively stable structure between the grains and pores. However, the distribution of particles and pores in backfilled loess (F-C, F-S) becomes denser, but the skeleton strength between the particles is destroyed, and the structure is more unstable. It was observed that the particles of backfilled loess are rounded with large pores filled with crushed fine particles, which results in weakness of the skeleton and cement strength among particles and pores and strengthening of the internal friction angle.

(3) By MIP tests and microscopic analysis, it can be concluded that some larger and medium pores in backfilled soils were destroyed and turned into smaller pores and micropores, and the cement strength between particles was damaged, which essentially affects the physical-mechanical properties of loess.

(4) Disturbance of backfilled loess leads to an obvious decrease of cohesion, and a slight increase in internal friction compared with natural loess. With increased backfilled time, the farming effect becomes prominent over time, the loess soil moisture content increases gradually, and both the cohesion and internal friction of the backfilled loess soil decrease to different degrees.

(5) Newly constructed farmland may not be suitable for planting crops in the first few years. Farmers can make the soil fertile by planting vigorous grasses or shrubs, like alfalfa and robinia pseudoacacia. At the same time, it is important to control the compactness of soil during construction.

Author Contributions: Conceptualization, S.Q. and S.G.; Data curation, L.M., B.Z., and S.G.; Formal analysis, L.M, S.Q., B.Z., S.G., and X.Y.; Funding acquisition, S.Q., S.G. and Q.H.; Investigation, L.M., S.Q., S.G., and Q.H.; Methodology, S.Q., S.G. and Q.H.; Project administration, S.Q., B.Z. and S.G.; Resources, S.Q.; Supervision, S.Q. and S.G.; Validation S.Q.; Writing—original draft, L.M.; Writing—review and editing, S.Q., S.G., Q.H., and X.Y. All authors have read and agreed to the published version of the manuscript.

Funding: This research was funded by the State Key Research Development Program of China under grant no. 2017YFD0800501, the Natural Science Foundation of China under grants nos. 41790442 and 41825018, and the Q.H.Strategic Priority Research Program of the Chinese Academy of Sciences under grant no. XDA23090402.

Acknowledgments: Special thanks to Zhang Bin at the China University of Geosciences, Beijing, and Yang Jijin at the Institute of Geology and Geophysics, Chinese Academy of Sciences, for their help with the experiments.

Conflicts of Interest: The authors declare no conflict of interest.

References

1. Liu, T.S. *Loess and Environment*; Science Press: Beijing, China, 1985; pp. 5–10.
2. Li, Y.H.; Du, G.M.; Liu, Y.S. Transforming the Loess Plateau of China. *Front. Agric. Sci. Eng.* **2016**, *3*, 181–185. [CrossRef]
3. Fu, B.J. Soil erosion and its control in the loess plateau of China. *Soil Use Manag.* **1989**, *5*, 76–82. [CrossRef]
4. Cai, Q.G. Soil erosion and management on the Loess Plateau. *J. Geogr. Sci.* **2001**, *11*, 53–70. [CrossRef]
5. Liu, T.S. *Loess Plateau & Agricultural Origin & Soil and Water Conservation*; Seismological Press: Beijing, China, 2004; pp. 1–15.
6. Su, Z.R.; Wang, J. Water conservation benefits' analysis of eight key national rehabilitation areas. *Soil Water Conserv. China* **1992**, *3*, 1–4.
7. Lü, Y.; Fu, B.; Feng, X.; Zeng, Y.; Liu, Y.; Chang, R.; Sun, G.; Wu, B. A policy-driven largescale ecological restoration: Quantifying ecosystem services changes in the Loess Plateau of China. *PLoS ONE* **2012**, *7*, e31782.
8. Chen, Y.P.; Zhang, Y. Sustainable Model of Rural Vitalization in Hilly and Gully Region on Loess Plateau. *Bull. Chin. Acad. Sci.* **2019**, *34*, 708–716. [CrossRef]
9. Xiao, J.F. Satellite evidence for significant biophysical consequences of the "Grain for Green" Program on the Loess Plateau in China. *J. Geophys. Res. Biogeosci.* **2014**, *119*, 2261–2275. [CrossRef]

10. Li, S.; Yang, S.N.; Liu, X.F.; Liu, Y.X.; Shi, M. NDVI-based analysis on the influence of climate change and human activities on vegetation restoration in the Shaanxi-Gansu-Ningxia Region, Central China. *Remote Sens.* **2015**, *7*, 11163–11182. [CrossRef]

11. He, C.X. How to develop modern agriculture in Yan 'an on the basis of the Gully Reclamation Project. *J. Yan'an Univ. (Soc. Sci.)* **2013**, *35*, 3.

12. Zhou, H.L.; Wang, W.Z.; Hao, Z.L. Gully Land Consolidation Benefits the People, Rebuild a Good South in North Shaanxi. Available online: http://www.igsnrr.ac.cn/xwzx/zhxw/201211/t20121116_3684668.html (accessed on 16 December 2015).

13. He, C.X. The situation, characteristics and effect of the Gully Reclamation Project in Yan'an. *J. Earth Environ.* **2015**, *6*, 59–64.

14. Liu, Q.; Wang, Y.; Zhang, J.; Chen, Y. Filling gullies to create farmland on the Loess Plateau. *Environ. Sci. Technol.* **2013**, *47*, 7589–7590. [CrossRef] [PubMed]

15. Liu, Z.P.; Shao, M.A.; Wang, Y.Q. Large-scale spatial interpolation of soil pH across the Loess Plateau, China. *Environ. Earth Sci.* **2013**, *69*, 2731–2741. [CrossRef]

16. Zhang, X.B.; Jin, Z. Gully land consolidation project in Yan'an is inheritance and development of wrap land dam project on the Loess Plateau. *J. Earth Environ.* **2015**, *6*, 261–264.

17. Jin, Z. The creation of farmland by gully filling on the Loess Plateau: A double-edged sword. *Environ. Sci. Technol.* **2014**, *48*, 883–884. [CrossRef] [PubMed]

18. Wei, H.A.; Wang, J.Y. Assessment of land consolidation suitability in Loess Hilly-gully Region in Yan'an City. *Areal Res. Dev.* **2013**, *3*, 129–132.

19. Jin, Z.; Guo, L.; Wang, Y.; Yu, Y.; Lin, H.; Chen, Y.; Chu, G.; Zhang, J.; Zhang, N. Valley reshaping and damming induce water table rise and soil salinization on the Chinese loess plateau. *Geoderma* **2019**, *339*, 115–125. [CrossRef]

20. Liu, Y.S.; Li, Y.R. Engineering philosophy and design scheme of gully land consolidation in Loess Plateau. *Trans. Chin. Soc. Agric. Eng.* **2017**, *33*, 1–9. [CrossRef]

21. Gao, G.R. Study of the Microstructures and the Collapse Mechanism in Loess Soil from Lanzhou. *J. Lanzhou Univ.* **1979**, *2*, 123–134.

22. Gao, G.R. Microstructure of loess soil in china relative to geologic environment. *Acta Geol. Sin.* **1984**, *58*, 265.

23. Gao, G.R. The distribution and geotechnical properties of loess soils, lateritic soils and clayey soils in China. *Eng. Geol.* **1996**, *42*, 95–104. [CrossRef]

24. Lei, X.Y.; Wang, S. Size of loess pores in relation to collapsibility. *Hydrogeol. Eng. Geol.* **1987**, *31*, 15–18.

25. Zhao, J.B.; Chen, Y. Study on porosity and collapsibility of loess. *J. Eng. Geol.* **1994**, *2*, 76–83.

26. Chen, Z.H.; Fang, X.W.; Zhu, Y.Q.; Qin, B.; Yao, Z.H. Research on meso-structures and their evolution laws of expansive soil and loess. *Rock Soil Mech.* **2009**, *30*, 1–11. [CrossRef]

27. Jiang, M.J.; Hu, H.J.; Peng, J.B.; Leroueil, S. Experimental study of two saturated natural soils and their saturated remolded soils under three consolidated undrained stress paths. *Front. Archit. Civ. Eng. China* **2011**, *5*, 225–238. [CrossRef]

28. Jiang, Z.K. Analysis of Stability and Slope Ratio Optimization of Excavated Slope of Gully Land Consolidation Project in the Loess Hilly Area: A Case Study of Nangou in Yan'an City. Master Thesis, Chang'an University, Xi'an, China, 2019.

29. Ministry of Water Resources of the People's Republic of China. *Standard for Soil Test method GB/T 50123-1999*; China Planning Press: Beijing, China, 1999.

30. Washburn, E.W. Note on a method of determining the distribution of pore sizes in a porous material. *Proc. Natl. Acad. Sci. USA* **1921**, *7*, 115–116. [CrossRef] [PubMed]

31. Li, T.L.; Fan, J.W.; Xi, Y.; Xie, X.; Hou, X.K. Analysis for effect of microstructure on SWCC of compacted loess. *J. Eng. Geol.* **2019**, *27*, 1019–1026. [CrossRef]

32. Ehlers, W.; Wendroth, O.; De Mol, F. Characterizing pore organization by soil physical parameters. In *Soil Struct*; CRC: Boca Raton, FL, USA, 1995; p. 257.

33. Gao, L.; Wang, B.; Li, S.; Wu, H.; Wu, X.; Liang, G.; Gong, D.; Zhang, X.; Cai, D. Degré, A. Soil wet aggregate distribution and pore size distribution under different tillage systems after 16 years in the Loess Plateau of China. *CATENA* **2019**, *173*, 38–47. [CrossRef]

34. Venkatesh, B.; Lakshman, N.; Purandara, B.K.; Reddy, V.B. Analysis of observed soil moisture patterns under different land covers in Western Ghats, India. *J. Hydrol.* **2011**, *397*, 281–294. [CrossRef]

35. Jia, Y.H.; Shao, M.A. Dynamics of deep soil moisture in response to vegetational restoration on the Loess Plateau of China. *J. Hydrol.* **2014**, *519*, 523–531. [CrossRef]

36. Deng, L.; Yan, W.; Zhang, Y.; Shangguan, Z. Severe depletion of soil moisture following land-use changes for ecological restoration: Evidence from northern China. *For. Ecol. Manag.* **2016**, *366*, 1–10. [CrossRef]

37. Frouz, J.; Livečková, M.; Albrechtová, J.; Chroňáková, A.; Cajthaml, T.; Pižl, V.; Háněl, L.; Starý, J.; Baldrian, P.; Lhotáková, Z.; et al. Is the effect of trees on soil properties mediated by soil fauna? A case study from post-mining sites. *Ecol. Manag.* **2013**, *309*, 87–95. [CrossRef]

38. Kravchenko, A.N.; Guber, A.K. Soil pores and their contributions to soil carbon processes. *Geoderma* **2016**, *287*, 31–39. [CrossRef]

39. Wang, H.; Yue, C.; Mao, Q.; Zhao, J.; Ciais, P.; Li, W.; Yu, Q.; Mu, X. Vegetation and species impacts on soil organic carbon sequestration following ecological restoration over the Loess Plateau, China. *Geoderma* **2020**, *371*, 114389. [CrossRef]

40. Lei, X.Y. Pore classification and collapsibility of loess in China. *Sci. China* **1987**, *12*, 1309–1318.

41. Osipov, V.I.; Sokolov, V.N. Factors and mechanism of loess collapsibility. In *Genesis and Properties of Collapsible Soils*; Derbyshire, E., Ed.; Proc. Workshop: Loughborough, UK, 1994; pp. 49–64.

42. Yang, Y.L. Study on collapsible mechanism of loess. *Sci. China* **1988**, *7*, 756–766.

43. Wei, Y.N.; Fan, W.; Yu, B.; Deng, L.S.; Wei, T.T. Characterization and evolution of three-dimensional microstructure of Malan loess. *CATENA* **2020**, *192*, 104585. [CrossRef]

44. Zhou, J.Y.; Gu, B.J.; Schlesinger, W.H.; Ju, X.T. Significant accumulation of nitrate in Chinese semi-humid croplands. *Sci. Rep. UK* **2016**, *6*, 25088. [CrossRef] [PubMed]

45. Song, X.L.; Gao, X.D.; Dyck, M.; Zhang, W.; Wu, P.T.; Yao, J.; Zhao, X.N. Soil water and root distribution of apple tree (Malus pumila mill) stands in relation to stand age and rainwater collection and infiltration system (RWCI) in a hilly region of the Loess Plateau, China. *Catena* **2018**, *170*, 324–334. [CrossRef]

46. Liu, Z.; Ma, P.; Zhai, B.; Zhou, J. Soil moisture decline and residual nitrate accumulation after converting cropland to apple orchard in a semiarid region: Evidence from the Loess Plateau. *Catena* **2019**, *181*, 104080. [CrossRef]

 *sustainability*

Article

Effectiveness of Polyacrylamide in Reducing Runoff and Soil Loss under Consecutive Rainfall Storms

Birhanu Kebede [1,2,*], Atsushi Tsunekawa [3], Nigussie Haregeweyn [4], Amrakh I. Mamedov [3], Mitsuru Tsubo [3], Ayele Almaw Fenta [3], Derege Tsegaye Meshesha [5], Tsugiyuki Masunaga [6], Enyew Adgo [5], Getu Abebe [1] and Mulatu Liyew Berihun [1,2]

[1] The United Graduate School of Agricultural Sciences, Tottori University, 4-101 Koyama-Minami, Tottori 680-8553, Japan; gabebe233@gmail.com (G.A.); mulatuliyew@yahoo.com (M.L.B.)

[2] Faculty of Civil and Water Resource Engineering, Bahir Dar Institute of Technology, Bahir Dar University, P.O. Box 26, Bahir Dar, Ethiopia

[3] Arid Land Research Center, Tottori University, 1390 Hamasaka, Tottori 680-0001, Japan; tsunekawa@tottori-u.ac.jp (A.T.); mamedov@tottori-u.ac.jp (A.I.M.); tsubo@tottori-u.ac.jp (M.T.); ayelealmaw@tottori-u.ac.jp (A.A.F.)

[4] International Platform for Dry Land Research and Education, Tottori University, 1390 Hamasaka, Tottori 680-0001, Japan; nigussie_haregeweyn@yahoo.com

[5] College of Agriculture and Environmental Sciences, Bahir Dar University, P.O. Box 1289, Bahir Dar, Ethiopia; deremesh@yahoo.com (D.T.M.); enyewadgo@gmail.com (E.A.)

[6] Faculty of Life and Environmental Science, Shimane University, Shimane Matsue 690-0823, Japan; masunaga@life.shimane-u.ac.jp

* Correspondence: birhanukg@gmail.com

Received: 3 January 2020; Accepted: 18 February 2020; Published: 20 February 2020

Abstract: The use of anionic polyacrylamide (PAM) as a soil conditioner could help prevent soil loss by water. In this study, we determined the effective granular PAM rate that best reduces runoff and soil loss from Oxisols. Furthermore, the effectiveness of the selected PAM rate was tested by applying it in a mixture with gypsum (G) or lime (L). The study was conducted in two phases: (i) Dry PAM rates of 0 (C), 20 kg ha^{-1} (P20), 40 kg ha^{-1} (P40), and 60 kg ha^{-1} (P60) were applied onto soil surface and run for six consecutive rainfall storms of 70 mm h^{-1} intensity for 1 h duration, and the effective PAM rate was selected; and (ii) G (4 t ha^{-1}) or L (2 t ha^{-1}) were applied alone or mixed with the selected PAM rate. The P20 was found to be effective in reducing runoff in the beginning while P40 and P60 were more effective starting from the third storm through the end of the consecutive storms, but with no statistically significant difference between P40 and P60. Hence, P40 was selected as the most suitable rate for the given test soil and rainfall pattern. On the other hand, the mixed application of P40 with G or L increased infiltration rate (IR) in the first two storms through improving soil solution viscosity. However, effectiveness of the mixtures had diminished by various degrees as rain progressed, as compared to P40 alone, which could be attributed to the rate and properties of G and L. In conclusion, the variation in effectiveness of PAM rates in reducing runoff with storm duration could indicate that the effective rates shall be selected based on the climatic region in that lower rates for the short rains or higher rates for elongated rains. Moreover, combined application of PAM with L could offer a good option to both fairly reduce soil erosion and improve land productivity especially in acidic soils like Oxisols, which requires further field verification.

Keywords: polyacrylamide; gypsum; lime; runoff; soil loss; dryland

---

1. Introduction

Soil erosion by water is the most threatening global problem causing adverse on- and off-site consequences, such as the depletion of soil fertility [1–3], siltation of downstream reservoirs [4–6], loss of vital ecosystem services, and associated economic costs [7]. Low soil fertility due to the removal of organic matter rich surface soil by erosion and exposure of lower soil layers causes nutrient deficits and physical hindrance to root growth, leading to reduced soil productivity [8]. However, this effect of erosion on soil productivity is a slow process and might not be noticed until crop production is no longer economically viable [9]. As a result, soil erosion has become a major threat to food security, particularly for developing countries where livelihoods predominantly depend on agriculture [2,10–15]. Hence, soil and water conservation are essential for sustaining food production and preserving the environment [16].

The major causes of soil erosion are the physical disintegration and dispersion of surface soil aggregates by the impact energy of raindrops [17,18] and the physico-chemical dispersion and migration of soil clays with the infiltrating water into the soil, thereby clogging the conducting pores [19–21], leading to crust formation, decreased infiltration, and increased runoff and soil loss [21–23]. Soil erosion can be prevented by using physical and biological measures, or through conventional management practices such as mulching, growing cover crops, etc. An alternative practice is the modification of soil properties through the application of chemical amendments to the soil, such as polyacrylamide (PAM) [22] to decrease aggregate disintegration.

PAM is a water-soluble, organic anionic polymer having a long molecule of identical atom chains held together by covalent bonds [24] that form bridges with the soil particles through cations in soil solution [25]. PAM has proved to be superior to other polymers in controlling erosion [26] and is also used to improve soil physical properties [25]. PAM remains effective in reducing soil loss by limiting the physical disintegration of aggregates caused due to water drop impact [27] by adsorption to the soil aggregates and increasing cohesion among soil particles [28], thus increasing the resistance of aggregates to the direct impact of raindrops or dragging force by runoff. Factors like soil characteristics, water quality, and PAM properties such as charge density and molecular weight play important roles in the adsorption of PAM [25]. PAM can stabilize an existing soil structure by preserving pervious pore structure during the surface seals formation [29], but is unable to remediate a poor soil structure [30]. PAM is infinitely soluble in water but dissolves very slowly [31]. It is more readily adsorbed by the water of a higher electrolyte concentration than by water with lower electrolyte concentration [32]. Nevertheless, the addition of dissolved PAM may have some negative effects, such as enhancing water viscosity at the beginning, which in turn could lead to a decrease in infiltration rate (IR) and increased runoff, although it may decrease soil erosion [28,33]. Furthermore, the high application rates required for effective erosion control and the large volume of water required for effective dissolution are the two major obstacles constraining the usage of PAM in agriculture [34].

Although anionic PAM is the most effective polymeric soil amendment to control erosion [35], its effectiveness can be enhanced by introducing a source of electrolyte that can create a cation bridge and help the polymer to adsorb to the soil [26]. The introduction of electrolyte (such as Ca^{2+}) at the soil surface reduces chemical dispersion and migration of clay particles by strengthening the bonds between primary soil particles, thus reducing seal formation [31,36]. Electrolytes are typically introduced in the form of gypsum and lime. The increase in the electrolyte concentration in soil replaces exchangeable Sodium (Na^+) ions from the exchange complex with dissolved Calcium (Ca^{2+}) ions. In addition, it decreases clay dispersion and surface sealing [36], enhances soil structure, increases infiltration rate (IR), and decreases runoff and sediment loss [37]. Furthermore, the effectiveness of PAM application depends primarily on the soil type and percentage of clay in the soil [35,38].

Many studies have been conducted to determine the effective rates of anionic PAM that can reduce runoff and soil loss through soil structure stabilization [39,40]. Most of these studies have shown that application rates of 10–20 kg ha^{-1} were effective in stabilizing the surface structure and decreasing runoff and soil erosion [40]. In earlier times, Gabriels et al. [41] found that applying 38 kg ha^{-1} of anionic PAM to soil surface resulted in increased IR and reduced runoff while other researchers [25,42,43] suggested the use of 20 kg ha^{-1} PAM as an effective and economical application rate. However, these effective PAM rates in most of the studies were determined using a maximum of three simulated rainfall storms (<250 mm rainfall), which may represent dry land regions.

For example, Lado et al. [39] evaluated the effectiveness of granular PAM at rates of 0, 25, 50, and 100 kg ha^{-1} to reduce post-fire erosion in a Calcic Regosol affected by different fire conditions using three consecutive rainfall storms of 80 mm depth each with an intensity of 47 mm h^{-1}. They found that the application of 50 kg ha^{-1} granular PAM increased runoff during the first storm due to increased viscosity of runoff. However, this rate was more effective in reducing soil loss during the three storms in unburnt and moderately burnt soils, with a total reduction of 42% and 34%, respectively. In addition, Abrol et al. [40] also evaluated the effect of the application rate of granular PAM (0, 5, 10, and 20 kg ha^{-1}) on IR as a function of cumulative rainfall in silt loam soil using 2-hour simulated rainstorm at rainfall intensity of 37 mm h^{-1}. That study showed that the application rate of 10 kg ha^{-1} was more effective in increasing IR and reducing erosion.

In both studies, the higher PAM rates, 100 kg ha^{-1} [39] and 20 kg ha^{-1} [38], were less effective, as compared to lower and medium rates. This could be that the amount of rainfall applied in these storms was insufficient to completely dissolve PAM at higher rates, as higher rates require a large volume of water for effective dissolution [34]. This suggests the importance of applying sufficient rainfall such that all the applied PAM gets fully dissolved to effectively act on soil aggregate stabilization and then monitoring the effect of PAM application rates on IR, runoff, and soil loss. PAM effectiveness is, therefore, believed to be strongly influenced by the rainfall pattern that a certain soil is subjected to and requires study under consecutive rainfall storms of high intensity. Hence, treating a test soil with different PAM rates, exposing it to consecutive rainfall storms of high intensity, which may represent humid and sub-humid regions, and determining the effective PAM rate that produces better results in terms of reducing runoff and soil loss is the best way forward. Moreover, applying PAM mixed with some source of electrolytes (e.g., gypsum and lime) at the effective rate and exposing it to consecutive rainfall storms will help us to understand whether the mixture of PAM and electrolytes can further enhance PAM effectiveness for a given test soil under consecutive storms.

The purpose of this study was, therefore, to evaluate the effectiveness of PAM rates (0, 20, 40, 60 kg ha^{-1}) in increasing IR and reducing runoff and soil loss when applied to a test soil (acidic Oxisol) and subjected to consecutive rainfall storms. The two main objectives were:

(i) To determine the effective granular PAM rate that increases IR and reduces runoff and soil loss under consecutive rainfall storms; and
(ii) To verify whether the effectiveness of the selected rate can further be enhanced by applying it mixed with gypsum (4 t ha^{-1}) or lime (2 t ha^{-1}) as a source of electrolyte.

2. Materials and Methods

2.1. Experimental Setup and Materials

This experiment was conducted at Arid Land Research Center (ALRC) of Tottori University during January–March 2019. The test soil used for this experiment was acidic clay red soil (similar to Oxisols by US Taxonomy or Acrisols and Luvisols by FAO classification), one of the widely abundant soils in Japan. Small runoff boxes of dimensions 50-cm length, 30-cm width, and 7-cm depth were filled with soil (Figure 1). The boxes have two outlets for collecting surface runoff and percolating (infiltration) water. The amendments used were environmentally friendly, non-toxic anionic PAM (Superfloc A-110, granular powder, 10–12% hydrolysis, and 12 Mg mole^{-1} molecular weight), gypsum, and lime. Similar to most previous studies, we were interested in granular application not only due to its durability but also because the granular method of application improves handling and reduces costs compared with that of the fluid (solution) application [16,37].

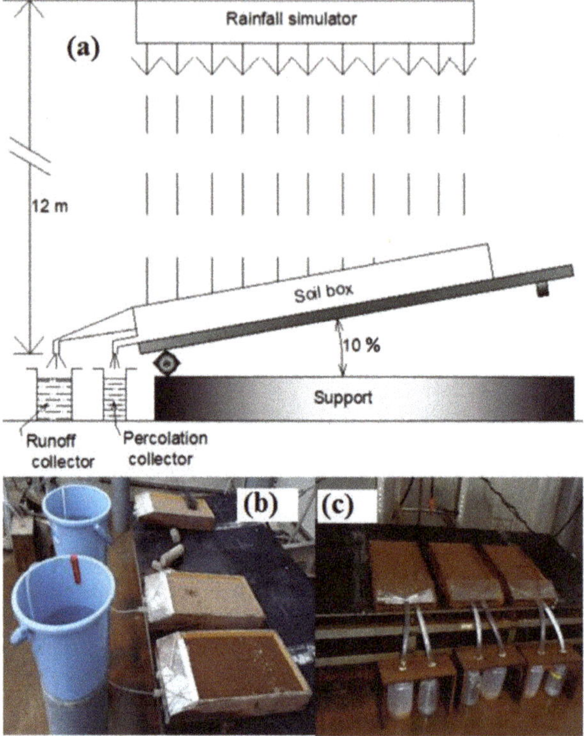

Figure 1. Experimental set-up showing (**a**) rainfall simulator, experimental soil box, and runoff and percolating water collector; (**b**) different treatments being saturated from below; and (**c**) runoff and percolating water collection during simulation.

The soil was air-dried, gently crushed, and sieved through a 5 mm sieve. Firstly, we packed gravel of 1–2 mm diameter up to a depth of 2 cm in the small runoff boxes to allow the percolation of infiltrated water. The sieved soil was then packed up to a depth of 3 cm in each box over the 2 cm layer of gravel, compacted to a bulk density of 1.2 g cm^{-3} using a wooden log, and finally, amendments were applied. PAM was applied onto the soil surface at a rate of 0 (C), 20 kg ha^{-1} (P20), 40 kg ha^{-1} (P40), and 60 kg ha^{-1} (P60) in granular form. The entire PAM rates were then exposed to the six consecutive storms, runoff and soil loss were measured, and effective PAM rate was selected. Later, gypsum (G) at a rate of 4 t ha^{-1}, and lime (L) at a rate of 2 t ha^{-1} were applied alone or in mixture with the effective PAM rate selected above. Each amendment was applied (uniformly distributed) over the soil surface by hand. For the combined treatments of PAM mixed with gypsum or lime, first PAM alone was applied followed by the application of gypsum or lime.

2.2. Rainfall Simulation Procedure

A drip-type rainfall simulator facility with raindrop fall-height of 12 m, raindrop diameter of 3 mm, and rain kinetic energy of 29 J^{-1} m^{-2} mm^{-1} [44] at the ALRC, Tottori University, Japan was used in this study. The simulation was conducted in two phases: The first phase using four PAM rates (0, 20, 40, and 60 kg ha^{-1}) and the second phase using gypsum, lime, and effective PAM rate mixed with gypsum or lime. Out of the four rates, the effective PAM rate was selected based on the simulation results from the first phase.

In both phases, after all amendments were applied, the boxes were placed over the simulator tray in a horizontal position and saturated from below for 10 to 15 minutes with deionized water in order to facilitate the immediate measurement of infiltrating water during the first simulation. After saturation, the boxes were air-dried for 24 h prior to the start of the simulation. The simulator tray was set to a slope of 10%, and three soil-packed runoff boxes, with treatments assigned randomly, were positioned side-by-side on the sloped platform to allow simultaneous testing of different treatments. Boxes were then exposed to six consecutive rainstorms using tap water (EC= 0.07 dS m^{-1}). Rainfall intensity was 70 mm h^{-1} lasting for 1 h rainfall duration, and time interval between two consecutive rainstorms was two days (48 h). Although the rainfall intensity of 70 mm h^{-1} lasting for 1 h may not be common under natural conditions, globally large-magnitude events are assumed to be a dominant cause of soil erosion [45,46]. Initially, runoff boxes were covered with plastic covers, and after the rain intensity was stabilized, we removed the plastic covers and started recording the time taken to initiate runoff (TRO), the infiltrated water that come to the box outlet as sub subsurface flow, and runoff (Figure 1).

During each storm, runoff and infiltrating water were collected at every 10-min interval, throughout the 1-h storm duration, using temperature resistant graduated plastic bottles placed underneath the outlet at the bottom of the box. At the end of the simulation, the 10-min volume of runoff and infiltrated water was measured using a graduated cylinder. The runoff volumes were oven-dried at 105 °C using the temperature resistant plastic bottles, and the weights of sediments in the runoff were determined. Then, the total soil loss of a treatment was found by summing up the amount of sediments obtained from the oven dried 10-min runoff volume in the six consecutive storms. It is worth noting that the transfer of splashed particles from one box to the other during rainfall simulations might affect the results. However, for the rain simulator and intensity used in this study, the splash erosion was reported to be minimal [47]. Therefore, splash from the runoff boxes, which could be directly related to the rain intensity or erosivity, was not measured. The cumulative runoff and cumulative soil loss were subjected to statistical analysis using IBM SPSS Statistics version 22 software. The simulation datasets (cumulative runoff and soil loss) were tested for normality and found to be significantly different from the normal distribution; hence parametric statistical tests could not be used for this study. Therefore, the differences among the median cumulative runoff and soil loss were subjected to analysis of variance using the Kruskal–Wallis non-parametric test [48] at a significance level of 0.95 ($\alpha = 0.05$).

2.3. Soil Physico-chemical Properties and Aggregate Stability Determination

Original soil properties and soil properties following the final rainfall storm were determined. Soil samples were collected from each treatment, dried, and used to determine the soil properties. Soil texture was determined for untreated soil samples using the hydrometer method [49]. The air-dried soil samples were used for the determination of electrical conductivity at 1:5 ratio, soil pH at 1:2.5 ratio, soil organic carbon with C/N coder apparatus, and cation exchange capacity (CEC) using atomic absorption spectroscopy (AAS) after extraction with ammonium acetate buffered at pH 7 (Table 1). The soil is characterized by clay texture (44.4%, 13.4%, and 42.2% for clay, silt, and sand, respectively) and with 3.6% organic matter content.

Table 1. Effect of treatments on soil properties. Different treatments are control (C), polyacrylamide (PAM) of 20 kg ha^{-1} (P20), 40 kg ha^{-1} (P40), 60 kg ha^{-1} (P60), PAM of 40 kg ha^{-1} + gypsum (4t ha^{-1}) (P+G), gypsum (4t ha^{-1}) (G), lime (2t ha^{-1}) (L), and PAM of 40 kg ha^{-1} + Lime (2t ha^{-1}) (P+L).

Treatments	pH	EC (dS m^{-1})	CEC (cmol$_c$ kg^{-1})	NH$_4$OA$_c$ Extractable Cations (cmol$_c^+$ kg^{-1})			
				Ca^{2+}	Mg^{2+}	K$^+$	Na$^+$
Untreated soil	4.84	0.05	16.35	10.83	4.40	0.71	0.41
C	4.86	0.05	14.69	9.52	4.20	0.62	0.35
P20	4.86	0.05	15.19	9.80	4.33	0.69	0.36
P40	5.03	0.05	15.15	9.88	4.29	0.59	0.39
P60	5.09	0.05	15.48	9.49	4.96	0.64	0.39
P40+G	5.11	0.08	17.78	11.03	6.00	0.57	0.18
G	4.85	0.12	16.18	10.93	4.50	0.56	0.19
L	6.75	0.37	19.23	13.52	4.80	0.67	0.24
P40+L	5.14	0.09	16.86	11.79	4.10	0.69	0.28

Aggregate (structure) stability with three replications (taken from upper 5 mm of the soil in each box) was determined using the modified high energy moisture characteristics (HEMC) method [38,50]. In this method, soil aggregates are wetted rapidly in a controlled manner, and a moisture content curve, at a matric potential range of 0–50 cm, corresponding to drainable pores of >60 µm, with small steps of 1–2 cm, was generated using a hanging water column. An index of aggregate stability or structural index (SI) was determined from differences among the water retention curves (differences in pore size distribution) of the treatments by using their specific water capacity curves. The volume of drainable pores and modal suction (matric potential at the peak of the specific water capacity curve corresponds to the most frequent pore size) were determined and SI was defined as the ratio of the volume of drainable pores to modal suction, and used to characterize soil aggregate and structure stability; the higher the value of SI, the higher the stability of samples [50].

3. Results

3.1. Effect of Treatments on Soil Properties

The effect of soil amendments on soil properties is presented in Table 1. Generally, the lime substantially increased the soil pH, EC, and CEC more than other treatments. However, this effect of lime was decreased when combined with PAM. Gypsum did not increase pH but slightly increased EC and CEC. This difference could probably be attributed to leaching of gypsum by the continued rainfall due to its high solubility.

3.2. Time to Runoff (TRO)

Time to runoff (TRO) for the different treatments is presented in Figure 2. During Storm 1, the shortest TRO i.e., 67% reduction was observed for higher PAM rates (P40 and P60), followed by 33% reduction for P20 and PAM associated treatments (i.e., P+G and P+L) while 33% increment was observed for gypsum treatment, compared with the control (Figure 2). For Storms 2 to 6, the runoff was initiated immediately after the start of rainfall.

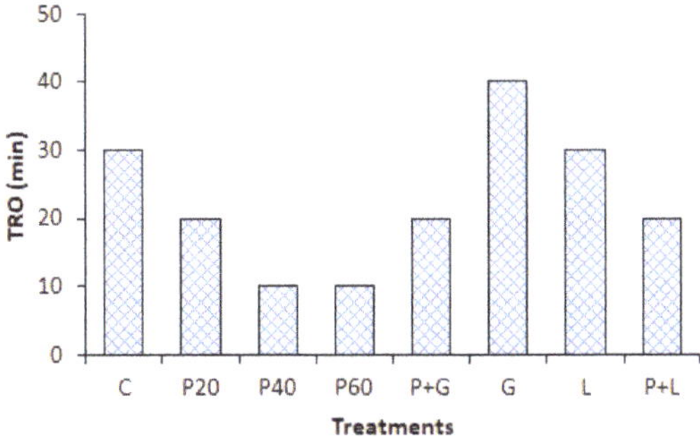

Figure 2. Effect of treatments on time to runoff (TRO) for control (C), PAM of 20 kg ha^{-1} (P20), 40 kg ha^{-1} (P40), 60 kg ha^{-1} (P60), PAM of 40 kg ha^{-1} + gypsum (4 t ha^{-1}) (P+G), gypsum (4 t ha^{-1}) (G), lime (2 t ha^{-1}) (L), and PAM of 40 kg ha^{-1} + Lime (2 t ha^{-1}) (P+L).

3.3. *Effect of Polyacrylamide (PAM) Rates on Infiltration Rate (IR), Runoff, and Soil Loss*

IR, runoff, and soil loss measured from the fine texture Oxisol treated with PAM rates of 0, 20, 40, and 60 kg ha^{-1} and subjected to six consecutive rainstorms separated by drying periods are presented in Figure 3. At the beginning of Storm 1, compared with the control, IR for all PAM rates reduced by 4%, 24%, and 46% for P20, P40, and P60, respectively. However, as the rainfall continued from Storms 2 to 6, IR for all PAM rates increased by 61–147%, 63–268%, and 20–338% for P20, P40, and P60, respectively. Furthermore, IR for higher PAM rates slightly decreased during Storm 2, compared with IR during Storm 1. However, IR sharply increased during Storms 3 and 4 but decreased again during Storm 5 and 6 for the higher rates. IR values from P20 continuously decreased throughout the consecutive storms, however, the final infiltration rate (FIR) during Storm 6 increased by 61%, 182%, and 229% for P20, P40, and P60, respectively, compared with the control (Figure 3).

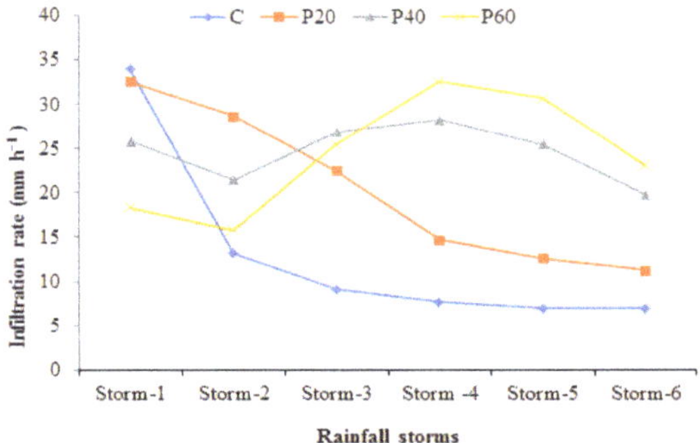

Figure 3. Effect of PAM rates on infiltration rate (mm h^{-1}). Each of the six storms had an average rainfall intensity of 70 mm h^{-1} for 1-h duration and 48-h drying period. The treatments were control (C), PAM of 20 kg ha^{-1} (P20), 40 kg ha^{-1} (P40), and 60 kg ha^{-1} (P60).

Following the reduction in IR at different PAM rates, when compared with the control, cumulative runoff increased during Storm 1 by 5%, 28%, and 54% for P20, P40, and P60 rates, respectively. However, cumulative runoff decreased for all the rates later in the consecutive storms depending on the changes in IR (Figure 4a). During Storms 2 to 6, cumulative runoff reduced by 7–27%, 15–29%, and 5–40% for P20, P40, and P60, respectively. The cumulative runoff increased in the order P60 > P40 > P20 for Storms 1 and 2 and in the order C > P20 > P60 > P40 for Storms 4 to 6. Nevertheless, runoff from control was the lowest during Storm 1, but increased sharply during Storm 2, and was the highest of all treatments throughout the consecutive Storms 2 to 6 (Figure 4a). The total runoff from all the six consecutive storms decreased in the order of C > P20 > P60 > P40 (Table 2).

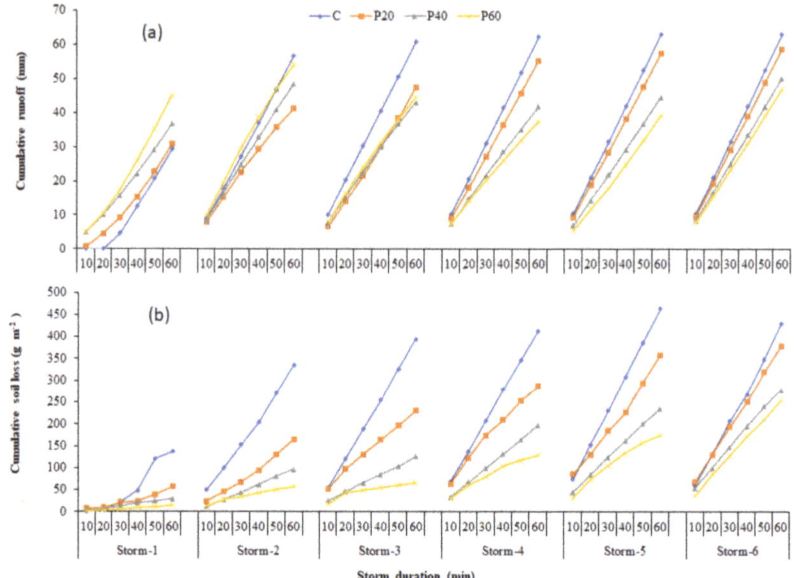

Figure 4. Effect of PAM rates on (**a**) cumulative runoff (mm) and (**b**) cumulative soil loss (g m^{-2}). Each of the six storms had an average rainfall intensity of 70 mm h^{-1} for 1-h duration and 48-h drying period interval. Data were collected at 10-min intervals in the 1-h rainfall duration. The treatments were control (C), PAM of 20 kg ha^{-1} (P20), 40 kg ha^{-1} (P40), and 60 kg ha^{-1} (P60).

Table 2. Effect of PAM rates on total runoff and total soil loss with respective percent reduction (%) compared with control and cumulative median runoff and soil loss from the six storms. Different letters following treatments indicate a significant difference between treatments using the Kruskal-Wallis test ($\alpha = 0.05$). The treatments were control (C), PAM of 20 kg ha^{-1} (P20), 40 kg ha^{-1} (P40), and 60 kg ha^{-1} (P60).

Treatments	Runoff			Soil Loss		
	Total Runoff (mm)	Reduction in Total Runoff (%)	Cumulative Median Runoff (mm)	Total Soil Loss (g m^{-2})	Reduction in Total Soil Loss (%)	Cumulative Median Soil Loss (g m^{-2})
C	336		62 [a]	2152		407 [a]
P20	292	13	52 [b]	1491	31	280 [b]
P40	265	21	45 [b]	919	57	159 [c]
P60	267	20	45 [b]	693	68	97 [c]

On the other hand, compared with the control, all PAM rates reduced cumulative soil loss ranging from 12–59%, 35–78%, and 41–90% for P20, P40, and P60 rates, respectively during Storms 1 to 6 (Figure 4b). For all the PAM rates, the maximum percentage reduction in soil loss was observed during the first and minimum was observed during the last storm. After six storms, the total soil loss decreased in the order C > P20 > P40 > P60 (Table 2). Despite the fact that all PAM rates reduced both runoff and soil loss, P40 and P60 reduced median cumulative soil loss significantly, compared with the control, however, there was no significant difference in the reduction caused by P40 and P60 treatments. (Table 2).

3.4. Effect of PAM, Gypsum, and Lime on Infiltration Rate (IR), Runoff, and Soil Loss

As P40 was determined as the effective rate in the first phase of the experiment, it was applied mixed with gypsum or lime at the same rate and subjected to the same number of storms as PAM treatments. During Storm 1, the highest IR was observed for gypsum treatment (41 mm h^{-1}) i.e., an increase of 21% compared with the control while the lowest was observed for P40 (26 mm h^{-1}) with reduction of 24%. During Storms 2 to 6, as the rain progressed, IR continued to decrease for all the treatments, except P40 in the order P+L > P+G > G or L > C (Figure 5). However, IR for P40 treatment sharply increased during Storm 3 and reached its maximum (28 mm h^{-1}, 268% increment) during Storm 4, however, it continuously dropped during Storms 5 and 6. Thus, for Storms 4 to 6, IR decreased in the order P40 > P+L > P+G > G or L > C, with respective final IR increment of 182%, 100%, 63%, 9%, and 4% (Figure 5). Due to the reduced IR during Storm 1, runoff from P40 increased by 28% during Storm 1 but decreased up to 33% in the subsequent storms. For other treatments, reduction in cumulative runoff ranged between 8–37% for P+L, 7–28% for P+G, 1–24% for gypsum, and 0–11% for lime (Figure 6a).

On the other hand, the amount of cumulative soil loss consistently increased in the consecutive storms in the order of P40 < P+L < P+G < G < L < C, with percentage reduction ranging between 35–78% for P40, 20–56% for P+L, 11–48% for P+G, 8–44% for gypsum, and 2–26% for lime, compared with the control (Figure 6b). At the end of six consecutive rainfall storms, both total runoff and soil loss increased in the order P40 < P+L < P+G < G < L < C (Table 3). The statistical analysis using Kruskal-Wallis test (at $\alpha = 0.05$) revealed that P40 and P+L treatments reduced median cumulative runoff and soil loss significantly, however, there was no significant difference in reduction between them while the runoff and soil loss from both gypsum and lime were not statistically different with that of the control (Table 3).

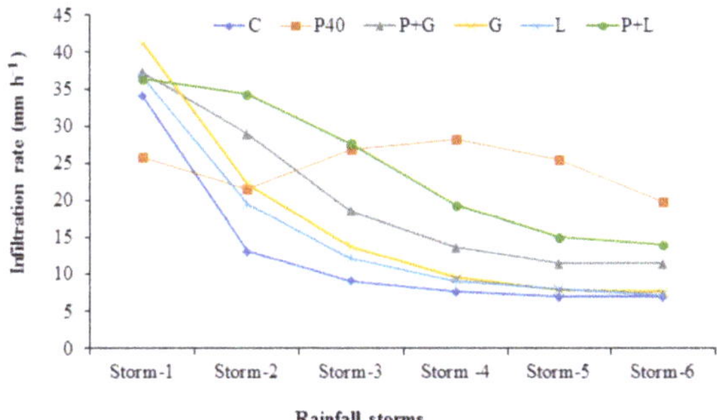

Figure 5. Effect of treatments on infiltration rate (mm h^{-1}). Each of the six storms had an average rainfall intensity of 70 mm h^{-1} for 1 h duration and 48 h drying period interval. Treatments were control (C), 40 kg ha^{-1} (P40), PAM of 40 kg ha^{-1} + Gypsum (4t ha^{-1}) (P+G), Gypsum (4t ha^{-1}) (G), lime (2t ha^{-1}) (L), and PAM of 40 kg ha^{-1} + Lime (2t ha^{-1}) (P+L).

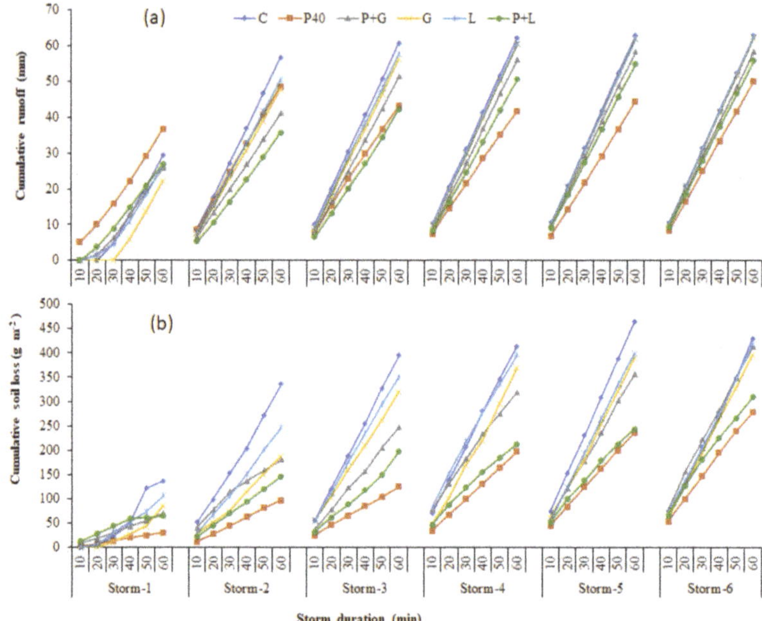

Figure 6. Effect of treatments on (**a**) cumulative runoff (mm) and (**b**) cumulative soil loss (g m^{-2}). Each of the six storms had an average rainfall intensity of 70 mm h^{-1} for 1 h duration and 48 h drying period interval. Treatments were control (C), 40 kg ha^{-1} (P40), PAM of 40 kg ha^{-1} + Gypsum (4t ha^{-1}) (P+G), Gypsum (4t ha^{-1}) (G), lime (2t ha^{-1}) (L), and PAM of 40 kg ha^{-1} + Lime (2t ha^{-1}) (P+L).

Table 3. Effect of treatments on total runoff and total soil loss with respective percent reduction (%) compared with the control and storm cumulative median runoff and soil loss from the six storms. Different letters following treatments indicate a significant difference between treatments using the Kruskal–Wallis test ($\alpha = 0.05$). Treatments were control (C), 40 kg ha^{-1} (P40), PAM of 40 kg ha^{-1} + gypsum (4t ha^{-1}) (P+G), gypsum (4t ha^{-1}) (G), lime (2t ha^{-1}) (L), and PAM of 40 kg ha^{-1} + Lime (2t ha^{-1}) (P+L).

Treatments	Runoff			Soil Loss		
	Total Runoff (mm)	Reduction in Total Runoff (%)	Cumulative Median Runoff (mm)	Total Soil Loss (g m^{-2})	Reduction in Total Soil Loss (%)	Cumulative Median Soil Loss (g m^{-2})
C	335.67		62.17 [a]	21.52		407 [a]
P40	265.04	21	44.75 [c]	9.19	57	159 [d]
P+G	291.44	13	54.88 [bc]	16.16	25	298 [bc]
G	313.33	7	58.00 [ab]	17.96	17	334 [ab]
L	322.44	4	59.92 [ab]	19.08	11	354 [ab]
P+L	266.94	20	47.42 [c]	11.53	46	188 [cd]

3.5. Effect of Aggregate and Structure Stability on Runoff and Soil Loss

Aggregate and structure stability, as described by SI, is an index that indicates the status of the soil aggregates in each treatment at the end of the experiment. After six consecutive rainfall storms, SI in P60 increased by 26% while it increased by 21% both in P40 and P+L compared with the control. The SI for all other treatments increased by less than 17% (Figure 7). In addition, both the total runoff and total soil loss showed a strong negative linear relationship with SI ($R^2 > 0.9$) (Figure 7).

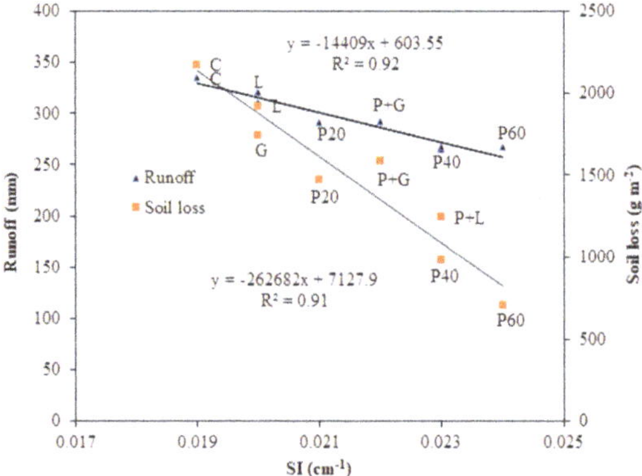

Figure 7. Relationship between structural index (SI) at the end of simulation with runoff (triangle) and soil loss (square).

4. Discussion

4.1. Effect of PAM Rates on TRO, IR, Runoff, and Soil Loss

The shorter TRO from PAM treatments, when compared with the control, was because of the high viscosity caused by the dissolution of PAM granules that led to a decrease in the IR and an increase in runoff [28,33]. Our result is consistent with the report by Lee et al. [51] who measured TRO for a silt loam soil amended with three granular PAM rates (0, 20, and 40 kg ha^{-1}), at a slope of 10%, 20%, and 40%. They found that the application of PAM (P20 and P40) decreased TRO for the 10% slope by an average of 5%.

During Storm 1, IR for all the PAM rates reduced, compared with the control (Figure 3). Nevertheless, IR from control was significantly reduced and became the lowest through Storms 2 to 6. This result is in agreement with observations by Inbar et al. [52], who used different granular PAM rates and consecutive rainstorms to mitigate post-fire soil erosion. They reported that the IR for control was higher during the first storm compared with the PAM treated setup. However, the effect of PAM treatments on IR and runoff in this study was not consistent. The application of granular PAM at higher rates (P40 and P60) decreased IR and increased runoff during Storms 1 and 2, compared with the control. However, IR for these rates substantially increased during Storms 3 and 4 while IR from lower rate (P20) decreased continuously through Storms 1 to 6. The mechanism responsible for decreasing IR and increasing runoff for the PAM treatments during the initial storm events was not the soil surface seal formation, as is the case with the control, but the large viscosity in the soil solution produced by the dissolution of PAM granules [52]. On the other hand, the reduction in IR from control during Storms 2 to 6 was due to the breakdown of aggregates by raindrop impact and subsequent seal formation [53].

On PAM dissolution during a rainfall event, the dissolved molecules are partially sorbed on the soil clay particles and improve aggregate stability, thereby increasing their resistance to detachment. Moreover, the non-sorbed segments of the molecules extend into the pores and drag the infiltrating water [40], decreasing the hydraulic conductivity of the soil and thus the infiltration rate [18]. However, the effect of viscosity disappeared after drying cycles, when the formation of bonds between PAM and soil particles was favored [39] and the positive effects of PAM applied at higher rates were enhanced after two raining and drying cycles that reversed the situation and sharply increased IR and decreased runoff later during Storms 3 and 4. The decrease in viscosity of the percolating solution together with the stabilization of aggregates through wetting and drying periods [52] led to an increase in the IR and a decrease in runoff during Storms 3 and 4 [18,28,33]. The IR, however, continuously declined during Storms 5 and 6, mainly due to washing out of PAM by runoff in the consecutive storms [52].

Furthermore, there was an increased reduction in IR with an increasing PAM rate at the beginning of the simulation; however, this reduction was reduced with increasing PAM rate at the latter stage of the simulation (Figure 3). This implies that higher PAM rates could either be applied as a split application or each of the first two rainfall storms could be applied in two or more applications with fair drying period intervals to minimize the excess runoff at the beginning.

Despite the higher runoff volume during Storms 1 and 2 as a result of high viscosity, higher PAM rates were effective in reducing soil loss throughout the consecutive storms. Similarly, Abrol et al. [40] reported that soil erosion decreases with an increase in viscosity of runoff in spite of the increase in runoff volume. Likewise, Inbar et al. [52] observed that increased viscosity of runoff reduces flow velocity and shear or drag forces that can detach soil particles. Furthermore, unlike the cumulative runoff and IR, cumulative soil losses for PAM treatments consistently decreased with increasing PAM rate but increased with time during the consecutive storm events. The reduction in soil loss with increasing PAM rate in this study is consistent with the results of Lado et al. [39], who reported that the relative viscosity of runoff increases with increasing granular PAM rate, thus decreasing runoff erosivity and hence soil loss. The consistent reduction in soil loss with increasing rates of PAM could also be attributed to the increased positive effect of PAM in preserving the soil aggregate structure [35] by increasing the soil erodibility resistance even during the last storm, which is in accordance with the reports by other researchers [27,54].

Similarly, the increase in soil loss for all PAM rates in the subsequent storms can be explained by the loss of effectiveness of PAM with time (e.g. dissolution, washing, and leaching), causing a decrease in IR and an increase in runoff, thus leading to increased soil loss during the last storms. The results from our study agree with the results of [55], who reported that the cumulative soil loss increases with increasing rainfall duration for each granular PAM rate because PAM effectiveness diminishes with time [39,51]. Furthermore, the lower PAM rate loses its effectiveness faster towards the end of the simulation when compared with the higher rates which could be due to the washout of PAM by runoff in the consecutive storms [52,54]. This leads to increased detachment and wash-in of the finer clay particles that plug the soil pores more at lower PAM rate (P20) because of the unavailability of sufficient PAM to protect the soil [25].

At the end of the simulation, it was observed that all the PAM rates were more effective at reducing total soil loss than total runoff, due to increased runoff observed at the beginning. This is in agreement with the observation made in a study by [18], who reported that the granular PAM application has less influence on runoff than soil loss. Furthermore, P40 and P60 reduced the median cumulative runoff and soil loss significantly, compared with the control, however, there was no significant difference in reduction when compared with each other, implying that P40 was the most suitable application rate (cost-effective) in reducing both runoff and soil loss from the given test soil under consecutive storms (Table 2). Nevertheless, higher final IR (Figure 3) and increased aggregate stability (Figure 3) after the last storm indicated that the PAM rates, especially the higher rates (P40 and P60) were still effective to reduce runoff and erosion for some more storms. The following figure (Figure 8) shows crusting of soil surfaces after six consecutive storms.

Figure 8. Crusts formed on the soil surface after six consecutive storms. Treatments were control (C), PAM of 20 kg ha^{-1} (P20), 40 kg ha^{-1} (P40), and 60 kg ha^{-1} (P60).

4.2. Effect of Gypsum, Lime, and Their Mixture with PAM on TRO, IR, Runoff, and Soil Loss

The reduction in TRO for PAM mixed with gypsum or lime treatments, as compared with the control, was attributed to the increased viscosity from the dissolution of PAM granules that reduced IR and increased runoff. On the other hand, the long TRO from gypsum could be related to the high solubility of gypsum that interacts with the soil faster. Our result agrees with the findings of [25], who compared the effects of phosphogypsum powder (PG) (5 t ha^{-1}) and granular PAM (20 kg ha^{-1}) on TRO, runoff, and erosion using a highly weathered Brazilian Alfisol. They reported that compared with control, the application of gypsum was effective in delaying runoff by increasing the TRO by 75%, whereas PAM and P+G decreased TRO by 16% and 85%, respectively. They attributed the longer TRO from gypsum treatment, which led to increased IR and delayed runoff, to changes in soil surface chemistry and the shorter TRO in PAM treatment to increased viscosity of the runoff.

The IR, runoff, and soil loss from gypsum, lime, and PAM mixed with gypsum or lime amendments varied during the consecutive rainfall storms. In the beginning, IR for P40 treatment was lower than that for other treatments and the control (Figure 5), due to increased soil solution viscosity, as discussed in Section 4.1. However, the effect of viscosity on the P40 solution started disappearing during Storm 3, when there was enough rainfall for the complete dissolution of the PAM and drying period for irreversible bonding or sorption to the soil aggregates [36]. This improved soil aggregate and structural stability increased IR and decreased cumulative runoff from P40 treatment substantially during Storms 4 to 6, compared with the control (Figure 5, Figure 6a). Despite the high runoff during Storms 1 and 2 from P40, cumulative soil loss from P40 was the lowest of all the treatments throughout the consecutive storms (Figure 6b), because the increased viscosity reduced the erosivity of the runoff at the beginning [52]. But when the viscosity decreased latter in the consecutive storms, PAM became adsorbed to and bound soil particles through cation bridging, thus, increasing aggregate stability and cohesion strength between the soil particles and decreasing soil erosion [56]. Thus, depending on the treatments and storm period, amendments affected surface soil aggregates and structural stability differently (e.g., resistance to disintegration and macro porosity).

Both gypsum and lime treatments increased IR, compared with the control (Figure 5). The higher IR from gypsum at the beginning could be due to the high solubility of gypsum, that it is likely to dissolve, release Ca^{2+} cations, and interact with the soil at a faster rate than lime [57]. On the other hand, the lower IR from lime implied the occurrence of higher dispersion from the lime treatment that sealed soil pores and decreased IR in the subsequent storms. As a result, runoff and soil loss from lime was higher than that from gypsum, and this difference in the two treatments could be explained by the varying effects of the two amendments on soil properties, mainly soil pH (Table 1).

Upon dissolution, lime dissociates into calcium and carbonate ions. The carbonate combines with hydrogen ions from the soil to form carbon dioxide and water. The removal of hydrogen ions from the soil in the form of water due to lime addition leads to an increase in negative charges in the soil while raising soil pH; hence the repulsive forces dominate between the soil particles and lead to dispersion [58–60] that seal soil pores and decrease IR. Our result is in agreement with the findings of de Castro and Celso [61] who studied the effect of gypsum and lime on clay dispersion and infiltration in Oxisols. They found that incorporating lime to the soil increased clay dispersion and reduced IR significantly than incorporating gypsum. Furthermore, the removal of hydrogen ions following the addition of lime to the soil in our study increased the pH of the Oxisols from 4.84 to 6.75 (Table 1). The increment in the soil pH of the Oxisols from lime addition might have also increased soil dispersion leading to a decrease in IR compared with the gypsum treatment that had no effect on the soil pH (Table 1). Similarly, Roloff [59] demonstrated the rise in pH values of oxidic soils leads to clay dispersion that seals pores in the soil surface and decreases IR. In addition, de Castro and Celso [61] have also highlighted a negative correlation between pH values and infiltration rate in acidic Oxisols, i.e., the IR was found to decrease significantly with increasing lime application rates.

In the case of P40 mixed with gypsum or lime treatments, increased cumulative runoff and cumulative soil loss was observed, compared with P40 alone treatments (Figure 6a,b). This can be explained by the increase in cations in the soil solution with the addition of gypsum or lime mixed with P40 promoted coiling of the dissolved PAM molecules and reduce the interaction of the charged functional groups with soil particles [39]. This reduces PAM effect on aggregate stabilization, increases surface seal formation, runoff and soil loss.

On the other hand, the presence of a higher concentration of electrolytes in P + G treatment, (4 t ha^{-1} of gypsum) compared with that in P + L (2 t ha^{-1} of lime), shortened the polymer chains and prevented the polymers from stretching and long-range bridging between the soil particles. This makes P + G less efficient in binding soil particles that are far apart and maintain soil aggregates, thus leading to increased soil erosion [18]. Hence, P + L was more effective than P + G in reducing both runoff and soil loss whereas both are less effective compared with P40 treatment. Furthermore, P + G and P + L increased IR and decreased cumulative runoff and cumulative soil loss, compared with treatments using gypsum or lime alone (Figure 6a,b). This implies that gypsum or lime was more effective for Oxisols when applied in mixture with PAM than when applied alone for Oxisols. Furthermore, the high clay content in the Oxisols could also be another factor that maintained the effectiveness of PAM alone treatment in the Oxisols [35].

The result of our study partially agrees with the study by Lepore et al. [37]. They reported that soil loss reduction in silt loam by P + G (58%) was lower than P + L (67%) while gypsum and lime reduced soil loss only by 18% and 30%, respectively. Our result was in agreement with that study in terms of highlighting that gypsum and lime were more effective when applied in mixture with PAM than when applied alone. Also, it showed that P + L was more effective than P + G in reducing soil loss. However, contrary to that study, our result showed that gypsum was less effective than lime when applied individually. The difference in the results between the two studies could be attributed to differences in soil properties (soil type, clay mineralogy), amendment rates applied [20,62], and their effect on aggregates and structural stability associated with treatments, rainfall duration, and wetting-drying condition in the two experiments. [18] also observed that the use of PAM mixed with gypsum increased the final infiltration rate and reduced runoff and wash erosion compared to application of gypsum alone in loamy sand and clay soils. Furthermore, [18] reported that application of 20 kg ha^{-1} PAM is the most effective in reducing soil losses for the silt loam and sandy clay soils compared with gypsum (2 t ha^{-1} and 4 t ha^{-1}) or 20 kg ha^{-1} PAM mixed with gypsum (2 t ha^{-1} and 4 t ha^{-1}), in spite of low IR and the resultant high runoff.

The SI of treatments after six consecutive storms also indicated that P60 increased SI by 26% while both P40 and P + L treatments increased SI by 21%, however, SI for other treatments remains lower than 17% (Figure 7), supporting the results of the effect of different treatments on runoff and soil loss. For all the eight treatments (control, PAM, G, L, and their combinations) there was a negative linear relationship between soil SI and soil loss or runoff with coefficient of determination ($R^2 > 0.9$) (Figure 7), showing that (i) both the runoff and soil loss were significantly affected by treatments and (ii) the relation between the SI and soil loss was affected more by the treatments than the relation between SI and runoff. Therefore, such an approach could be used for the evaluation of the effect of amendments on soil status to resist erosion under consecutive rainfall storms.

5. Conclusions

In this study, we tested the effectiveness of different PAM rates in reducing runoff and soil loss under consecutive rainfall storms and selected the effective PAM rate for fine-textured Oxisols. The application of PAM at a rate of 20 kg ha^{-1} (i.e., P20) was more effective in reducing runoff during the first two storms and higher rates (i.e., P40 and P60) were more effective towards the end of the consecutive storms. However, the effectiveness of PAM in reducing soil loss increased with increasing PAM rate but diminished with time over the entire duration of the simulation. Furthermore, the application of PAM leads to a better reduction in soil loss compared to runoff. Our results also revealed that reductions in runoff and soil loss by P40 and P60 were not statistically different. Hence, by taking into consideration the price and application cost, P40 was selected as the most appropriate application rate for the given test soil. The selected PAM rate was further tested by applying it mixed with sources of electrolytes (gypsum and lime), but the concurrent application of P40 with gypsum or lime was less effective in reducing runoff and soil loss than P40 alone treatments. However, as only one application rate of gypsum and lime was considered, the result from this study may not be conclusive and further tests using multiple rates of gypsum or lime mixed with P40 and wetting-drying experimental conditions are recommended. Furthermore, the split application of P40 and evaluation of its effectiveness in reducing runoff against its gross application may provide additional knowledge about the effective PAM application rate.

Author Contributions: Conceptualization, B.K. and N.H.; methodology, B.K., N.H., A.I.M., A.A.F., A.T. and T.M.; software, B.K.; validation, A.T., N.H., and M.T.; formal analysis, B.K.; investigation, B.K.; resources, A.T. and N.H.; data curation, B.K., G.A., M.L.B.; writing—original draft preparation, B.K.; writing—review and editing, N.H., A.I.M., A.A.F., D.T.M., E.A.; visualization, B.K.; supervision, A.T., N.H., M.T., T.M. and A.I.M.; project administration, A.T., N.H.; funding acquisition, A.T. All authors have read and agreed to the published version of the manuscript.

Funding: This research was funded by the Science and Technology Research Partnership for Sustainable Development (SATREPS)—Development of a Next-Generation Sustainable Land Management (SLM) Framework to Combat Desertification project, Grant Number JPMJSA1601, Japan Science and Technology Agency (JST)/Japan International Cooperation Agency (JICA).

Acknowledgments: The authors thank the Science and Technology Research Partnership for Sustainable Development (SATREPS) and Japan Science and Technology Agency (JST)/Japan International Cooperation Agency (JICA). We also would like to thank the KEMIRA OYJ company for providing anionic polyacrylamide, SUPERFLOC A-110, free of charge.

References

1. Haregeweyn, N.; Poesen, J.; Nyssen, J.; Govers, G.; Verstraeten, G.; Vente, J.; Deckers, J.; Moeyersons, J.; Mitiku, H. Sediment yield variability in Northern Ethiopia: A quantitative analysis of its controlling factors. *Catena* **2008**, *75*, 65–76. [CrossRef]

2. Lal, R. Erosion-crop productivity relationships for soils of Africa. *Soil Sci. Soc. Am. J.* **1995**, *59*, 661–667. [CrossRef]

3. Okoba, B.O.; Sterk, G. Catchment-level evaluation of farmers' estimates of soil erosion and crop yield in the Central Highlands of Kenya. *Land Degrad. Dev.* **2010**, *21*, 388–400. [CrossRef]

4. Haregeweyn, N.; Poesen, J.; Nyssen, J.; De Wit, J.; Haile, H.; Govers, G. Reservoirs in Tigray: Characteristics and sediment deposition problems. *Land Degrad. Dev.* **2006**, *17*, 211–230. [CrossRef]

5. Tamene, L.; Abegaz, A.; Aynekulu, E.; Woldearegay, K.; Vlek, P.L. Estimating sediment yield risk of reservoirs in northern Ethiopia using expert knowledge and semi-quantitative approaches. *Lakes Reserv. Res. Manag.* **2011**, *16*, 293–305. [CrossRef]

6. Vanmaercke, M.; Poesen, J.; Maetens, W.; de Vente, J.; Verstraeten, G. Sediment yield as a desertification risk indicator. *Sci. Total Environ.* **2011**, *409*, 1715–1725. [CrossRef]

7. Kirui, O.; Mirzabaev, A. Economics of land degradation in Eastern Africa. Available online: https://www.researchgate.net/publication/265612178_Economics_of_Land_Degradation_in_Eastern_Africa (accessed on 3 January 2020).

8. Dimotta, A.; Lazzari, M.; Cozzi, M.; Romano, S. *Soil Erosion Modelling on Arable Lands and Soil Types in Basilicata, Southern Italy. Proceedings of the Computational Science and Its Applications – ICCSA 2017: 17th International Conference, Trieste, Italy, 3–6 July 2017*; Gervasi, O., Murgante, B., Misra, S., Borruso, G., Torre, C.M., Rocha, A.M.A.C., Taniar, D., Apduhan, B.O., Stankova, E., Cuzzocrea, A., Eds.; Springer: Berlin, Germany, 2017; pp. 57–72.

9. Dimotta, A.; Cozzi, M.; Romano, S.; Lazzari, M. *Soil Loss, Productivity and Cropland Values GIS-based Analysis and Trends in the Basilicata Region (southern Italy) from 1980 to 2013, Proceedings of the International Conference on Computational Science and Its Applications 2016, Beijing, China, 4–7 July 2016*; Gervasi, O., Murgante, B., Misra, S., Borruso, G., Torre, C.M., Rocha, A.M.A.C., Taniar, D., Apduhan, B.O., Stankova, E., Cuzzocrea, A., Eds.; Springer: Berlin, Germany, 2016; pp. 29–45.

10. Ebabu, K.; Tsunekawa, A.; Haregeweyn, N.; Adgo, E.; Meshesha, D.T.; Aklog, D.; Masunaga, T.; Tsubo, M.; Sultan, D.; Fenta, A.A.; et al. Analyzing the variability of sediment yield: A case study from paired watersheds in the Upper Blue Nile basin, Ethiopia. *Geomorphology* **2018**, *303*, 446–455. [CrossRef]

11. Ebabu, K.; Tsunekawa, A.; Haregeweyn, N.; Adgo, E.; Meshesha, D.T.; Aklog, D.; Masunaga, T.; Tsubo, M.; Sultan, D.; Fenta, A.A.; et al. Effects of land use and sustainable land management practices on runoff and soil loss in the Upper Blue Nile basin, Ethiopia. *Sci. Total Environ.* **2019**, *648*, 1462–1475. [CrossRef]

12. Fenta, A.A.; Yasuda, H.; Shimizu, K.; Haregeweyn, N.; Negussie, A. Dynamics of soil erosion as influenced by watershed management practices: A case study of the Agula watershed in the semi-arid highlands of northern Ethiopia. *Environ. Manag.* **2016**, *58*, 889–905. [CrossRef]

13. Fenta, A.A.; Yasuda, H.; Shimizu, K.; Haregeweyn, N.; Kawai, T.; Sultan, D.; Ebabu, K.; Belay, A.S. Spatial distribution and temporal trends of rainfall and erosivity in the Eastern Africa region. *Hydrol. Process.* **2017**, *31*, 4555–4567. [CrossRef]

14. Fenta, A.A.; Tsunekawa, A.; Haregeweyn, N.; Poesen, J.; Tsubo, M.; Borrelli, P.; Panagos, P.; Vanmaercke, M.; Broeckx, J.; Yasuda, H.; et al. Land susceptibility to water and wind erosion risks in the East Africa region. *Sci. Total Environ.* **2019**, *703*, 135016. [CrossRef] [PubMed]

15. Haregeweyn, N.; Tsunekawa, A.; Poesen, J.; Tsubo, M.; Meshesha, D.T.; Fenta, A.A.; Nyssen, J.; Adgo, E. Comprehensive assessment of soil erosion risk for better land use planning in river basins: Case study of the Upper Blue Nile River. *Sci. Total Environ.* **2017**, *574*, 95–108. [CrossRef] [PubMed]

16. Graber, E.R.; Fine, P.; Levy, G.J. Soil stabilization in semiarid and arid land agriculture. *J. Mater. Civ. Eng.* **2006**, *18*, 190–205. [CrossRef]

17. Mamedov, A.I.; Shainberg, I.; Levy, G.J. Irrigation with effluent water: Effect of rain energy on soil infiltration. *Soil Sci. Soc. Am. J.* **2000**, *64*, 732–737. [CrossRef]

18. Yu, J.; Lei, T.; Shainberg, I.; Mamedov, A.I.; Levy, G.J. Infiltration and erosion in soils treated with dry PAM and gypsum. *Soil Sci. Soc. Am. J.* **2003**, *67*, 630–636. [CrossRef]

19. Warrington, D.N.; Mamedov, A.I.; Bhardwaj, A.K.; Levy, G.J. Primary particle size distribution of eroded material affected by degree of aggregate slaking and seal development. *Eur. J. Soil Sci.* **2009**, *60*, 84–93. [CrossRef]

20. Teo, J.A.; Ray, C.; El-Swaf, S.A. Screening of polymers on selected Hawaii soils for erosion reduction and particle settling. *Hydrol. Process.* **2006**, *20*, 109–125. [CrossRef]

21. Mamedov, A.I.; Levy, G.J. Soil erosion-runoff relationships in cultivated lands: Insights from laboratory studies. *Eur. J. Soil Sci.* **2019**, *70*, 686–696. [CrossRef]

22. Flanagan, D.C.; Chaudhari, K.; Norton, L.D. Polyacrylamide soil amendment effects on runoff and sediment yield on steep slopes: Part I. Simulated rainfall conditions. *Trans. Asae* **2002**, *45*, 1327–1337.

23. Wakindiki, I.I.C.; Ben-Hur, M. Soil mineralogy and texture effects on crust micromorphology, infiltration, and erosion. *Soil Sci. Soc. Am. J.* **2002**, *66*, 897–905. [CrossRef]

24. Seybold, C.A. Polyacrylamide review: Soil conditioning and environmental fate. *Commun. Soil Sci. Plant Anal.* **1994**, *25*, 2171–2185. [CrossRef]

25. Cochrane, B.H.W.; Reichert, J.M.; Eltz, F.L.F.; Norton, L.D. Controlling soil erosion and runoff with polyacrylamide and phosphogypsum on subtropical soil. *Trans. Asae* **2005**, *48*, 149–154. [CrossRef]

26. Peterson, J.R.; Flanagan, D.C.; Tishmack, J.K. PAM application method and electrolyte source effects on plot–scale runoff and erosion. *Trans. Asae* **2002**, *45*, 1859. [CrossRef]

27. Sepaskhah, A.R.; Bazrafshan-Jahromi, A.R. Controlling runoff and erosion in sloping land with polyacrylamide under a rainfall simulator. *Biosyst. Eng.* **2006**, *93*, 469–474. [CrossRef]

28. Sojka, R.E.; Bjorneberg, D.L.; Entry, J.A.; Lentz, R.D.; Orts, W.J. Polyacrylamide (PAM) in agriculture and environmental land management. *Adv. Agron.* **2007**, *92*, 75–162.

29. Mamedov, A.I.; Beckmann, S.; Huang, C.; Levy, G.J. Aggregate stability as affected by polyacrylamide molecular weight, soil texture, and water quality. *Soil Sci. Soc. Am. J.* **2007**, *71*, 1909–1918. [CrossRef]

30. Lentz, R.D.; Sojka, R.E. Field results using polyacrylamide to manage furrow erosion and infiltration. *Soil Sci.* **1994**, *158*, 274–282. [CrossRef]

31. Kumar, A.; Saha, A. Effect of polyacrylamide and gypsum on surface runoff, sediment yield and nutrient losses from steep slopes. *Agric. Water Manag.* **2011**, *98*, 999–1004. [CrossRef]

32. Shainberg, I. Soil crusting in South America. In *Soil Crusting: Chemical and Physical Processes. Proceedings of the 1st International Symposium on Soil Crusting. Advances in Soil Science*; Sumner, M.E., Stewart, B.A., Eds.; Lewis Publishers: Boca Raton, FL, USA, 1992; p. 3353.

33. Ajwa, H.A.; Trout, T.J. Polyacrylamide and water quality effects on infiltration in sandy loam soils. *Soil Sci. Soc. Am. J.* **2006**, *70*, 643–650. [CrossRef]

34. Petersen, A.L.; Thompson, A.M.; Baxter, C.A.; Norman, J.M.; Roa-Espinosa, A. A new polyacrylamide (PAM) formulation for reducing erosion and phosphorus loss in rainfed agriculture. *Trans. Asae* **2007**, *50*, 2091–2101. [CrossRef]

35. Tumsavas, Z.; Kara, A. The effect of polyacrylamide (PAM) applications on infiltration, runoff and soil losses under simulated rainfall conditions. *Afr. J. Biotechnol.* **2011**, *10*, 2894–2903.

36. Mamedov, A.I.; Shainberg, I.; Wagner, L.E.; Warrington, D.N.; Levy, G.J. Infiltration and erosion in soils treated with dry PAM of two molecular weights and phosphogypsum. *Aust. J. Soil Res.* **2009**, *47*, 788–795. [CrossRef]

37. Lepore, B.J.; Thompson, A.M.; Petersen, A.L. Impact of polyacrylamide delivery method with lime or gypsum for soil and nutrient stabilization. *J. Soil Water Conserv.* **2009**, *64*, 223–231. [CrossRef]

38. Mamedov, A.I.; Huang, C.H.; Aliev, F.A.; Levy, G.J. Aggregate Stability and Water Retention Near Saturation Characteristics as Affected by Soil Texture, Aggregate Size and Polyacrylamide Application. *Land Degrad. Dev.* **2017**, *28*, 543–552. [CrossRef]

39. Lado, M.; Inbar, A.; Sternberg, M.; Ben-Hur, M. Effectiveness of granular polyacrylamide to reduce soil erosion during consecutive rainstorms in a calcic regosol exposed to different fire conditions. *Land Degrad. Dev.* **2015**, *27*, 1453–1462. [CrossRef]

40. Abrol, V.; Shainberg, I.; Lado, M.; Ben-Hur, M. Efficacy of dry granular anionic polyacrylamide (PAM) on infiltration, runoff and erosion. *Eur. J. Soil Sci.* **2013**, *64*, 699–705. [CrossRef]

41. Gabriels, D.M.; Moldenhauer, W.C.; Kirkham, D. Infiltration, hydraulic conductivity and resistance to water drop impact of clod bed as affected by chemical treatment. *Soil Sci. Soc. Am. Proc.* **1973**, *37*, 364–367. [CrossRef]

42. Smith, H.; Levy, G.J.; Shainberg, I. Water drop energy and soil amendments: Effect on infiltration and erosion. *Soil Sci. Soc. Am. J.* **1990**, *54*, 1084–1087. [CrossRef]

43. Shainberg, I.; Warrington, D.; Rengasamy, P. Effect of soil conditioner and gypsum application on rain infiltration and erosion. *Soil Sci.* **1990**, *149*, 301–307. [CrossRef]

44. Meshesha, D.T.; Tsunekawa, A.; Tsubo, M.; Haregeweyn, N.; Tegegne, F. Evaluation of kinetic energy and erosivity potential of simulated rainfall using Laser Precipitation Monitor. *Catena* **2014**, *137*, 237–243.

45. Bagarello, V.; Ferro, V.; Keesstra, S.; Comino, J.R.; Pulido, M.; Cerdà, A. Testing simple scaling in soil erosion processes at plot scale. *Catena* **2018**, *167*, 171–180. [CrossRef]

46. Zhou, P.H.; Wang, Z.L. Soil erosion storm rainfall standard in the Loess Plateau. *Bull. Soil Water Conserv.* **1987**, *7*, 38–44.

47. Abd Elbasit, M.A.M.; Yasuda, H.; Salmi, A.; Anyoji, H. Characterization of rainfall generated by dripper-type rainfall simulator using piezoelectric transducer and its impact on splash soil erosion. *Earth Surf. Process. Landf.* **2010**, *35*, 466–475. [CrossRef]

48. Kruskal, W.H.; Wallis, W.A. Use of ranks in one-criterion analysis of variance. *J. Am. Stat. Assoc.* **1952**, *47*, 583–621. [CrossRef]

49. Gee, G.W.; Bauder, J.W. Particle size analysis. In *Methods of Soil Analysis: Part 1. Physical and Mineralogical Methods*; American Society of Agronomy: Madison, WI, USA, 1986; pp. 383–411.

50. Levy, G.J.; Mamedov, A.I. High-energy-moisture-characteristic aggregate stability as a predictor for seal formation. *Soil Sci. Soc. Am. J.* **2002**, *66*, 1603–1609. [CrossRef]

51. Lee, S.S.; Gantzer, C.J.; Thompson, A.L.; Anderson, S.H. Polyacrylamide efficacy for reducing soil erosion and runoff as influenced by slope. *J. Soil Water Conserv.* **2011**, *66*, 172–177. [CrossRef]

52. Inbar, A.; Ben-Hur, M.; Sternberg, M.; Lado, M. Using polyacrylamide to mitigate post-fire soil erosion. *Geoderma* **2014**, *239*, 107–114. [CrossRef]

53. Ben-Hur, M. Seal formation effects on soil infiltration and runoff in arid and semiarid regions under rainfall and sprinkler irrigation conditions. In *Climatic changes and water resources in the Middle East and North Africa*; Zereini, F., Hötzl, H., Eds.; Springer: Berlin, Germany, 2008; pp. 429–452.

54. Sepaskhah, A.R.; Mahdi-Hosseinabadi, Z. Effect of polyacrylamide on the erodibility factor of a loam soil. *Biosyst. Eng.* **2008**, *99*, 598–603. [CrossRef]
55. Li, Y.; Shao, M.; Horton, R. Effect of polyacrylamide applications on soil hydraulic characteristics and sediment yield of sloping land. *Proc. Env. Sci.* **2011**, *11*, 763–773. [CrossRef]
56. Roa-Espinosa, A.; Bubenzer, G.D.; Miyashita, E.S. Sediment runoff control on construction sites using four application methods of polyacrylamide mix. In Proceedings of the National Conference on Tools for Urban Water Resource Management and Protection, Chicago, IL, USA, 7–10 February 2000; pp. 278–283.
57. Bennett, J.; Cattle, S. Viability of lime and gypsum use in mitigating sodicity in an irrigated Vertosol. In Proceedings of the 19th World Congress of Soil Science, Brisbane, Australia, 1–6 August 2010.
58. Bennett, J.; Cattle, S. Viability of lime and gypsum use in mitigating sodicity in an irrigated Vertosol. Available online: https://www.iuss.org/19th%20WCSS/Symposium/pdf/2431.pdf (accessed on 3 January 2020).
59. Roloff, G. Strength of low and variable charge soils. Ph.D. Thesis, University of Minnesota, Minneapolis and Saint Paul, Minnesota, USA, June 1987; p. 114.
60. Tama, K.; El-Swaify, S.A. Charge, colloidal and structural stability interrelationships for Oxic soils. In *Modification of Soils Structure*; Emerson, W.W., et al., Eds.; John Wiley & Sons: New York, NY, USA, 1978.
61. De Castro, C.F. Effects of Liming on Characteristics of a Brazilian Oxisol at Three Levels of Organic Matter as Related to Erosion. Ph.D. Thesis, The Ohio State University, Columbus, OH, USA, 1988.
62. Roth, C.H.; Pavan, M.A. Effects of lime and gypsum on clay dispersion and infiltration in samples of a Brazilian Oxisol. *Geoderma* **1991**, *48*, 351–361. [CrossRef]

Article

New Soil, Old Plants, and Ubiquitous Microbes: Evaluating the Potential of Incipient Basaltic Soil to Support Native Plant Growth and Influence Belowground Soil Microbial Community Composition

Aditi Sengupta [1,*,†], Priyanka Kushwaha [2], Antonia Jim [3], Peter A. Troch [1,4] and Raina Maier [2]

[1] Biosphere 2, University of Arizona, Tucson, AZ 85721, USA; patroch@hwr.arizona.edu
[2] Department of Environmental Science, University of Arizona, Tucson, AZ 85721, USA; pkushwaha@email.arizona.edu (P.K.); rmaier@ag.arizona.edu (R.M.)
[3] Department of Environment and Sustainability, Fort Lewis College, Durango, CO 81301, USA; aljim1@fortlewis.edu
[4] Department of Hydrology and Atmospheric Sciences, University of Arizona, Tucson, AZ 85721, USA
* Correspondence: aditi.sengupta@pnnl.gov
† Author's current address: Earth and Biological Sciences Directorate, Pacific Northwest National Laboratory, 902 Battelle Boulevard, MSIN J4-18, Richland, Washington, DC 99352, USA.

Received: 22 March 2020; Accepted: 18 May 2020; Published: 21 May 2020

Abstract: The plant–microbe–soil nexus is critical in maintaining biogeochemical balance of the biosphere. However, soil loss and land degradation are occurring at alarmingly high rates, with soil loss exceeding soil formation rates. This necessitates evaluating marginal soils for their capacity to support and sustain plant growth. In a greenhouse study, we evaluated the capacity of marginal incipient basaltic parent material to support native plant growth and the associated variation in soil microbial community dynamics. Three plant species, native to the Southwestern Arizona-Sonora region, were tested with three soil treatments, including basaltic parent material, parent material amended with 20% compost, and potting soil. The parent material with and without compost supported 15%, 40%, and 70% germination of Common Bean (*Phaseolus vulgaris* L. 'Tarahumara Norteño'), Mesquite (*Prosopis pubescens* Benth), and Panic Grass (*Panicum Sonorum* Beal), respectively, though germination was lower than in the potting soil. Plant growth was also sustained over the 30 day period, with plants in parent material (with and without amendment) reaching 50% height compared to those in the potting soil. A 16S rRNA gene amplicon sequencing approach showed *Proteobacteria* to be the most abundant phyla in both parent material and potting soil, followed by *Actinobacteria*. The potting soil showed *Gammaproteobacteria* (19.6%) to be the second most abundant class, but its abundance was reduced in the soil + plants treatment (5.6%–9.6%). Within the basalt soil type, *Alphaproteobacteria* (42.7%) and *Actinobacteria* (16.3%) had a higher abundance in the evaluated bean plant species. Microbial community composition had strong correlations with soil characteristics, but not plant attributes within a given soil material. Predictive functional potential capacity of the communities revealed chemoheterotrophy as the most abundant metabolism within the parent material, while photoheterotrophy and anoxygenic photoautotrophy were prevalent in the potting soil. These results show that marginal incipient basaltic soil, both with and without compost amendments, can support native plant species growth, and non-linear associations may exist between plant–marginal soil–microbial interactions.

Keywords: marginal soil; land degradation; endemic plant species; soil microbes

1. Introduction

Soils provide a wide range of ecosystem services and are central to sustaining life in the biosphere. These include supporting plant growth and sustenance, ensuring food security, and modulating biogeochemical cycles [1]. Additionally, soils hold anthropological significance, as civilizations have developed and flourished around their ability to harness soils' power to grow crops. However, global soil loss is occurring at unprecedented rates with depletion rates twenty times faster than formation in the United States alone [2,3]. Soil loss and eventual land degradation are exacerbated by anthropogenic activities including urbanization, population growth, machine-intensive agricultural practices, conversion of forest land to agricultural land, and land-use practices such as mining. Degraded land quality, in turn, negatively impacts food production, livelihoods, and ecosystem services [4].

The process of rock to soil formation and, therefore, landscape development occurs over hundreds of years, with a predicted 500–1000 years needed to form 2.5 cm of top soil [1]. This imbalance between rates of soil formation and depletion is unsustainable. Soil conservation therefore stands out as a necessity if humans are to secure food, fiber, economy, and health. The Land Degradation Neutrality (LDN) Program adopted by the United Nations Convention to Combat Desertification (UNCCD) recognized that land conservation requires restoring degraded land and soil to achieve a "degradation-neutral world" [5]. A key concept of the neutrality framework is to improve the productive potential of land/soil that is already degraded [6], with calls for restoration and rehabilitation mechanisms. We propose exploring the capacity of marginal soils as one approach to revegetate marginal soils and increase ecosystem service of degraded lands both in terms of regulating services (e.g., climate regulation, hydrological regulation, regulation of soil erosion) provisioning services (e.g., crops, livestock, fuel, fiber), and cultural services (local heritage, recreation, tourism, education) [7].

Seed germination is a critical step in plant establishment and needs detailed evaluation in marginal soils with constrained physicochemical conditions. Previous studies of contaminated mine tailings have shown successful germination of native *Atriplex lentiformis* and *Bouteloua dactyloides* seeds within the first week in mine tailings amended with compost [8–10]. Previous studies have also demonstrated use of plant-growth-promoting bacteria to establish native plants in Klondyke mine tailings in Arizona [11] and aid phytoremediation of contaminated soils [12]. Another study showed a seed mixture of 15 native seeds sown on degraded soils on the Falkland island, of which three species, *Elymus magellanicus*, *Poa flabellate*, and *Poa alopecarus*, successfully germinated during a revegetation trial [13]. Therefore, plant-specific establishment rates may vary, and, along with plant diversity, tissue chemistry, and root traits [14], will impact soil characteristics and plant–soil–microbe interactions in marginal soils. From a land-sustainability standpoint, the concept of using native seeds specifically sourced from the areas close to the parent material is environmentally sustainable. For example, native seed use preserves the local biodiversity and ensures that a diverse plant gene pool is present, thereby reducing disease vulnerability [15]. Moreover, using locally sourced seeds can reduce environmental and economic costs incurred in producing and transporting seeds from distant locations, especially for subsistence farmers who cannot rely on industry-scale operations [16,17].

Native plants have also been reported to act as ecological resource islands by improving soil chemical and microbial properties of adjacent rhizosphere soil [18] and can therefore improve local land productivity. Land degradation reduces soil microbial biomass and microbial activity [19], with reports of significant decrease in beneficial microorganisms and increase in pathogenic ones in degraded soils [20]. Another study [21] found higher bacterial richness and diversity in restored soils and soils under native vegetation in comparison to degraded soils. As invisible engineers of terrestrial ecosystems, soil microorganisms contribute to soil structure, are involved in biogeochemical cycling of nutrients [22], decompose organic matter, impact plant diversity and productivity, and play a critical role in soil fertility [23].

The aboveground–belowground links between plant and microbes are especially crucial in marginal systems' characteristic of nutrient-poor soils and inferior soil structure. Soil biota is one of the five soil forming factors proposed by Hans Jenny [24]. In marginal soils lacking higher life forms, microbes are therefore the first biotic component to contribute to soil stabilization and structure. Hence, evaluating microbial community establishment and response in marginal soils is critical. Empirical knowledge of these linkages can then be potentially applied to enhance marginal soils' capacity to support plant growth, e.g., by developing microbe-mediated amendment technologies suited to low-productivity landscapes.

Marginal soils can include incipient soils, soils affected by mining processes and fire, over-used agricultural lands, urban vacant plots, and primary succession ecosystems [25,26]. Soils in these landscapes have poor quality, have a persistent lack of plant growth, and are characterized by a lack of stable soil aggregates, organic carbon forms, and essential plant nutrients [26,27]. Restoring the productive capacity of marginal land is critical to the mission of the LDN Framework of the UNCDD taskforce [5]. Our study uses basaltic material as a marginal soil medium and is novel in its concept, since a literature survey revealed that such material, despite being rich in nutrients, has not been evaluated for plant growth purposes. Marginal soil productivity and efficacy in supporting plant growth are therefore a potential way of compensating for global soil loss and supporting sustainable land management.

Landscape evolution of marginal incipient soils includes spatio-temporal interactions spanning coupled hydrobiogeochemical processes. Field studies fall short of capturing these critical process responses during the initial stages of development, including soil stabilization and plant establishment [28]. In this study, we evaluated the capacity of basaltic soil material to support native plant growth and the associated variation in soil microbial community dynamics following a month-long greenhouse experiment. The incipient soil is basaltic crushed tephra sourced from Merriam crater in northern Arizona, and is being extensively studied at the Landscape Evolution Observatory (LEO) housed at Biosphere 2 in University of Arizona to understand coupled hydrobiogeochemical processes of landscape evolution [28,29]. The timeframe of the greenhouse study was selected to test for seeds that could germinate and plants that could grow within a month, keeping in mind the larger scope of this research at the LEO hillslopes, which see two–three month-long rainfall episodes each year and are in the process of being vegetated. The seeds selected for this study were native to the Arizona-Sonora region, reflected those of plants with different rooting patterns, could withstand high water availability as expected to be present in the LEO hillslopes, and could withstand high Arizona summer temperatures. This lithogenic basalt parent material is oligotrophic, with low carbon $(9.33 \times 10^{-5}$ μg/mg), nitrogen $(4.33 \times 10^{-6}$ μg/mg) [30], and phosphorus content (0.003 μg/mg) [31], but has been shown to harbor microbial life [32]. The soil texture of LEO basalt material is loamy sandy (% sand: 84.6, % silt: 12.2, and % clay: 3.2). The soil was contrasted with a potting soil treatment, which served as a positive control, as well as a basalt soil amended with compost treatment, which served as a comparative mid-point between the marginal basalt soil and the nutrient-rich potting soil.

The objectives of this study were to (i) evaluate germination capacity and plant growth capacity of the basaltic parent material for three seed types, (ii) compare growth of the different species in the parent material with and without compost amendment as compared to potting soil, and (iii) assess soil microbial community composition and functional potential between treatments. The seeds used in this study were sourced from the seed banks of Native Seed Search [33] and were native to the Arizona/Sonora region of Arizona. We hypothesized that for each plant species, the highest germination, plant height, and biomass will be observed in the potting soil, followed by compost-amended basalt soil, with the lowest plant attributes observed in the basalt soil. These differences would also be visible as distinct soil characteristics at the end of the experiment. Additionally, we hypothesized that the soil microbial community for each plant species would be significantly different between the soil treatments, resulting in different predictive functional profiles of the microbial communities.

2. Materials and Methods

2.1. Experimental Design

A greenhouse pot experiment was conducted with three seed types: Panic Grass (*Panicum Sonorum* Beal), Mesquite (*Prosopis pubescens* Benth), and Common Bean (*Phaseolus vulgaris* L. 'Tarahumara Norteño'), as well as three soil materials: Basaltic parent material from the LEO experiment (LPM), LEO parent material + 20% w/w commercially available compost (LPMC), and commercially available (Miracle Gro® Potting Mix) potting soil (PS). Percent moisture was measured after drying field-moist soils at 105 °C for 24 hrs. The total carbon (TC), organic carbon (TOC), and nitrogen (TN) concentrations (U.S. EPA method 415.3) were measured using a TOC-L Series total organic carbon and nitrogen analyzer, equipped with a TOC-LCSH autosampler (Shimadzu, Kyoto, Japan) [28,34]. The TC, TOC, and TN concentrations in the compost material were 138 ± 7 μg g^{-1}, 120 ± 5 μg g^{-1}, and 3.0 ± 0.56 μg g^{-1}, respectively, while the potting soil recorded $294 \pm$ μg g^{-1}, 213 ± 3.0 μg g^{-1}, and 13.34 ± 1.0 μg g^{-1}, respectively. Compost was added to evaluate the effect of natural amendment on LPM's capacity to support plant growth while potting soil served as a positive control. Each seed-type–soil material combination (n = 27) was set up in randomized triplicate one-gallon plastic pots with 2.1 kg LPM, 2.1 kg (1.68 kg LPM + 0.42 kg compost) LPMC, and 0.6 kg PS, and ten seeds sown per pot. The drip irrigation system was controlled at three one-minute events a day at 8:30 a.m., 12:00 p.m., and 4:30 p.m., respectively, for seven days a week, totaling 720 mL/day/pot of tap water and pots monitored for 30 days (24 June–24 July 2017). The lowest and highest temperature range for the 30 days were 19–23 °C and 31–44 °C, respectively.

The one-month timeframe was chosen for multiple reasons. First, it was primarily driven by the ability to study seeds that have short germination times, eventually leading up to fast-growing plants. This would allow initial responses of plant–soil–microbe interaction in the nascent stages of germination and growth in an incipient soil to be captured. Moreover, to inform the larger scope of the research, seeds were chosen that fit the research motivation at the LEO in Biosphere 2, which sees two–three month-long rainfall episodes in a year and, therefore, needs plants that can germinate within a month and withstand the water exposure at the hillslopes. This also informed our irrigation regime, which was designed to not be water-constrained. This allowed observations to be made with the underlying fact that the only constraints provided to the seeds/plants were the marginal soil characteristics.

The soil treatments were chosen to first evaluate the plant growth capacity of the incipient basalt parent material. Next, a literature review suggested that marginal soils support plant growth when treated with amendments, which motivated us to add a compost-amendment treatment. The decision to go with 20% compost was arrived at after evaluating literature [8–10]. Potting soil was chosen as a positive control to validate our experimental design and evaluate whether the seeds we were using would indeed germinate and grow in the one-month-long summer experiment. The treatments therefore provided contrasting soil environments, especially beneficial to test the initial plant–soil–microbe interactions and responses in the initial stages of plant growth, and to differing levels of soil fertility (C, N) and moisture.

2.2. Seed Information

Detailed seed information and sowing are provided in a preliminary publication by Jim and Sengupta [35]. Panic grass (*Panicum Sonorum*) was cultivated in the Arizona and Sonora region four thousand years ago. *Prosopis pubescens*, or Mesquite, grows in arid and semi-arid environments, such as deserts, woodlands, floodplains, grasslands, and shrublands, and has deep tap roots with leaves adapted to reduce water loss. *Tarahumara Norteño* bean (also known as the common bean), originated in the Tarahumara area of the Chihuahua region in Mexico and has been widely cultivated by Native American farmers throughout the Southwest [33,36]. Common beans are known to grow in semi-tropical regions and host nitrogen-fixing bacteria in their root nodules. The three seed types typically grow in arid to semi-arid environments, are heat tolerant, can retain water, and have short germination periods. The grass and bean seeds were sourced from Native Seeds, Tucson, AZ and mesquite seeds were sourced from Desert Nursery, Phoenix, AZ.

The plant species were selected primarily keeping in mind the larger goal of providing robust preliminary data to the research team at the Landscape Evolution Observatory, where the goal is to study the hydrobiogoechemical trajectory of landscape evolution from incipient soil to a vegetated landscape. The criteria were: (i) The plants would germinate fast and grow considerably during the one-month study period, (ii) the plants would represent a range of root systems (for example, Mesquite for tap roots, Panic Grass for shallow fibrous root, and Common Bean for a mix of adventitious and basal root systems), and (iii) the plants would be native to the Arizona-Sonora region primarily so that they could tolerate the heat and would be able to grow on the basaltic hillslopes of the Landscape Evolution Observatory.

2.3. Plant and Soil Measurements

Percent germination was calculated after ten days. One plant per pot was marked and monitored continuously, with height measurements taken once every week. At the end of the experiment, the marked plant was harvested for wet and dry aboveground biomass measurement as per protocol outlined in Jim and Sengupta [35]. Bulk density (BD) cores were collected using metallic cores of height 2.9 cm and diameter 5.3 cm. Bulk soil samples close to the roots were also collected for geochemical and microbiological analyses, stored on ice, and brought back to the lab for processing. Half of the bulk samples were air-dried and sieved for pH, carbon, and nitrogen measurements, while the other half was frozen at −80 °C.

2.4. Soil Microbial Community Analysis

Soil microbial DNA from the frozen soils was extracted using a modified extraction protocol [8,32] of the FastDNA™ SPIN Kit for Soil from MP Biomedicals with an additional 5.5 M guanidine thiocyanate wash step to remove humic acids, and was sent for sequencing to the University of Arizona Genetics Core (UAGC; Tucson, AZ, USA). Sequence data were analyzed as per protocols highlighted in Sengupta et al. 2019 [32]. Briefly, paired-end sequencing (2 × 150 bp) was performed on the bacterial and archaeal 16S rRNA gene V4 (515F-806R primers) hypervariable region using the Illumina MiSeq platform (Illumina, CA, USA) [37]. All of the sequencing procedures, including the construction of the Illumina sequencing library, were performed using the protocol previously published [38] with modifications [39]. Illumina MiSeq v2 (300 bp) chemistry was used for sequencing and was performed on the Illumina MiSeq (SN M02149 with the MiSeq Control Software v 2.5.0.5) at the UAGC following their standard protocols. The UAGC provided standard Illumina quality control, base-calling, demultiplexing, adaptor removal, and conversion to FastQ format. Raw sequence data were submitted to NCBI's Sequence Read Archive SUB4001574, ProjectID PRJNA464263.

Paired-end sequence merging, barcode removal, quality filtering, singleton-sequence removal, chimera checking and removal, and open-reference Operational Taxonomic Unit (OTU) picking were conducted using default parameters unless otherwise specified in QIIME v 1.9.1 [37]. A minimum overlap of 20 bases was specified for joining the paired reads to give an average sequence length of 253 base pairs. A summary of the sequences, post-merging and quality filtering, was performed using mothur (v 1.25) [39], OTU picking was done using UCLUST [40], and sequence alignment was performed with PyNAST [41]. Clustering was done with Greengenes database at 97% sequence similarity [42], chimera were removed with Chimera Slayer [43], taxonomy was assigned with RDP Classifier [44], tree building was completed with FastTree [45], and comparative diversity calculations were done with UniFrac [46]. OTUs that were observed only once after chimera filtering were removed. All data files generated from the QIIME workflow were imported into the *R* environment program [47] for alpha and beta diversity estimation and visualization using *Phyloseq* [48].

2.5. Functional Annotation of Sequence Data

The 16S rRNA amplicon gene sequences were used to infer metabolic traits from phylogeny using the tool Functional Annotation of Prokaryotic Taxa (FAPROTAX) [49]. FAPROTAX facilitates interpretation of microbial functional profiles from 16S rRNA bacterial and archaeal sequence data based on available literature of cultured representatives. Briefly, an OTU is associated with a particular metabolic function if all cultured representatives of that OTU are reported to exhibit that function. For example, if all cultured representatives of a genus have been identified as nitrifiers, FAPROTAX assumes all uncultured members to be nitrifiers as well. This approach was well suited to our study given the complexity of soil environments, and a large portion of soil microbes remain uncultured.

2.6. Data Analyses

Percent germination, plant height, aboveground wet and dry biomass, pH, bulk density, carbon (total, organic, and inorganic), total nitrogen, and moisture content were statistically analyzed using JMP® 13.0. Significant differences of mean were determined by one-way analysis of variance (ANOVA) ($P < 0.05$), followed by pairwise comparisons of the means of each soil material/plant group using Tukey's Honest Significance Difference test with significance levels at $P < 0.05$. Sequences were evaluated for alpha (Richness and Shannon's Index), and beta diversity metrics (Bray–Curtis) were analyzed with Principal Coordinate analysis (PCoA). To visualize predicted functional differences between the soil types, a heatmap was generated using the *gplots* [50] package in R. Additionally, a hierarchical cluster was generated using the average linkage method. To determine the dissimilarity in the microbial community structures and the soil physicochemical properties together with plant measurements, a non-parametric multivariate analysis of variance (PERMANOVA) was performed based on the Bray–Curtis dissimilarity. Additionally, Mantel tests were performed to determine correlations between the soil microbial communities and the environmental factors. PERMANOVA and Mantel's test analyses were carried out in R (v.3.5) using the package *vegan* [51] and *ade4* [52], respectively. For PERMANOVA analyses, there was no replication. For LEO soil material and potting soil, there were three and four samples, respectively. To perform the PERMANOVA as well as Mantel's test, n = 999 iterations were used. ANOVA analysis was conducted using relative abundance of OTUs classified as specific functional guilds using FAPROTAX. The F-value was used to evaluate the ratio of mean square values of the samples separated into the two soil types.

3. Results and Discussion

3.1. Germination, Plant Height and Aboveground Biomass

Plant characteristics including germination, plant height, and aboveground biomass were monitored for 30 days in a greenhouse (Figure 1). The results are presented in Jim and Sengupta [35] with additional statistical testing between all treatments and interpretation discussed here.

Briefly, all three soil treatments supported germination, with the maximum percent germination observed in the potting soil and the overall highest germination of grasses across treatments. Malfunctioning irrigation drips discovered towards the end of the experiment in one replicate pot each of parent material and parent material amended with compost may explain the low germination rates for the bean plants in basalt soil and basalt amended with compost.

Figure 1. Greenhouse experimental set-up on Day 10.

As noted in Jim and Sengupta [35], significantly higher plant growth was observed in the potting soil as compared to basalt soil and basalt soil amended with compost. The bean and grass plants grew twice the height (averages of 40 and 54 cm, respectively) in the potting soil as compared to basalt with and without amendments, while the slowest growing plant was bean in parent material. Significant aboveground biomass accumulation was observed for bean and grass plants in the potting soil as compared to the parent material with and without amendment, with the lowest biomass accumulation in mesquite across soil treatments and no significant differences between the soil treatments [35]. However, it is likely that a period of 30 days may not be sufficient for compost benefits to be visible owing to the slow release of nutrients [53]. Therefore, a longer growth period may be necessary for compost-amendment benefits to be observable.

This provides a positive indication of LEO basalt material's ability to support germination and promote plant growth (Figure 2). Additionally, these results show that locally sourced native seeds can prove to be an effective strategy in vegetating incipient soils that are localized around the native plant source. Moreover, the rapid germination of seeds in the LEO basaltic material can stabilize the soil and facilitate ecosystem succession, as highlighted by Smith et al. [13], who reported rapid revegetation of degraded land by incorporating native plant species.

Figure 2. Growth observed in basalt parent material for all three seed types.

3.2. Physico-Chemical Properties of the Studied Soil Materials

Physico-chemical data of the treatments as well as irrigation water, parent material, and potting soil are presented in Supplementary Table S1. Soil pH differed between treatments, with pH significantly higher in the parent material and parent material amended with compost as compared to potting soil. A decrease in pH was observed in the basaltic parent material (pH = 9.8) in the presence of plant roots (pH = 9.4) and compost amendment (pH = 8.5). Bulk density of the potting soil and treatments was significantly lower (average of different plant types = 0.78 g cm^{-3}) than the parent material (average of different plant types = 1.83 g cm^{-3}), and even within the parent material and treatments, significant differences in mean bulk density were also observed. Potting soil and treatments were held on average 50% moisture and were significantly higher than the parent material soil with or without compost. Carbon and nitrogen estimates were also significantly higher in the potting soil (average of different plant types = 314 µg mg^{-1}) than the parent material and parent material with compost treatments (average of different plant types = 14.24 µg mg^{-1}). The marginal soil characteristic of the parent material (lack of structure, high bulk density, low moisture holding capacity, and low nutrients), as compared to the potting soil, primarily separates the treatments into the two broad groups, as observed in the principal component plot showing variation in soil samples and irrigation water (Figure 3). The average moisture content of the soils ranged from 50% in potting soil, 3.33% in basalt soil amended with compost, and 5.33% in basalt soil.

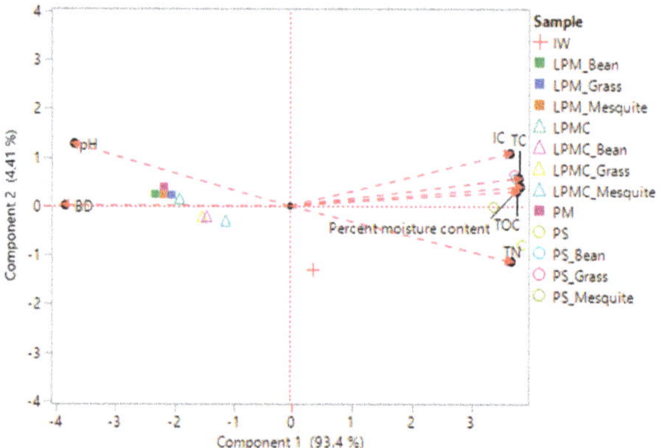

Figure 3. Variation observed in the physico-chemical characteristics of triplicate soil–plant treatments and water samples after 30 days, plotted as principle component of variation superimposed by variable loadings. Points are averages of three replicates. Abbreviations: LPM (LEO Parent Material), LPMC (LEO Parent Material + Compost), PS (Potting Soil), and the plant species follow the soil material type, IW (Irrigation Water), TC (Total Carbon), TOC (Total Organic Carbon), IC (Total Inorganic Carbon), TN (Total Nitrogen), and BD (Bulk Density).

For the LEO parent material, significant differences were not observed for the bare soil or those with plants, while parent material amended with compost had a significant increase in carbon (total, organic, inorganic) and nitrogen concentrations for each plant type. It is interesting to note that while plant germination and growth differences did not differ significantly between basalt material with or without compost, the soil characteristics between the two treatments differed significantly. Therefore, soil characteristics likely start to change in these incipient soils within a month when amended with compost, though the effects are not parallelly observable in aboveground plant growth dynamics, suggesting that abiotic shifts in soil characteristics may not have linear observable shifts on plant growth.

3.3. Composition of Microbial Communities

High-throughput 16S rRNA gene amplicon sequencing was performed to evaluate soil microbial community composition. A total of 5955 Operational Taxonomic Unit (OTUs) were obtained across basalt parent material samples and potting soil samples. The DNA extracts from compost-amended LEO soil samples failed to amplify and are therefore absent from the analyses. Despite treating the compost samples with a 5.5 M guanidine thiocyanate (GTC) wash to remove PCR-interfering humic matter [54] and attempting to amplify the samples twice, the compost sample DNA extracts failed to amplify. It is likely that the DNA was coextracted with PCR-inhibiting material (either humic acid and/or ions that bind to DNA) in these samples. Therefore, for this section of the results, the samples were separated into the two broad soil treatments: Parent material and potting soil.

A total of 27 phyla were observed across the samples (Supplementary Figure S1). The top ten phyla detected were *Acidobacteria* (0.8%–5.7%), *Actinobacteria* (2.6%–26%), *Bacteroidetes* (2.1%–11.1%), *Chloroflexi* (0.6%–2.5%), *Cyanobacteria* (1.1%–3.7%), *Firmicutes* (1.4%–7.9%), *Gemmatimonadetes* (0.7%–2.8%), *Planctomycetes* (0.7%–5.0%), *Proteobacteria* (43.9%–60.7%), and *Verrucomicrobia* (1.5%–16.8%). *Proteobacteria* were the most abundant phyla in both soil types. The potting soil showed *Actinobacteria* (26%) to be the second most abundant phyla, but its abundance was reduced in the soil + plus plant (11.1%–14.9%). Within the potting soil samples, an increase in Verrucomicrobial OTUs was also observed in the soils with plants (5.8%–11.3%) from 1.5% in bare soil. The abundance of *Verrucomicrobia* (Mesquite > Grass > Bean) showed a similar relative abundance pattern in the basalt soil to that in potting soil, which could be indicative of plant-specific microbial community response.

A total of 71 classes were observed across the samples (Supplementary Figure S2). The top ten classes detected were [*Pedosphaerae*] (0.8%–5.3%), [*Saprospirae*] (0.7%–9.1%), *Alphaproteobacteria* (22.2%–42.7%), *Actinobacteria* (1.1%–16.3%), *Betaproteobacteria* (2.3%–14.4%), *Bacilli* (1.6%–7.6%), *Deltaproteobacteria* (0.2%–4.9%), Gammaproteobacteria (2.5%–19.6%), *Opitutae* (0.6%–9.8%), and *Thermoleophilia* (0.3%–6.2%) (Figure 4). The *Alphaproteobacteria* were the most abundant class in both soil types. The potting soil showed *Gammaproteobacteria* (19.6%) to be the second most abundant class, but its abundance was reduced in the soil plus plant (5.6%–9.6%). In addition, a decrease in *Acidobacteriia, Actinoobacteria, Betaproteobacteria,* and *Thermophilia* OTUs was observed in potting soils with plants. Within the potting soil samples, OTUs of [*Pedosphaerae*], [*Saprospirae*], [*Spartobacteria*], *Acidobacteria-6, Alphaproteobacteria, Chloroplast, Cytophagia, Deltaproteobacteria, Opitutae, Phycisphaerae,* and *Planctomycetia* were observed to increase in the soils with plants. The observed shift in microbial community abundances is likely a result of plant establishment in the potting soils. The number of different classes in the bean plant was the lowest compared to grass and mesquite in both basalt and potting soil, suggesting that plant-specific microbial community establishment during plant growth.

Furthermore, within the basalt soil type, the bean plant soil had low abundance of *Acidobacteria* (0.8%), *Armatimonadetes* (0.3%), *Chloroflexi* (0.6%), *Gemmatimonadetes* (0.7%), and *Planctomycetes* (0.7%) when compared to the grass and mesquite plant soils. However, within the basalt soil type, *Alphaproteobacteria* (42.7%) and *Actinobacteria* (16.3%) had a higher abundance in the bean plant, which was much higher than in grass (27.4%, 1.8%) and mesquite (22.2%, 1.1%). The bean plant is a legume and is therefore known to have nitrogen fixation mediated by nitrogen-fixing bacteria in the plant root nodules [55,56]. Interestingly, over the recent years, multiple studies have highlighted the role of many nitrogen-fixing actinobacterial taxa, as highlighted in a review by Gtari et al., 2012 [57], as well as the filamentous nature of actinobacterial groups promoting successful colonization of oligotrophic land surfaces [58].

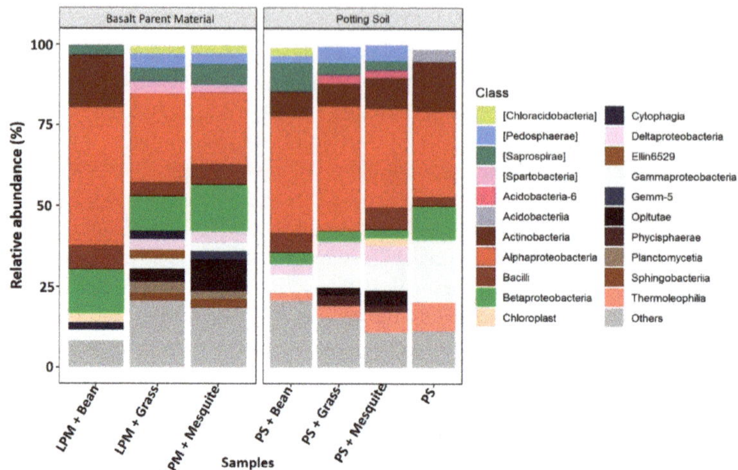

Figure 4. Relative proportions of bacterial community at the class level. Classes representing more than 2% in all samples are summarized, and the remaining are indicated as others in the figure. Abbreviations: LPM (LEO Parent Material), PS (Potting Soil), and the plant species follow the soil material type. The sample without plant species name is a control sample.

Out of a total of 343 identified different genera, 191 were unclassified genera. The top ten genera in the basalt soil were Agromyces (0.94%–10.4%), Bacillus (5.76%–10.0%), Balneimonas (4.1%–7.328%), Bdellovibrio (1.0%–4.04%), Flavisolibacter (1.9%–4.2%), Methyloversatilis (1.9%–13.9%), Opitutus (0.63%–6.5%), Sediminibacterium (2.2%–5.6%), Sinorhizobium (24.6%), and Sphingopyxis (3.0%–8.5%). The top ten genera identified in the potting soil were Bacillus (0.89%–4.41%), Devosia (6.12%–10.74%), Kaistobacter (1.3%–20.1%), Luteibacter (0.1%–17.3%), Mycobacterium (1.3%–6.4%), Opitutus (3.9%–10.5%), Rhodoplanes (12.0%–30.2%), Sediminibacterium (0.4%–5.0%), Sphaerisporangium (2.3%–4.2%), and Streptomyces (3.9%–9.4%). The genera Alicyclobacillus (0.2%–3.6%) and Sphaerisporangium (2.34%–4.2%) were only detected in the potting soil, whereas the basalt soil had the following unique genera: Agromyces (0.94%–10.4%), Balneimonas (3%–7.3%), Methyloversatilis (1.9%–13.5%), Pleomorphomonas (0.5%–2.1%), and Renibacterium (0.13%–3.0%).

The DNA extracted from the LEO basalt parent material without plants did not have a high enough number of sequences to pass the sequence analysis quality control in this study. However, as a standalone analysis for visual comparison (Supplementary Figure S3), the parent material community consists of 62% *Bacteroidetes*, followed by 16% *Firmicutes*, and less than 10% *Proteobacteria*. This suggests a dramatic month-long shift in the soil microbial community composition of the basalt soil. Since the seeds were not sterilized before planting, it cannot be ruled out that the seeds may have contributed to the increase in microbial community diversity. We posit that a combination of the parent material, dust and irrigation water input, and seeds provided a heterogeneous initial composition, and the root exudates further altered the availability and quality of carbon compounds that increased community richness and introduced an abundance of new taxa in the incipient basalt material.

A comparison of species richness and Shannon's diversity index after rarefaction showed minor differences in the composition of the microbial community (Figure 5). Richness in potting soil samples had a wider range from 394–728 compared to parent material samples, 475–672. Potting soil plus bean had the highest richness of 728, while potting soil plus mesquite had the lowest richness of 394. Shannon's indices in LEO parent material samples ranged from 4.8–5.8 versus 4.7–5.7 in potting soils. The richness and Shannon's index of microbial communities were not significantly different between the soil types and plant types. Additionally, the percent richness across samples ranged from 10%–18.4%, whereas % Shannon's index ranged from 12.7%–15.6% (Supplemental Table S2).

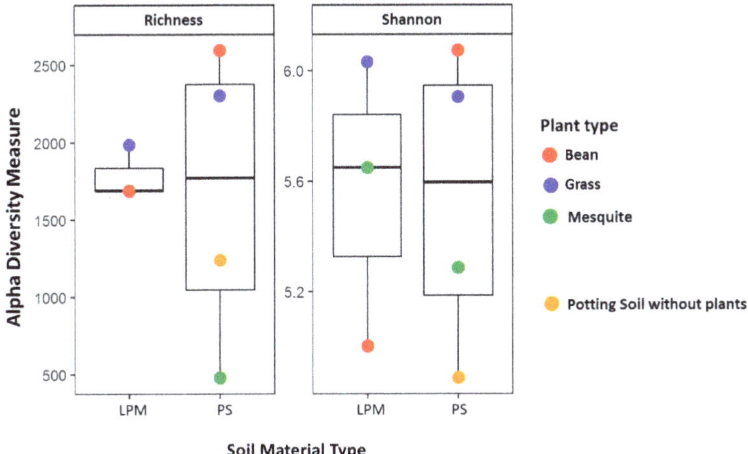

Figure 5. Alpha diversity in the studied soils represented by richness and Shannon's index. The plant types—bean, grass, and mesquite—are indicated by the colors red, blue, and green. The potting soil control (without plant) is represented by the color yellow. The soil material is indicated as LPM (LEO Parent Material) and PS (Potting Soil).

A principal coordinate analysis (PCoA) plot of the Bray–Curtis dissimilarity distance matrix demonstrated the differences in microbial communities between the soil types (Figure 6). The first two PCoA axes together explained 65.3% of the variation in the microbial communities. The samples from each soil type grouped distinctly separate from the control sample of potting soil plus plant samples. Additionally, PERMANOVA analysis revealed that microbial communities in LEO parent material and potting soil were significantly different ($P = 0.03$) with a R^2 of 0.47, suggesting 47% of the variation in the microbial communities to be explained by the soil material type.

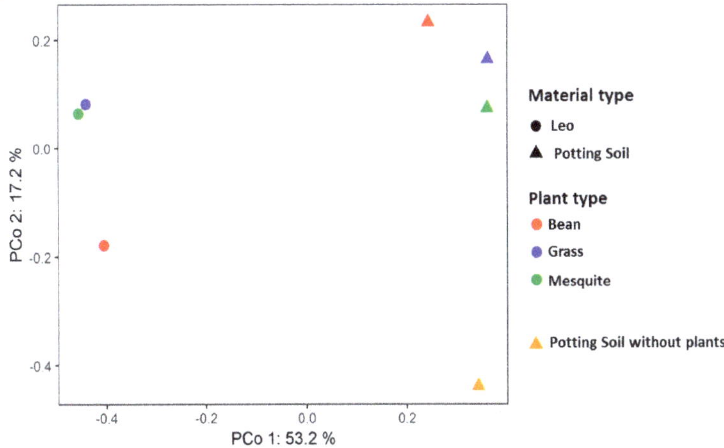

Figure 6. A Principle Coordinate Analysis (PCoA) plot based on the Bray–Curtis distance matrix showing similarity of the bacterial and archaeal community in the studied soil material type. The first two PCoA axes together explain 65.3% of the variation in the microbial community based on the soil material type. The non-parametric multivariate analysis of variance (PERMANOVA) of soil type was significantly different ($P = 0.03$).

3.4. Microbial Communities and Environmental Variables

The PERMANOVA and Mantel test were carried out to decipher linkages between soil microbial communities and environmental variables, and between soil microbial communities and plant attributes (Table 1). PERMANOVA as well as the Mantel test demonstrated that the majority of the soil factors influenced the microbial communities. Among these soil factors, pH, TC, TIC, TOC, TN, moisture content, bulk density, and soil material type exhibited significant ($P < 0.05$) associations with microbial communities. The R^2 values for the environmental variables ranged from 0.42–0.47, representing the 42%–47% contribution of the variables towards the microbial community. These soil factors that exhibited significant correlations had strong correlation values (r) ranging from 0.69 (TIC) to 0.91 (pH). The soil material type also had a strong significant r value of 0.87, thus indicating that the microbial community shifts were significantly correlated with the soil material type. Plant attributes did not show significant interactions with the soil microbial communities for the duration of the experiment.

Table 1. Correlations between the environmental variables and plant attributes with the Bray–Curtis dissimilarity index using a non-parametric multivariate analysis of variance (PERMANOVA) and Mantel test.

	PERMANOVA		Mantel Test	
Environmental variables	R^2	*P*	r	*P*
pH	0.46	**0.002**	0.91	**0.002**
TC	0.47	**0.007**	0.87	**0.012**
TN	0.45	**0.02**	0.9	**0.002**
TOC	0.47	**0.002**	0.85	**0.006**
TIC	0.42	**0.02**	0.69	**0.023**
Moisture content	0.47	**0.018**	0.86	**0.039**
Bulk density	0.43	**0.03**	0.81	**0.017**
Soil material type	0.47	**0.035**	0.87	**0.034**
Plant attributes	R^2	*P*	r	*P*
Germination rate	0.15	0.474	0.02	0.413
Plant height	0.21	0.232	0.10	0.310
Wet aboveground biomass	0.17	0.429	−0.16	0.660
Dry aboveground biomass	0.17	0.416	−0.14	0.605
Plant type	0.39	0.799	0.00	0.452

Mantel coefficient (r) ranges from −1 to +1. A value of −1 indicates strong negative correlation, a value of +1 indicates strong positive correlation, and 0 means no correlation. Significant values ($P < 0.05$) are in bold.

An observable difference is present in aboveground plant attributes and belowground microbial community structure shifts. Therefore, as with the abiotic changes in the soil, belowground biotic changes (in this case, microbial community structure) had non-linear associations with plant establishment and growth for the initial time of growth captured in our study. This time-lag between aboveground and belowground shift may be a critical point when evaluating and predicting temporal feedbacks between plant–soil–microbe interactions [59]. These results are useful in informing future directions of our study, where plant establishment and growth in the basalt material can be studied over a longer duration, followed by periodic sampling of soil and plant material to temporally identify linear versus non-linear associations between plant–soil–microbe interactions.

3.5. Predicted Functional Potential of Microbial Community

The OTUs were utilized to predict the functional potential of the soil microbial community in LEO and potting soil samples. Using FAPROTAX analyses, 59 predicted functions were identified. A subset of 24 functions were further evaluated for their relative profiles in the different soil samples (Figure 7). These functions were chosen to reflect broad carbon fixation/utilization mechanisms (autotrophy versus heterotrophy), and nitrogen cycling metabolisms. Like the microbial community differences, the LEO and potting soils also demonstrated differences in their functional potential. Chemoheterotrophy was the most abundant identified function within LEO soils, while photoheterotrophy and anoxygenic photoautotrophy were higher in the potting soils.

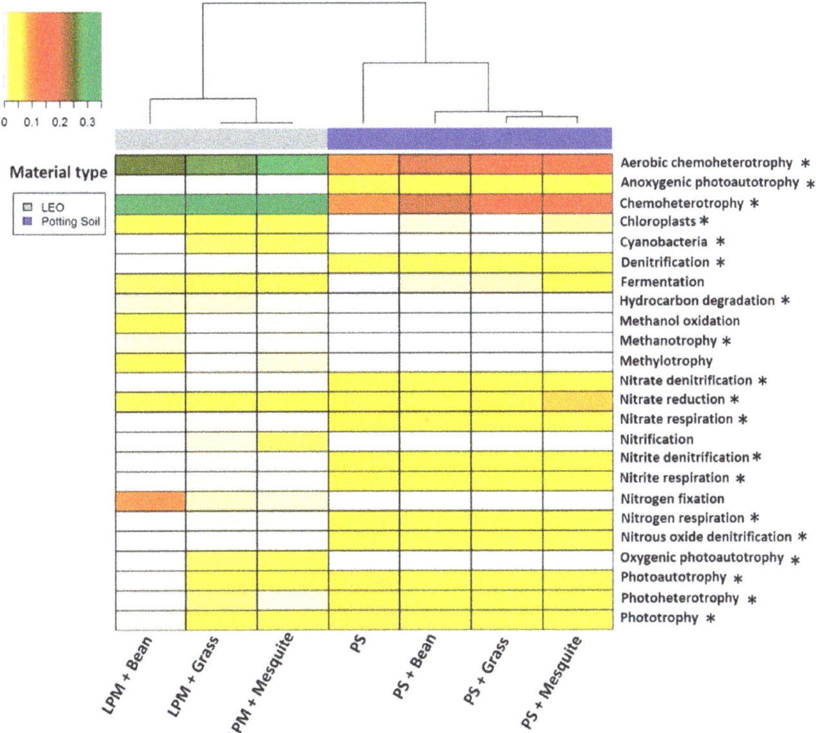

Figure 7. A heatmap showing relative abundance of the selected predicted functional potential of the microbial community using FAPROTAX analysis in the studied soil materials. The relative abundance ranges from 0 to 0.35 and is represented by white (0–0.025), red (0.026–0.1), yellow (0.11–0.21), and green (0.21–0.35), respectively. Abbreviations: LPM (LEO Parent Material), PS (Potting Soil), and the plant species follow the soil material type. As per the Mantel test, nineteen functions (represented by *) were significantly different ($P < 0.05$) between the soil types with the exception of fermentation, methanol oxidation, methylotrophy, nitrification, and nitrogen fixation.

The majority of the 24 functions were significantly different between the soil types (Table 2) with the exception of fermentation, methanol oxidation, methylotrophy, nitrification, and nitrogen fixation. The Mantel test showed a strong correlation of the predicted functions with soil type material (r = 0.89, P = 0.026). The hierarchical cluster grouped the samples according to their soil types, indicating the soil-type-specific functional capacity. The grass and mesquite samples in both LEO parent material and potting soil clustered closely compared to the bean. The basalt soil plus bean sample had the highest nitrogen fixation capacity among all samples. In contrast, the potting soils did not have high abundance of nitrogen fixation, but presented higher abundance of other nitrogen cycling pathways (denitrification, reduction, and respiration) as compared to the basalt material soils. The function of fermentation was predicted to be higher in the basalt material, which suggests that the basalt material may constitute fermentators that fix carbon [60,61]. While we did not observe high autotrophic predictions for the incipient basalt material, the presence of microbes in this low-carbon soil and the overall increase in organic carbon concentration is evidence of carbon-fixation mechanisms at play.

Furthermore, the higher predictive heterotrophic activity in basalt soils could be attributed to the likely diminished autotrophic activity of the *Cyanobacteria* during plant colonization, resulting in the availability of fixed plant carbon for heterotrophs [62]. Additionally, we propose a second hypothesis that heterotrophy rates far exceed autotrophy rates in these low-carbon environments. It may be likely that as soon as autotrophs are able to fix carbon and produce by-products, the heterotrophs utilize the fixed carbon and proliferate. The FAPROTAX results should be treated with caution, as they do not confirm the presence/absence of in situ microbial metabolisms. However, these results do serve as a starting point to generate further hypotheses of soil–microbe interactions and metabolic strategies potentially at play in low-carbon versus nutrient-rich soils.

Table 2. Analysis of variance showing statistical differences in predicted Functional Annotation of Prokaryotic Taxa (FAPROTAX) functions between soil types.

Predicted Functions	F	P
Aerobic chemoheterotrophy	28.6	**0.003**
Anoxygenic photoautotrophy	87.8	**0.0002**
Chemoheterotrophy	40.2	**0.001**
Chloroplasts	26.5	**0.004**
Cyanobacteria	7.6	**0.040**
Denitrification	84.1	**0.0003**
Fermentation	4.3	0.094
Hydrocarbon degradation	38.1	**0.002**
Methanol oxidation	2.8	0.157
Methanotrophy	10.7	**0.022**
Methylotrophy	3.9	0.106
Nitrate denitrification	84.1	**0.0003**
Nitrate reduction	12.0	**0.018**
Nitrate respiration	82.8	**0.0003**
Nitrification	4.2	0.096
Nitrite denitrification	84.1	**0.0003**
Nitrite respiration	84.4	**0.0003**
Nitrogen fixation	2.4	0.185
Nitrogen respiration	82.8	**0.0003**
Nitrous oxide denitrification	84.1	**0.0003**
Oxygenic photoautotrophy	7.6	**0.040**
Photoautotrophy	20.3	**0.006**
Photoheterotrophy	45.1	**0.001**
Phototrophy	7.9	**0.037**

An F-value closer to 1 indicates that the means of the two groups are equal. A higher F-value indicates that the means of the two groups are not identical. Significant values ($P < 0.05$) are in bold.

4. Conclusions

The ability to use marginal soils is crucial to the framework of Land Degradation Neutrality. In this study, we evaluated whether marginal incipient basaltic soil will be able to support plant growth under ideal growth conditions. Here, plants native to the geographical area and climate of a marginal incipient basaltic soil material, with and without organic amendment, were evaluated for their germination and growth attributes. Comparisons of soil microbial communities and plant attributes were made between the incipient soil and a commercially available potting soil. The results show that marginal incipient basaltic soil can support native plant growth and that distinct soil microbial communities develop in these soils alongside plant establishment. Furthermore, nonlinear associations between abiotic shifts in soil characteristics, microbial community compositional changes, and plant growth parameters exhibited.

A direct outcome of this study is being applied to current experiments being conducted at the Landscape Evolution Observatory to establish plants and monitor the co-evolving hydrobiogeochemical signatures in the incipient landscapes. As a future direction, we propose detailed experiments with longer growth periods and different combinations of marginal soils with and without amendments to evaluate the capacity of such soils to support and sustain plant growth. Furthermore, additional cleaning steps to obtain clean DNA from the basalt plus compost samples may be evaluated. With LEO experiments, the use of compost may also be considered to aid vegetation establishment in the hillslopes. Additionally, an approach of sourcing and using native seeds can be a collaborative exercise between scientists and social scientists to involve and evaluate native cultures' knowledge base to answer modern-day sustainability issues.

Supplementary Materials: The following are available online at http://www.mdpi.com/2071-1050/12/10/4209/s1, Supplementary Table S1 Raw data (soil characterization and plant attributes). Supplementary Table S2: Alpha diversity measurements represented by richness and Shannon's diversity index and the % richness and Shannon's index differences across samples. Supplementary Figure S1. Relative abundance of microbial community at the phyla level. Abbr. LPM (LEO Parent Material), PS (Potting Soil). Supplementary Figure S2: Relative abundance of bacterial and archaeal community at the class level. A total of 71 classes were identified. Abbreviations: LPM (LEO Parent Material), PS (Potting Soil), and the plant species follow the soil material type. Sample without a plant species name is a control sample. Supplementary Figure S3. Relative abundance of the incipient basaltic parent material.

Author Contributions: Conceptualization, A.S.; Methodology, A.S. and A.J.; Formal Analysis, A.S., A.J., and P.K.; Investigation, A.S., P.K., and A.J.; Resources, P.A.T. and R.M.; Data Curation, A.S. and P.K.; Writing—Original Draft Preparation, A.S. and P.K.; Writing—Review and Editing, A.S., A.J., and P.A.T.; Visualization, A.S., P.K., and A.J.; Supervision, A.S.; Project Administration, A.S.; Funding Acquisition, P.A.T. and R.M. All authors have read and agreed to the published version of the manuscript.

Funding: This research was funded by the National Science Foundation through a Center for Integrated Access Networks (CIAN) grant #EEC-0812072 and Research Experience for Undergraduates (REU) site award #EEC-1359163 and #EEC-165910.

Acknowledgments: A.J. would like to acknowledge Integrated Optics for Undergraduate Native American (IOU-NA) REU Program directors Amee Hennig, Alison Hoff-Lohmeier, and Emily Lynch for the opportunity to carry out this research. A.S. and A.J. would like to thank Karen Serrano and Edward Hunt for analyzing soil chemical characteristics, as well as Emalee Eisenhauer, Roopkamal Kaur, Katarena Matos, Genesis Matos, Antonio Alveres Meira-Neto, and Scott Alexander White for their support during the experimental set-up and sampling. A.S. and A.J. also thank Katerina Dontsova and Kevin E. Bonine for facilitating the REU program at the University of Arizona.

Conflicts of Interest: "The authors declare no conflict of interest." "The funders had no role in the design of the study; in the collection, analyses, or interpretation of data; in the writing of the manuscript, or in the decision to publish the results".

References

1. Weil, R.R.; Brady, N.C. *The Nature and Properties of Soils*, 15th ed.; Pearson: London, UK, 2017; ISBN 0133254488.
2. Pimentel, D. Soil erosion: A food and environmental threat. *Environ. Dev. Sustain.* **2006**, *8*, 119–137. [CrossRef]
3. O'geen, A.T.; Schwankl, L.J. *University of California Davis, Division of Agriculture and Natural Resources, Publication 8196*; University of California: Davis, CA, USA, 2006.
4. World Health Organization. Land Degradation and Desertification. 2012. Available online: http://www.who.int/globalchange/ecosystems/desert/en/ (accessed on 6 April 2018).
5. UNCCD/Science-Policy Interface. *Land in Balance the Scientific Conceptual Framework for Land Degradation Neutrality Conceptual Framework for Land Degradation Neutrality (LDN)*; United Nations Convention to Combat Desertification: Bonn, Germany, 2016.
6. Orr, B.J.; Cowie, A.L.; Sanchez, V.M.C.; Chasek, P.; Crossman, N.D.; Erlewein, A.; Louwagie, G.; Maron, M.; Metternicht, G.I.; Minelli, S.A.E.; et al. *Scientific Conceptual Framework for Land Degradation Neutrality: A Report of the Science-Policy Interface*; United Nations Covention to Combat Desertification: Bonn, Germany, 2017.
7. Montoya-Tangarife, C.; de la Barrera, F.; Salazar, A.; Inostroza, L. Monitoring the effects of land cover change on the supply of ecosystem services in an urban region: A study of Santiago-Valparaíso, Chile. *PLoS ONE* **2017**, *12*, e0188117. [CrossRef] [PubMed]
8. Valentín-Vargas, A.; Root, R.A.; Neilson, J.W.; Chorover, J.; Maier, R.M. Environmental factors influencing the structural dynamics of soil microbial communities during assisted phytostabilization of acid-generating mine tailings: A mesocosm experiment. *Sci. Total Environ.* **2014**, *500–501*, 314–324. [CrossRef] [PubMed]

9. Honeker, L.K.; Neilson, J.W.; Root, R.A.; Gil-Loaiza, J.; Chorover, J.; Maier, R.M. Bacterial rhizoplane colonization patterns of buchloe dactyloides growing in metalliferous mine tailings reflect plant status and biogeochemical conditions. *Microb. Ecol.* **2017**, *74*, 853–867. [CrossRef] [PubMed]

10. Valentín-Vargas, A.; Neilson, J.W.; Root, R.A.; Chorover, J.; Maier, R.M. Treatment impacts on temporal microbial community dynamics during phytostabilization of acid-generating mine tailings in semiarid regions. *Sci. Total Environ.* **2018**, *618*, 357–368. [CrossRef]

11. Grandlic, C.J.; Mendez, M.O.; Chorover, J.; Machado, B.; Maier, R.M. Plant growth-promoting bacteria for phytostabilization of mine tailings. *Environ. Sci. Technol.* **2008**, *42*, 2079–2084. [CrossRef] [PubMed]

12. Mendez, M.O.; Maier, R.M. Phytoremediation of mine tailings in temperate and arid environments. *Rev. Environ. Sci. Biotechnol.* **2008**, *7*, 47–59. [CrossRef]

13. Smith, S.W.; Ross, K.; Karlsson, S.; Bond, B.; Upson, R.; Davey, A. Going native, going local: Revegetating eroded soils on the Falkland Islands using native seeds and farmland waste. *Restor. Ecol.* **2017**, *26*, 134–144. [CrossRef]

14. Gould, I.J.; Quinton, J.N.; Weigelt, A.; De Deyn, G.B.; Bardgett, R.D. Plant diversity and root traits benefit physical properties key to soil function in grasslands. *Ecol. Lett.* **2016**, *19*, 1140–1149. [CrossRef]

15. Debenport, S.J.; Assigbetse, K.; Bayala, R.; Chapuis-Lardy, L.; Dick, R.P.; McSpadden Gardener, B.B. Shifting populations in the root-zone microbiome of millet associated with enhanced crop productivity in the Sahel. *Appl. Environ. Microbiol.* **2015**. [CrossRef] [PubMed]

16. Tittonell, P.; Scopel, E.; Andrieu, N.; Posthumus, H.; Mapfumo, P.; Corbeels, M.; van Halsema, G.E.; Lahmar, R.; Lugandu, S.; Rakotoarisoa, J.; et al. Agroecology-based aggradation-conservation agriculture (ABACO): Targeting innovations to combat soil degradation and food insecurity in semi-arid Africa. *Field Crops Res.* **2012**, *132*, 168–174. [CrossRef]

17. Abd-Elmabod, S.; Bakr, N.; Muñoz-Rojas, M.; Pereira, P.; Zhang, Z.; Cerdà, A.; Jordán, A.; Mansour, H.; De la Rosa, D.; Jones, L. Assessment of soil suitability for improvement of soil factors and agricultural management. *Sustainability* **2019**, *11*, 1588. [CrossRef]

18. Hernandez, R.R.; Debenport, S.J.; Leewis, M.C.C.E.; Ndoye, F.; Nkenmogne, K.I.E.; Soumare, A.; Thuita, M.; Gueye, M.; Miambi, E.; Chapuis-Lardy, L.; et al. The native shrub, Piliostigma reticulatum, as an ecological "resource island" for mango trees in the Sahel. *Agric. Ecosyst. Environ.* **2015**, *204*, 51–61. [CrossRef]

19. Nunes, J.S.; Araujo, A.S.F.; Nunes, L.A.P.L.; Lima, L.M.; Carniero, R.F.V.; Salviano, A.A.C.; Tsai, S.M. Impact of land degradation on soil microbial biomass and activity in Northeast Brazil. *Pedosphere* **2012**, *22*, 88–95. [CrossRef]

20. Zhang, H.; Wang, R.; Chen, S.; Qi, G.; He, Z.; Zhao, X. Microbial taxa and functional genes shift in degraded soil with bacterial wilt. *Sci. Rep.* **2017**, *7*, 39911. [CrossRef] [PubMed]

21. Araújo, A.S.F.; Borges, C.D.; Tsai, S.M.; Cesarz, S.; Eisenhauer, N. Soil bacterial diversity in degraded and restored lands of Northeast Brazil. *Antonie Van Leeuwenhoek* **2014**, *106*, 891–899. [CrossRef]

22. van der Heijden, M.G.A.; Bardgett, R.D.; van Straalen, N.M. The unseen majority: Soil microbes as drivers of plant diversity and productivity in terrestrial ecosystems. *Ecol. Lett.* **2008**, *11*, 296–310. [CrossRef]

23. Delgado-Baquerizo, M.; Maestre, F.T.; Reich, P.B.; Jeffries, T.C.; Gaitan, J.J.; Encinar, D.; Berdugo, M.; Campbell, C.D.; Singh, B.K. Microbial diversity drives multifunctionality in terrestrial ecosystems. *Nat. Commun.* **2016**, *7*, 10541. [CrossRef]

24. Jenny, H.K. *Factors of Soil Formation: A System of Quantitative Pedology*; Tata McGraw Hill: New York, NY, USA, 1941.

25. Dauber, J.; Brown, C.; Fernando, A.L.; Finnan, J.; Krasuska, E.; Ponitka, J.; Styles, D.; Thrän, D.; Jan, K.; Groenigen, V.; et al. Bioenergy from "surplus" land: Environmental and socio-economic implications. *BioRisk* **2012**, *7*, 5–50. [CrossRef]

26. Mendez, M.O.; Neilson, J.W.; Maier, R.M. Characterization of a bacterial community in an abandoned semiarid lead-zinc mine tailing site. *Appl. Environ. Microbiol.* **2008**, *74*, 3899–3907. [CrossRef]

27. Ciria, C.; Sanz, M.; Carrasco, J.; Ciria, P. Identification of arable marginal lands under rainfed conditions for bioenergy purposes in Spain. *Sustainability* **2019**, *11*, 1833. [CrossRef]

28. Volkmann, T.H.M.; Sengupta, A.; Pangle, L.A.; Dontsova, K.; Barron-Gafford, G.A.; Harman, C.J.; Niu, G.-Y.; Abramson, N.; Meira-Neto, A.A.; Wang, Y.; et al. Controlled experiments of hillslope coevolution at the biosphere 2 landscape evolution observatory: Toward prediction of coupled hydrological, biogeochemical, and ecological change. In *Hydrology of Artificial and Controlled Experiments*; Intech Open Limiited: London, UK, 2017; pp. 25–74.

29. Sengupta, A.; Pangle, L.A.; Volkmann, T.H.M.; Dontsova, K.; Troch, P.A.; Meira-neto, A.A.; Neilson, J.W.; Hunt, E.A.; Chorover, J.; Zeng, X.; et al. Advancing understanding of hydrological and biogeochemical interactions in evolving landscapes through controlled experimentation at the landscape evolution observatory. In *Terrestrial Ecosystems Research Infrastructure: Challenges and Opportunities*; Chabbi, A., Loescher, H.W., Eds.; CRC Press, Taylor and Francis Group: Boca Raton, FL, USA, 2017; pp. 83–118.

30. Pangle, L.A.; DeLong, S.B.; Abramson, N.; Adams, J.; Barron-Gafford, G.A.; Breshears, D.D.; Brooks, P.D.; Chorover, J.; Dietrich, W.E.; Dontsova, K.; et al. The landscape evolution observatory: A large-scale controllable infrastructure to study coupled Earth-surface processes. *Geomorphology* **2015**, *244*, 190–203. [CrossRef]

31. Pohlmann, M.; Dontsova, K.; Root, R.; Ruiz, J.; Troch, P.; Chorover, J. Pore water chemistry reveals gradients in mineral transformation across a model basaltic hillslope. *Geochem. Geophys Geosystems* **2016**, *17*, 2054–2069. [CrossRef]

32. Sengupta, A.; Stegen, J.C.; Nielson, J.W.; Meira-Neto, A.; Wang, Y.; Troch, P.A.; Chorover, J.; Maier, R.M. Assessment of microbial community patterns under incipient conditions in a basalt soil system. *J. Geophys. Res. Biogeosci.* **2019**, *124*, 941–958. [CrossRef]

33. Native Seeds/SEARCH—Home. Available online: https://www.nativeseeds.org/ (accessed on 29 August 2017).

34. Sengupta, A.; Wang, Y.; Meira Neto, A.A.; Matos, K.A.; Dontsova, K.; Root, R.; Neilson, J.W.; Maier, R.M.; Chorover, J.; Troch, P.A. Soil lysimeter excavation for coupled hydrological, geochemical, and microbiological investigations. *J. Vis. Exp.* **2016**, *2016*. [CrossRef] [PubMed]

35. Jim, A.; Sengupta, A. Assessing the ability of incipient basaltic soil to support plants native to Southwestern United States. *J. Undergrad. Res.* **2018**, *IX*, 30–33.

36. Duke, J.A. *Phaseolus vulgaris* L. In *Handbook of Energy Crops*. Available online: https://www.hort.purdue.edu/newcrop/duke_energy/Phaseolus_vulgaris.html (accessed on 15 July 2017).

37. Caporaso, J.G.; Lauber, C.L.; Walters, W.A.; Berg-Lyons, D.; Huntley, J.; Fierer, N.; Owens, S.M.; Betley, J.; Fraser, L.; Bauer, M.; et al. Ultra-high-throughput microbial community analysis on the Illumina HiSeq and MiSeq platforms. *ISME J.* **2012**, *6*, 1621–1624. [CrossRef]

38. Caporaso, J.G.; Kuczynski, J.; Stombaugh, J.; Bittinger, K.; Bushman, F.D.; Costello, E.K.; Fierer, N.; Peña, A.G.; Goodrich, J.K.; Gordon, J.I.; et al. QIIME allows analysis of high-throughput community sequencing data. *Nat. Methods* **2010**, *7*, 335–336. [CrossRef] [PubMed]

39. Laubitz, D.; Harrison, C.A.; Midura-Kiela, M.T.; Ramalingam, R.; Larmonier, C.B.; Chase, J.H.; Caporaso, J.G.; Besselsen, D.G.; Ghishan, F.K.; Kiela, P.R. Reduced epithelial Na+/H+ exchange drives gut microbial dysbiosis and promotes inflammatory response in T cell-mediated murine colitis. *PLoS ONE* **2016**, *11*, e0152044. [CrossRef]

40. Edgar, R.C. Search and clustering orders of magnitude faster than BLAST. *Bioinformatics* **2010**, *26*, 2460–2461. [CrossRef]

41. Caporaso, J.G.; Bittinger, K.; Bushman, F.D.; DeSantis, T.Z.; Andersen, G.L.; Knight, R. PyNAST: A flexible tool for aligning sequences to a template alignment. *Bioinformatics* **2010**, *26*, 266–267. [CrossRef] [PubMed]

42. DeSantis, T.Z.; Hugenholtz, P.; Larsen, N.; Rojas, M.; Brodie, E.L.; Keller, K.; Huber, T.; Dalevi, D.; Hu, P.; Andersen, G.L. Greengenes, a chimera-checked 16S rRNA gene database and workbench compatible with ARB. *Appl. Environ. Microbiol.* **2006**, *72*, 5069–5072. [CrossRef]

43. Haas, B.J.; Gevers, D.; Earl, A.M.; Feldgarden, M.; Ward, D.V.; Giannoukos, G.; Ciulla, D.; Tabbaa, D.; Highlander, S.K.; Sodergren, E.; et al. Chimeric 16S rRNA sequence formation and detection in Sanger and 454-pyrosequenced PCR amplicons. *Genome Res.* **2011**, *21*, 494–504. [CrossRef]

44. Wang, Q.; Garrity, G.M.; Tiedje, J.M.; Cole, J.R. Naive Bayesian classifier for rapid assignment of rRNA sequences into the new bacterial taxonomy. *Appl. Environ. Microbiol.* **2007**, *73*, 5261–5267. [CrossRef] [PubMed]

45. Price, M.N.; Dehal, P.S.; Arkin, A.P. FastTree 2—Approximately maximum-likelihood trees for large alignments. *PLoS ONE* **2010**, *5*, e9490. [CrossRef] [PubMed]

46. Lozupone, C.; Knight, R. UniFrac: A new phylogenetic method for comparing microbial communities. *Appl. Environ. Microbiol.* **2005**, *71*, 8228–8235. [CrossRef]

47. R Core Team. *R: A Language and Environment for Statistical Computing*; R Foundation for Statistical Computing: Vienna, Austria, 2014; Available online: https://www.R-project.org/ (accessed on 21 May 2020).

48. McMurdie, P.J.; Holmes, S. Phyloseq: An R package for reproducible interactive analysis and graphics of microbiome census data. *PLoS ONE* **2013**, *8*, e61217. [CrossRef]

49. Louca, S.; Parfrey, L.W.; Doebeli, M. Decoupling function and taxonomy in the global ocean microbiome. *Science* **2016**, *353*, 1272–1277. [CrossRef]

50. Warenes, G.R.; Bolker, B.; Bonebakker, L.; Gentleman, R.; Huber, W.; Liaw, A.; Lumley, T.; Maechler, M.; Magnusson, A.; Moeller, S.; et al. *gplots: Various R Programming Tools for Plotting Data*; R Package Version; 2016. Available online: https://rdrr.io/cran/gplots/ (accessed on 21 May 2020).

51. Oksanen, J. *Multivariate Analysis of Ecological Communities in R: Vegan tutorial*; 2015; pp. 1–40. Available online: https://www.mooreecology.com/uploads/2/4/2/1/24213970/vegantutor.pdf (accessed on 21 May 2020).

52. Dray, S.; Dufor, A.-B.; Thioulouse, J. Analysis of Ecological Data: Exploratory and Euclidean Methods in Environmental Sciences. 2018. Available online: https://rdrr.io/cran/ade4/ (accessed on 21 May 2020).

53. Tilman, D.; Cassman, K.G.; Matson, P.A.; Naylor, R.; Polasky, S. Agricultural sustainability and intensive production practices. *Nature* **2002**, *418*, 671–677. [CrossRef]

54. Solís-Domínguez, F.A.; Valentín-Vargas, A.; Chorover, J.; Maier, R.M. Effect of arbuscular mycorrhizal fungi on plant biomass and the rhizosphere microbial community structure of mesquite grown in acidic lead/zinc mine tailings. *Sci. Total Environ.* **2011**, *409*, 1009–1016. [CrossRef]

55. Martinez-Romero, E.; Segovia, L.; Mercante, F.M.; Franco, A.A.; Graham, P.; Pardo, M.A. Rhizobium tropici, a novel species nodulating phaseolus vulgaris l. beans and leucaena sp. trees. *Int. J. Syst. Bacteriol.* **1991**, *41*, 417–426. [CrossRef] [PubMed]

56. Martínez-Romero, E. Diversity of rhizobium-phaseolus vulgaris symbiosis: Overview and perspectives. *Plant Soil* **2003**, *252*, 11–23. [CrossRef]

57. Gtari, M.; Ghodhbane-Gtari, F.; Nouioui, I.; Beauchemin, N.; Tisa, L.S. Phylogenetic perspectives of nitrogen-fixing actinobacteria. *Arch. Microbiol.* **2012**, *194*, 3–11. [CrossRef]

58. Neilson, J.W.; Quade, J.; Ortiz, M.; Nelson, W.M.; Legatzki, A.; Tian, F.; LaComb, M.; Betancourt, J.L.; Wing, R.A.; Soderlund, C.A.; et al. Life at the hyperarid margin: Novel bacterial diversity in arid soils of the Atacama Desert, Chile. *Extremophiles* **2012**, *16*, 553–566. [CrossRef]

59. Kardol, P.; De Deyn, G.B.; Laliberté, E.; Mariotte, P.; Hawkes, C.V. Biotic plant-soil feedbacks across temporal scales. *J. Ecol.* **2013**, *101*, 309–315. [CrossRef]

60. Kelly, L.C.; Cockell, C.S.; Herrera-Belaroussi, A.; Piceno, Y.; Andersen, G.; DeSantis, T.; Brodie, E.; Thorsteinsson, T.; Marteinsson, V.; Poly, F.; et al. Bacterial diversity of terrestrial crystalline volcanic rocks, Iceland. *Microb. Ecol.* **2011**, *62*, 69–79. [CrossRef]

61. Akob, D.M.; Küsel, K. Where microorganisms meet rocks in the Earth's Critical Zone. *Biogeosciences* **2011**, *8*, 3531–3543. [CrossRef]

62. Knelman, J.E.; Legg, T.M.; O'Neill, S.P.; Ashenberger, C.L.W.; Gonzalez, A.G.; Cleveland, C.C.; Nemergut, D.R. Bacterial community structure and function change in association with colonizer plants during early primary succession in a glacier forefield. *Soil Biol. Biochem.* **2012**, *46*, 172–180. [CrossRef]

Article

Effect of Soil Microbiome from Church Forest in the Northwest Ethiopian Highlands on the Growth of *Olea europaea* and *Albizia gummifera* Seedlings under Glasshouse Conditions

Getu Abebe [1,2,*], Atsushi Tsunekawa [3], Nigussie Haregeweyn [4], Takeshi Taniguchi [3], Menale Wondie [2], Enyew Adgo [5], Tsugiyuki Masunaga [6], Mitsuru Tsubo [3], Kindiye Ebabu [3,5], Amrakh Mamedov [3] and Derege Tsegaye Meshesha [5]

[1] The United Graduate School of Agricultural Sciences, Tottori University, 4-101 Koyama-minami, Tottori 680-8553, Japan

[2] Amhara Agricultural Research Institute, Forestry Research Department, P.O. Box 527 Bahir Dar, Ethiopia; menalewondie@yahoo.com

[3] Arid Land Research Center, Tottori University, 1390 Hamasaka, Tottori 680-0001, Japan; tsunekawa@tottori-u.ac.jp (A.T.); takeshi@alrc.tottori-u.ac.jp (T.T.); tsubo@tottori-u.ac.jp (M.T.); kebabu@tottori-u.ac.jp (K.E.); amrakh03@yahoo.com (A.M.)

[4] International Platform for Dryland Research and Education, Tottori University, 1390 Hamasaka, Tottori 680-0001, Japan; nigussie_haregeweyn@yahoo.com

[5] College of Agriculture and Environmental Sciences, Bahir Dar University, P.O. Box 1289 Bahir Dar, Ethiopia; enyewadgo@gmail.com (E.A.); deremesh@yahoo.com (D.T.M.)

[6] Faculty of Life and Environmental Science Shimane University, Shimane Matsue 690-082, Japan; masunaga@life.shimane-u.ac.jp

[*] Correspondence: gabebe233@gmail.com or d17a4114b@edu.tottori-u.ac.jp

Received: 3 April 2020; Accepted: 13 June 2020; Published: 18 June 2020

Abstract: Loss of beneficial microbes and lack of native inoculum have hindered reforestation efforts in the severely-degraded lands worldwide. This is a particularly pressing problem for Ethiopia owing to centuries-old unsustainable agricultural practices. This study aimed to evaluate the inoculum potential of soils from church forest in the northwest highlands of Ethiopia and its effect on seedling growth of two selected native tree species (*Olea europaea* and *Albizia gummifera*) under a glasshouse environment. Seedlings germinated in a seed chamber were transplanted into pots containing sterilized and/or non-sterilized soils collected from under the canopy of three dominant church forest trees: *Albizia gummifera* (AG), *Croton macrostachyus* (CM), and *Juniperus procera* (JP) as well as from adjacent degraded land (DL). A total of 128 pots (2 plant species × 4 soil origins × 2 soil treatments × 8 replicates) were arranged in a factorial design. Overall, seedlings grown in AG, CM, and JP soils showed a higher plant performance and survival rate, as a result of higher soil microbial abundance and diversity, than those grown in DL soils. The results showed significantly higher plant height, root collar diameter, shoot, and total mass for seedlings grown in non-sterilized forest soils than those grown in sterilized soils. Furthermore, the bacterial relative abundance of *Acidobacteria*, *Actinobacteria*, and *Nitrospirae* was significantly higher in the non-sterilized forest soils AG, CM, and JP (r^2 = 0.6–0.8, $p < 0.001$). Soil pH had a strong effect on abundance of the bacterial community in the church forest soils. More specifically, this study further demonstrated that the effect of soil microbiome was noticeable on the performance of *Olea* seedlings grown in the soil from CM. This suggests that the soils from remnant church forests, particularly from the canopy under CM, can serve as a good soil origin, which possibly would promote the native tree seedling growth and survival in degraded lands.

Keywords: arid regions; bacteria; degraded land; fungi; ITS; microbial community; restoration; 16S rRNA

---

1. Introduction

Land degradation is a major global problem affecting all terrestrial biomes in arid and humid regions [1,2]. Human activities, such as deforestation, overgrazing, and improper agricultural practices are the main factors causing land degradation, all of which significantly reduce environmental quality, and socio-economic sustainability [3–6]. Land degradation also causes deterioration of soil communities and negatively influences ecosystem function [7]. Studies have revealed the significant role of soil microbes in ecosystem functioning [8,9]. However, the loss of beneficial microbes, including fungi and bacteria, adversely affects the recovery potential of a degraded ecosystem [10].

Symbiotic relationships between the roots of higher plants and microbes (fungi and bacteria) strongly influence plant survival, growth, and ecosystem properties [11]. These beneficial microbes can enhance soil nutrient supply, drought tolerance, and pathogen resistance [12] of the host plant. The interaction of soil microbes with plant roots and organic matter can improve soil aeration and resistance to slaking and erosion by enhancing soil aggregation and structural stability; the microorganisms influence soil aggregation via chemical stabilization, and the organic matter contributes a cementing effect [13]. However, these functional roles of soil microbes are limited in degraded ecosystems because of a low level of microbial diversity, poor vegetation cover, high soil disturbance, and severe erosion rates [10,14,15]. Consequently, in many regions of the world, various methods of ecological restoration are necessary to rectify degraded ecosystems [16]. For instance, soil and water conservation practices [17], including afforestation and exclosure establishment [2,18], have been experimented with within Ethiopia. However, achieving restored ecosystem function through re-establishment of native tree species [19] has proved challenging due to the lack of a native soil microbial community in degraded lands (DLs) [20]. Indeed, soil microbes in combination with plant species play a crucial role in restoring DL [8,15]. In the case of the Ethiopian highlands, very little information exists about the source of native inoculum for the successful restoration of lost microbial community functions in degraded landscapes.

Small patches of natural forest, called "church forests," exist around Ethiopian Orthodox Tewahedo churches and monasteries, and these constitute the last remnants of the original forest cover, having been conserved for more than a century. Church forests are located in a matrix of intensively degraded agricultural landscapes [21,22]. Apart from their social and spiritual value, church forests are obvious and important foci of biodiversity and act as a source of seeds and germplasm for native flora [21–24]. However, there is a lack of studies characterizing the role of microbial communities or evaluating the inoculum potential and the effects of microbes from church forests on the early stages of native tree establishment.

Olea europaea L. subsp. *cuspidata* and *Albizia gummifera* are among the most important native tree species of Ethiopia. These two tree species were selected for this study based on their social and ecological importance and their limited survival and regeneration ability in degraded lands of the Ethiopian highlands [25,26]. *Olea europaea* subsp. *cuspidata* (Wall. ex DC.) is a late-successional evergreen tree species found in dry Afromontane forest between 1250 and 3100 m a.s.l. [27]. The species is hardy and drought resistant once established, even in poor soils; adult trees are commonly 15–25 m high [28]. *Albizia gummifera* (J.F.Gmel.) C.A.Sm., is a deciduous tree species; it can reach up to 15 m height and occurs in semi-humid and humid highland forests between 1400 and 2500 m a.s.l. [28]. It often co-exists with *Olea europaea* and *Juniperus procera* [29]. Despite *Olea europaea* and *Albizia gummifera* are among the suitable native tree species supposed to restoring degraded lands in the highlands of Ethiopia, no or very limited information is available on the growth performance and survival rate of their seedlings in soils from conserved forest and degraded lands. This study was, therefore, designed to: (1) assess the soil microbial diversity in remnant church forest and surrounding degraded land, (2) evaluate the effect of soil microbiome, from under the canopy of church forest, on early growth and survival rate of seedlings of *Olea europaea* and *Albizia gummifera* under glasshouse conditions, and (3) evaluate the association between soil microbial and chemical properties in relation to plant growth.

2. Materials and Methods

2.1. Site Description

The study was conducted using soil sampled from Ethiopian Orthodox Tewahedo Church, Laguna St. Giorgis forest (Figure 1). Laguna St. Giorgis forest is a remnant forest around a church built in 1500 A.D. It is located at 11°39′21″ N and 37°30′36″ E at an altitude of 2100 m a.s.l. The current forest covers 5.25 ha. The mean annual rainfall ranges from 895 to 2037 mm (Figure 2a). The mean annual temperature range is 17–31 °C [30]. The vegetation type of the area is *Albizia–Juniperus-Croton*-dominated dry Afromontane forest [31]. Leptosols and Regosols are the major soil types in the study area [32].

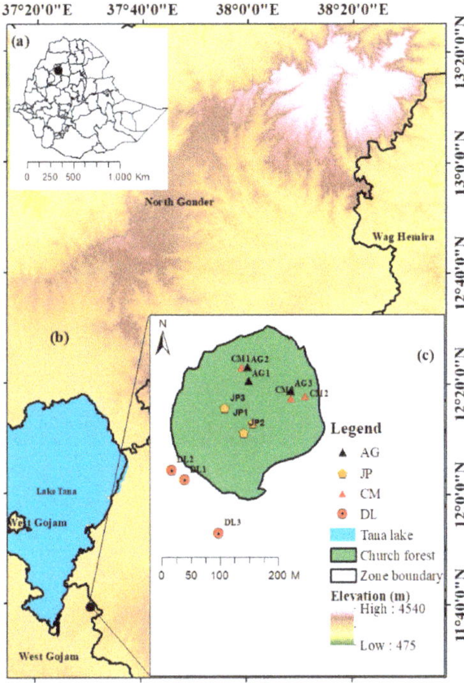

Figure 1. Location of the study area. The study site is indicated by the black circle (**a**,**b**). *Albizia gummifera* (AG), *Croton macrostachyus* (CM), *Juniperus procera* (JP) are tree species in the church forest, and DL is adjacent degraded land (**c**).

2.2. Field Soil Sampling

Soil samples were collected under the canopy of three predominant native tree species: *Albizia gummifera* (AG), *Croton macrostachyus* (CM), and *Juniperus procera* (JP), in the church forest (Figure 1c), as well as from adjacent degraded land (DL). For each tree species and DL, three replicate soil samples were collected from the top 0 to 20 cm soil depth using a ruler and a hand shovel measuring around 3.0 kg of soil in plastic bags. Soil samples from a similar source were mixed to obtain a composite inoculum. The samples were prepared at the soil laboratory of Bahir Dar University, Ethiopia, and transported to Japan for the experiment: half of the total samples for each soil origin were sterilized using gamma-rays (30–60 kGy; [33]) to evaluate the effects of soil microbes. For each soil origin, a 2 g sample was taken to store at −80 °C for downstream DNA extraction to evaluate soil microbiome before the Glasshouse (GH) experiment.

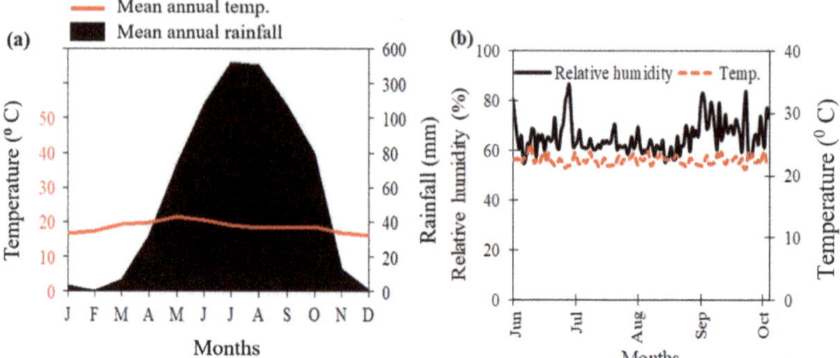

Figure 2. (**a**) Climate characteristics of the study area from 1962 to 2017 [30,34] and (**b**) glasshouse atmospheric conditions during the experiment. Climate diagram of the study site (**a**) is based on Walter and Lieth climate diagrams [35].

2.3. Experimental Design

A pot experiment was carried out in a glasshouse at Tottori University, Arid Land Research Center, Japan. The temperature in the glasshouse was in the range of 21–25 °C with an average relative humidity of 65.7% (Figure 2b).

Surface-sterilized seeds of *Olea* and *Albizia* were germinated, in May 2018, using autoclaved vermiculite (15 min at 121 °C) in a seed germination chamber (MLR-351H, SANYO, Tokyo, Japan); temperature and relative humidity of the chamber was 25 °C and 60–70%, respectively. *Albizia* seeds were germinated in a week, while *Olea* seeds took 2 weeks after sowing in the chamber. The seedlings were transplanted to the glass pots (4 cm diameter and 14 cm height each) at the depth of 2–3 cm containing either sterilized or non-sterilized soil (100 g for samples from forest and 125 g for samples from degraded land). The pots were kept at room conditions (25 °C and 12 h light) for 2 weeks.

A total of 128 pots (2 plant species × 4 soil origins × 2 soil treatments × 8 replicates) were transferred to the glasshouse for further monitoring and evaluation. Pots were maintained at 15% moisture content throughout the experiment period (5 months) and were weighed every 3 days. Every 2 weeks, 10 mL of sterilized distilled water was added to the pots containing non-sterilized soil and, to control contamination, 10 mL of 1:50 antibiotic (penicillin/streptomycin/amphotericin B solution [36] and sterilized distilled water) solution was added to the pots containing sterilized soil.

2.4. Growth and Survival Data

Seedling survival was recorded monthly until the end of the GH experiment. Plant height (cm) and root collar diameter (RCD, mm) were measured using a ruler and digital caliper, respectively, at the end of the GH experiment. Root to shoot (R/S) ratio was calculated by dividing root dry weight by shoot dry weight.

2.5. Soil Analysis

Soil samples collected from field (before GH experiment) and pots (after the GH experiment) were air-dried, passed through a 2-mm sieve and analyzed for selected soil parameters: pH, organic carbon, total nitrogen, available phosphorus (P), aggregate–structure stability, and texture (particle size). Soil pH was measured in 1:2.5 soil: water suspension. Soil organic carbon (SOC) and total nitrogen (TN) were determined using a CN corder (Macro Corder JM1000CN, J-Science Lab, Kyoto, Japan). Soil available P was extracted with a solution of 1.0 M ammonium fluoride and 2.5 M hydrochloric acid (Bray II method; Bray and Kurtz [37]). Phosphate was detected by absorption spectrometry (UV–140–02; Shimadzu, Kyoto, Japan) using the molybdenum blue colorimetric procedure. Soil aggregate–structure stability (SAS) was determined using the modified high energy moisture characteristics method [38], and particle size was analyzed by the hydrometer method [39]. In addition, soil moisture content was gravimetrically measured after drying moist samples in an oven at 105 °C for 24 h [39].

2.6. Plant and Soil Sample Collection

Seedlings were harvested at 5 months after transplantation (Figures S2 and S3). At harvest, for each seedling: (i) the shoots (stem and leaves) were separated and weighed, (ii) after breaking the glass pots, the roots were carefully separated, gently washed with tap water, and weighed in fresh. Shoots and roots were then dried at 60 °C for 72 h and weighed to estimate the shoot, root and total biomass. From each pot, 2 g of soil samples were collected to store at −80 °C for further molecular microbiome characterization after the GH experiment.

2.7. Soil DNA Extraction, Polymerase Chain Reaction (PCR) Amplification, and Sequencing

For each sample, 0.25 g of soil was extracted using DNeasy PowerSoil® DNA kit (Qiagen, Germany) following the manufacturer's protocol. The extracted DNA solution was diluted 10 times and PCR was carried out using BIO-RAD T100™ Thermal Cycler (Bio-Rad Laboratories, Hercules, CA). For each sample, two replicates were amplified in a 20 μL (total volume) reaction mixture, containing 1 μL of template DNA (10 ng/μL), 7.0 μL sterilized distilled water, 10 μL of 2 × Q5 High-Fidelity DNA Polymerase (New England Biolabs Inc., Ipswich, MA, USA), and 1.0 μL for each primer.

The V4 region in the 16S rRNA gene was amplified using primers S-D-Arch-0519-a-S-15 (CAGCMGCCGCGGTAA) and S-D-Bact-0785-a-A-21(GACTACHVGGGTATCTAATCC) for bacteria [40]. The internal transcribed spacer (ITS) region was amplified with the primers ITS1F_KYO2 (TAGAGGAAGTAAAAGTCGTAA) and ITS2_KYO2 (TTYRCTRCGTTCTTCATC) for fungi [41]. Each forward primer was tagged with the Ion Torrent specific adapters and Ion Xpress barcode to distinguish the origin of samples. The expected band size for 16S rRNA and ITS primers was 350 bp and 360 bp, respectively.

The PCR thermal cycling conditions were: initial denaturation at 98 °C for 30 s, followed by 35 cycles of 98 °C for 10 s, 53 °C for 30 s, and 72 °C for 1 min, followed by a 2 min final extension at 72 °C. The PCR products were checked by agarose gel electrophoresis, and the replicates were composited to make one mix for purification. The purified DNA concentration was quantified for each PCR product with the Qubit dsDNA HS Assay Kit on a Qubit fluorometer 2.0 (Invitrogen, Carlsbad, CA, USA). Then, pooled DNA distribution and size were checked using Agilent 2100 (Agilent Technologies, Inc. Santa Clara, CA, USA) after putting an equimolar amount of amplified DNA into a tube. Following this, DNA sequencing by Ion Torrent Personal Genome Machine (Life Technology, Inc. Carlsbad, CA, USA) was performed as described by [41].

2.8. Sequence Data Processing

Sequence data processing was performed as previously described in Tian et al. [42]. Quality sequences were gained using the Quantitative Insight into Microbial Ecology (QIIME) 1.8.0 pipeline [43] after removing sequences shorter than 200 bp for bacteria and 360 bp for fungi, and sequences with expected errors predicted by Phred (Q) scores greater than 0.8. ITSx were used to precisely pick the ITS region for fungal DNA [44]. Successful sequences (Figure S1) were clustered into operational taxonomic units (OTUs) at 97% similarity level using USEARCH [45]. Then, sequencing chimeras were checked and removed using UCHIME [46]. Taxonomy of each OTU was assigned to fungal and bacterial taxa using UNITE [47] and SILVA (SILVA 128 QIIME release) databases, respectively.

2.9. Statistical Analysis

Statistical analyses were performed in R [48] using the interference implemented RStudio (version 1.1.383). The data were checked for normal distribution before analysis using the Shapiro–Wilk test, and non-normally distributed data were log and square-root transformed. The effects of soil from the four origins (DL, AG, CM, and JP) with soil treatment on plant height, R/S ratio, mass, pH, soil organic carbon (SOC), total nitrogen (TN), carbon/nitrogen (C/N) ratio, available P, and SAS were analyzed with a general linear model procedure. Three-way ANOVA was used to check the effect of species on pooled plant and soil properties, then the two species (*Olea* and *Albizia*) separately and soil origins (with and without soil treatment) were fixed factors. Two-way ANOVA was used to test the interaction between soil origins and soil treatments on plant growth indices and soil properties. Differences in means across soil origins and soil treatments were analyzed with Tukey's HSD test using the R package Agricolae [49]. Survival was measured using the Kaplan–Meier procedure; survival curves were compared statistically using log-rank test (LogranKA package) and Cox-regression (survival package) survival analysis was used to test the interaction between soil origins and soil treatments [50].

Rarefied OTU tables were used to analyze soil microbial community composition and diversity Indices (Shannon (H′) and Simpson (D)) and number of OTUs observed. The effects of soil origin between treatments on H′, D, and number of OTUs observed were non-parametrically determined (Kruskal–Wallis test). To test the effect of soil origins and soil treatments on microbial communities, PerMANOVA (1000 permutations) was performed by using the Adonis function in the vegan package of R [51]. Non-metric multidimensional scaling based on Bray–Curtis dissimilarity was used to visualize the results, and the relationships between soil properties, and microbial communities were tested using the envfit function in the vegan package of R [51].

3. Results

3.1. Soil Microbial Community

3.1.1. Soil Bacterial Community Composition, Abundance, and Relationship with Soil Properties

Acidobacteria, Actinobacteria, Crenarchaeota, Nitrospirae and *Proteobacteria* were the most abundant bacterial phyla in the original forest soils (Figure 3a). Whereas, *Actinobacteria, Gammatimonadetes,* and *Proteobacteria* were the most dominant bacterial phyla in the DL soil (Figure 3a). For both plant species, the relative abundance of *Proteobacteria* was higher in sterilized soils (47% and 53%, respectively) than in non-sterilized ones (28% and 29%, respectively); in contrast, the relative abundance of *Actinobacteria* was higher in non-sterilized soils (Figure 3b,c). Also, in non-sterilized soils, the abundance of *Acidobacteria* was higher in DL, CM, and JP for *Olea* seedlings (Figure 3b) and in all soil origins for *Albizia* seedlings (Figure 3c).

The values of indices (H′ and D), and the number of OTUs observed for bacteria under *Olea* (Figure 4a,e,i), and *Albizia* (Figure 4c,g,k) seedlings were significantly higher in non-sterilized than in sterilized soils. The bacterial community significantly varied among soil origins (PerMANOVA; $F = 13.63$, $p < 0.001$) and between sterilize versus non-sterilize soil treatments ($F = 3.27$, $p < 0.001$) and seedling types ($F = 16.79$, $p < 0.001$). The relative abundance of *Acidobacteria, Actinobacteria,* and *Nitrospirae* was significantly correlated with bacterial community in non-sterilized forest soils ($r^2 = 0.63$, $r^2 = 0.77$, $r^2 = 0.79$, respectively; $p < 0.001$). *Proteobacteria* were strongly correlated with bacterial community in sterilized soils ($r^2 = 0.63$, $p < 0.001$), in particular for the forest soils; for the DL soil, the bacterial communities were less related to *Acidobacteria, Nitrospirae,* and *Proteobacteria*. In addition, the bacterial communities of DL were grouped in a separate cluster distant from the forest soils (Figure 5a). Among different soil properties, soil pH was strongly correlated ($r^2 = 0.6$, $p < 0.001$) with the bacterial community (Figure 5a), whereas SOC, TN, the C/N ratio, and available P did not show a strong correlation with the bacterial community.

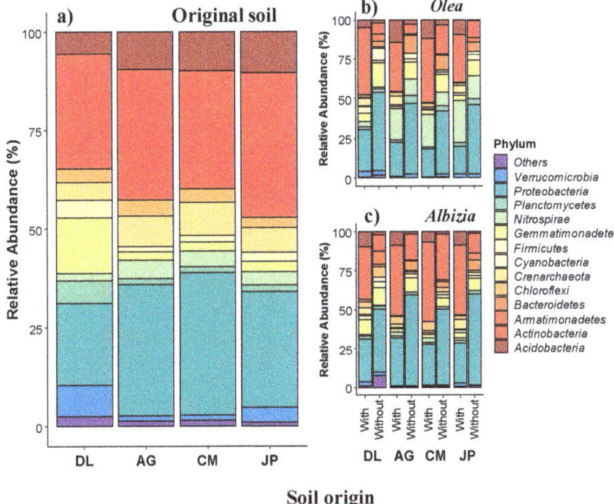

Figure 3. Relative abundances of bacteria at the phylum level in the soil samples; before GH experiment (**a**) (original soil), and after GH experiment for *Olea* (**b**) and *Albizia* (**c**) seedlings with treatment (non-sterilized soil) and without treatment (sterilized soil). The values shown are means (n = 3 for soil before GH experiment and n = 8 for soil after GH experiment). DL, AG, CM, and JP represent degraded land, *Albizia gummifera*, *Croton macrostachyus*, and *Juniperus procera*, respectively.

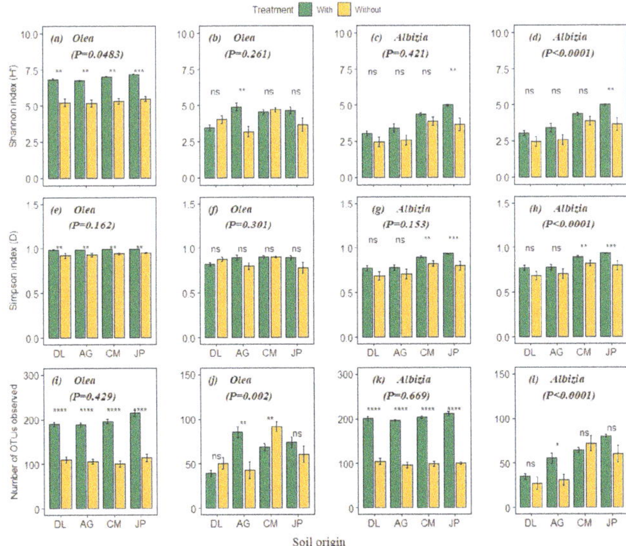

Figure 4. Shannon (H′) index (**a–d**), Simpson (D) index (**e–h**), and number of OTUs (**i–l**) observed for bacteria (columns 1 and 3) and fungi (columns 2 and 4) for *Olea* and *Albizia* seedlings with (non-sterilized; green bars) and without (sterilized; yellow bars) treatments. DL, AG, CM, and JP stand for different soil origins from degraded land, or from beneath *Albizia* gummifera, Croton macrostachyus, and Juniperus procera, respectively. Asterisks indicate statistically significant differences between seedlings in non-sterilized (with treatment) and sterilized (without treatment) soil (* $p \leq 0.05$; ** $p \leq 0.01$; *** $p \leq 0.001$; **** $p \leq 0.0001$; and ns, not significant) and *p*-value (Kruskal test among soil origins). Values are shown as mean ± standard error (n = 8).

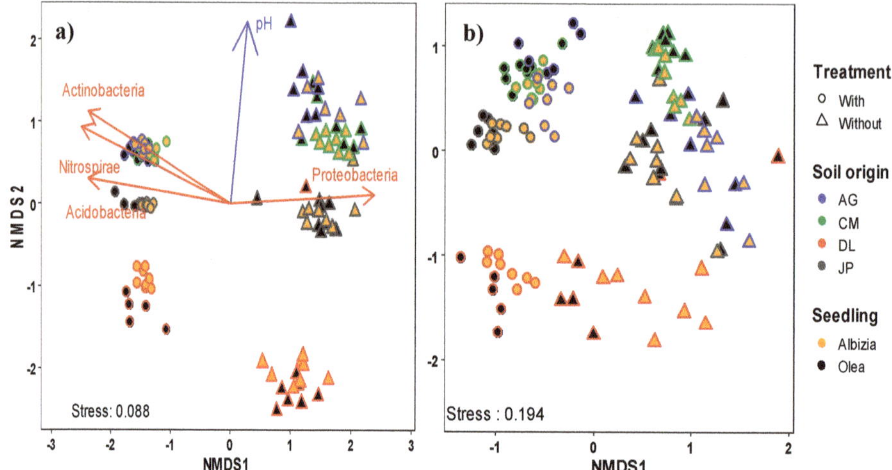

Figure 5. Non-metric multidimensional scaling based on Bray–Curtis dissimilarities for the (**a**) bacterial and (**b**) fungal communities. Shapes of symbols represent soil treatments, colors of symbols represent soil origins (AG, CM, DL, and JP), and fills with the symbols represent *Albizia* and *Olea* seedlings. Arrows indicate significant correlations and the lengths of arrows are relative to the strength of the correlation. DL, AG, CM, and JP stands for soil from degraded land or from beneath *Albizia gummifera, Croton macrostachyus, Juniperus procera*, respectively.

3.1.2. Soil Fungal Community Composition, Abundance, and Relationship with Soil Properties

In the soils before GH experiment, *Ascomycota* (48%, 70%, 72%, and 68%) and *Basidiomycota* (21%, 16%, 11% and 17%) were the most abundant fungi phyla, respectively in DL, AG, CM, and JP (Figure 6a). In addition, a relatively higher abundance of *Glomeromycota* was found in DL soils. Similarly, for soils after GH experiment, *Ascomycota* and *Basidiomycota* were the most abundant fungal phyla under *Olea* and *Albizia* seedlings (Figure 6b,c). *Ascomycota* had a higher relative abundance for *Olea* seedlings in non-sterilized (89%, 93%, 93%, and 76%) than in sterilized (81%, 80%, 78%, and 78%) DL, AG, CM, and JP soils, respectively. However, the abundance of *Basidiomycota* was relatively higher in sterilized (11%, 14%, and 18%) than non-sterilized (3%, 5%, and 5%) of DL, AG, and CM soils, respectively (Figure 4b). The relative abundance of *Ascomycota* under *Albizia* seedlings in DL, AG, CM, and JP were 80%, 88%, 80%, and 83%; and 62%, 62%, 75%, and 74% for seedlings in non-sterilized and sterilized soils, respectively (Figure 6c).

Significant differences in H′ for the fungal community were found between soil treatments for *Olea* (Figure 4b) and *Albizia* seedlings (Figure 4d) in AG and JP soils, respectively. In addition, significant variation in D was found in CM and JP soils for *Albizia* seedlings (Figure 4h) between non-sterilized and sterilized soil. The number of OTUs observed for fungi were significantly higher under seedlings with non-sterilized AG soil for both *Olea* and *Albizia* (Figure 4j,l) than with sterilized soil, but the case was the opposite for CM soil for *Olea* (Figure 4j). However, the DL soil had significantly fewer OTUs for fungi than the forest soils (AG, CM, and JP).

PerMANOVA results indicated that the fungal community significantly differed among soil origins (F = 12.71, $p < 0.001$) and between soil treatments (F = 2.70, $p < 0.001$) and seedling types (F = 8.40, $p < 0.001$). The relative abundance of *Ascomycota* and *Basidiomycota* had a weak correlation ($r^2 < 0.5$, $p < 0.001$ with the fungal community. There was no correlation between the soil properties (pH, SOC, TN, C/N ratio, and available P) and the fungal community (Figure 5b).

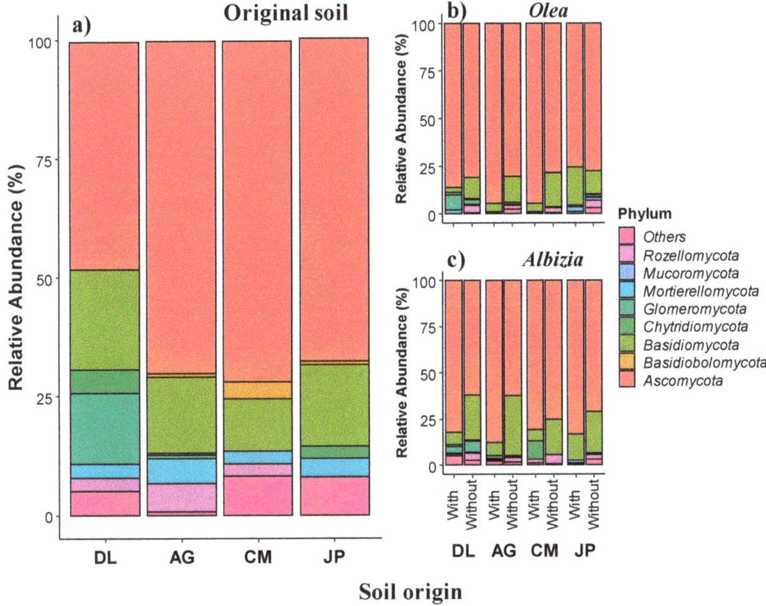

Figure 6. Relative abundances of fungi at the phylum level in the soil samples: before GH experiment (**a**) (original soil), and after GH experiment for *Olea* (**b**) and *Albizia* (**c**) seedlings with treatment (non-sterilized soil) and without treatment (sterilized soil). The values shown are means (n = 3 for soil before GH experiment and n = 8 for soil after GH experiment). DL, AG, CM, and JP represent degraded land, *Albizia gummifera*, *Croton macrostachyus*, and *Juniperus procera*, respectively.

3.2. Effect of Soil Origins and Soil Treatments on Plant Growth and Survival

The factors including species, soil origins and soil treatments were all considered in the model for analysis of variance (Table S1). The results indicated that most of plant characteristics and soil properties are related with species, soil origins, soil treatments, and their interactions. The results of the three-way ANOVA demonstrated that species, and the interactions between species and soil origins, soil treatments and soil origins, and species, soil treatments and soil origins were the most important factors affecting plant characteristics excluding the R/S ratio (Table S1). The soil origins also significantly affected plant characteristics ($p < 0.05$; Table S1). Except for root mass, soil treatments substantially influenced plant characteristics. However, the interaction of species and soil treatments significantly determined survival rate, shoot and total mass of plants ($p < 0.05$; Table S1).

After separating per species, the two-way ANOVA indicated that soil treatments, soil origins, and their interaction had a significant effect on plant height, RCD, shoot mass, total mass, and survival rate of *Olea* and *Albizia* seedlings (Table S1). *Olea* seedlings grown in non-sterilized AG, CM, and JP soils showed significantly higher plant height (Figure 7a), total mass (Figure 7i), RCD (Figure S4a), and shoot mass (Figure S4c), than seedlings grown in sterilized soils. For the same plant characteristics, there was no significant difference between sterilized and non-sterilized DL soils (Figure 7a,c,e). Root mass did not vary between soil treatments, except in CM soil, where it was significantly increased more than double in non-sterilized soil (Figure S4e). In contrast, the R/S ratio for *Olea* seedlings grown in sterilized AG, CM, and JP soils was significantly higher than seedlings grown in the non-sterilized soils (Figure 7e). For *Albizia* seedlings, the RCD (Figure S4b) and shoot mass (Figure S4d) significantly varied in DL and CM non-sterilized soils than seedlings in sterilized soils. The shoot mass of *Albizia* seedlings were significantly higher for both treatments of AG and non-sterilized CM soil than the other treatments (Figure S4d). Root mass (Figure S4f) was significantly higher in sterilized AG and JP soils, significantly lower in CM soil, and was not significantly different in DL soil.

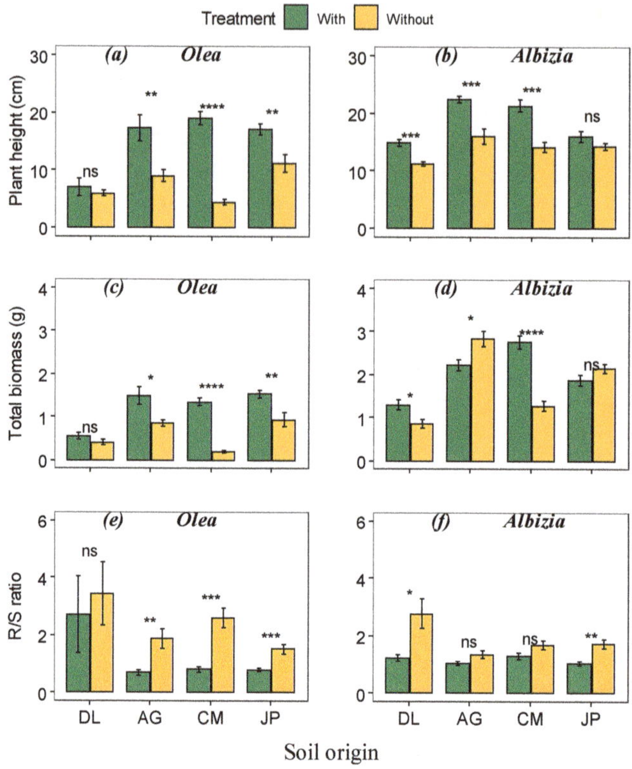

Figure 7. Effects of soil from different origins on plant height (**a,b**), total biomass (**c,d**), and root to shoot (R/S) ratio (**e,f**) of *Olea* and *Albizia* seedlings, respectively, with treatment (non-sterilized soil) and without treatment (sterilized soil). DL, AG, CM, and JP stand for soil origins from degraded land, or from beneath *Albizia gummifera*, *Croton macrostachyus*, and *Juniperus procera*, respectively. Asterisks indicate statistically significant differences between seedlings with treatment (in non-sterilized soil) and without treatment (in sterilized soil): * $p \leq 0.05$; ** $p \leq 0.01$; *** $p \leq 0.001$; **** $p \leq 0.0001$; and ns, not significant). Values are mean ± standard error (n = 8).

Survival analysis showed no significant differences in survival rate between *Olea* seedlings with non-sterilized and sterilized soil ($\chi^2(1) = 3.32$, $p = 0.061$), although the survival rate for seedlings with sterilized soil (97%) was slightly higher than for those in non-sterilized soil (95%). Survival rate significantly differed among soil origins for *Olea* seedlings ($\chi^2(3) = 17.64$, $p < 0.001$). Survival of *Albizia* seedlings was significantly affected by the soil treatments (Table S1) and soil origins (Table S1). *Albizia* seedlings survived better with the sterilized treatment (100%) than with the non-sterilized treatment (97%). The survival rate in AG, CM, and JP soils was found to be similar and significantly higher than in DL (92%). The interaction between soil treatments and soil origins had no significant effect on survival of either *Olea* or *Albizia* seedlings (Table S1; $\chi^2(3) = 3.32$, $p = 0.06$ and $\chi^2(3) = 6.03$, $p = 0.42$, respectively).

3.3. Soil Chemical and Physical Properties

Soil pH, TN content and C/N ratio were strongly dependent on species ($p < 0.05$). Soil origins ($p < 0.05$) also influenced pH, SOC, TN and available P, but not the C/N ratio. Except for available P, soil treatments did not affect soil characteristics. In addition, the interaction between species and soil origins, and species, soil origin and soil treatments affected SOC and TN contents ($p < 0.05$; Table S1). Available P and SOC content were also dependent on the interaction of species and soil treatments, and soil origins and soil treatments, respectively.

Significant differences ($p < 0.05$) in soil pH between original soil and the soils after the GH experiment were observed for both species and soil treatments in AG and JP soils (Tables 1 and 2). SOC content for *Olea* and *Albizia* seedlings varied significantly between before (original soil) and after GH experiments in both soil treatments of AG and DL, respectively. Whereas SOC content only varied significantly in DL (sterilized) and JP (non-sterilized) soils, respectively for *Olea* and *Albizia* seedlings. The TN contents in DL soil for both species and soil treatments were significantly different between original soil and the soils after the GH experiments. There was significant variation in the C/N ratio of DL (sterilized) and AG in both soil treatments between original soil and the soils after the GH experiments. However, available P only significantly varied in non-sterilized AG soil of *Albizia* seedlings.

Table 1. Major soil characteristics of degraded land and church forest soils (n = 3).

Soil	pH	MC (%)	SOC (%)	TN (%)	C/N	Avail. P (mg kg^{-1})	Sand (%)	Silt (%)	Clay (%)	Texture Class (USDA) [a]
DL	5.6 (0.2) [b]	11.7 (1.1) [c]	6.0 (0.2) [b]	0.8 (0.1) [b]	10.2 (0.3) [b]	29.4 (1.1) [b]	68	11	21	Sandy clay loam
AG	6.1 (0.1) [ab]	26.7 (2.0) [a]	12.0 (1.2) [a]	1.0 (0.1) [a]	11.8 (1.8) [a]	88.2 (5.9) [a]	68	12	20	Sandy loam
CM	6.5 (0.2) [ab]	23.3 (1.5) [ab]	8.1 (1.1) [ab]	0.7 (0.1) [b]	11.3 (0.4) [a]	39.8 (4.2) [b]	64	17	19	Sandy loam
JP	7.0 (0.3) [a]	17.0 (1.6) [bc]	8.7 (0.4) [ab]	0.8 (0.1) [b]	10.8 (0.1) [b]	76.3 (5.1) [a]	67	14	19	Sandy loam

Different lowercase letters indicate significant differences among soil origins [values are mean (±standard error)]; Tukey HSD, $p < 0.05$, n = 3. Soil in degraded land (DL) and under mature trees of *Albizia gummifera* (AG), *Croton macrostachyus* (CM), *Juniperus procera* (JP). MC, moisture content at sampling. SOC, soil organic carbon. TN, total nitrogen. C/N, carbon/nitrogen ratio. Avail.P, available phosphorus. [a] Soil texture class is according to [52].

Table 2. Soil characteristics of *Olea* and *Albizia* seedlings inoculated with and without four different soil origins at the end of the 5 months experiment (DL, soil from degraded land; AG/CM/JP, soil from beneath *A. gummifera*, *C. macrostachyus*, and *J. procera*, respectively).

Species	Soil Origin	pH with	pH without	SOC (%) with	SOC (%) without	TN (%) with	TN (%) without	C/N with	C/N without	Avail.P (mg kg^{-1}) with	Avail.P without	SAS (SI, cm^{-1}) with	SAS without
Olea	DL	5.9 (0.1) cA	5.9 (0.1) cA	4.5 −1.3	7.9 −0.8	0.4 −0.1	0.7 −0.1	11 −0.2	11 −0.1	28.4 (0.6) d	30.7 (0.4) d	0.02 (0.0) c	0.01 (0.0) c
		ns	ns	ns	*	*	*	ns	ns	ns	ns		
	AG	6.7 (0.1) aB	7.2 (0.1) aA	6.7 −1.8	10.1 −1.5	0.6 −0.2	0.92 −0.1	11.1 −0.5	11.4 −0.2	88 (3.3) a	86.1 (4.7) a	0.05 (0.0) aA	0.04 (0.0) aB
		*	*	*	*	*	ns	ns	ns	ns	ns		
	CM	6.6 (0.1) aA	6.6 (0.1) aA	6.3 −1.4	5.7 −1.5	0.6 −0.1	0.5 −0.1	 −0.2	10.9 −0.2	42.2 (0.7) c	42.5 (0.7) c	0.04 (0.0) bA	0.03 (0.0) bB
		ns	ns	ns	ns	ns	ns	ns	ns	ns	ns		
	JP	6.3 (0.1) bA	5.9 (0.1) bB	8.3 −1.7	6.8 −0.7	0.7 −0.1	0.62 −0.1	11.3 −0.1	10.9 −0.1	77.9 (2.2) b	74.8 (2.0) b	0.04 (0.0) b	0.03 (0.0) b
		*	*	ns	ns	ns	ns	ns	ns	ns	ns		

Table 2. *Cont.*

Species	Soil Origin	pH		SOC (%)		TN (%)		C/N		Avail.P (mg kg⁻¹)		SAS (SI, cm⁻¹)	
		with	without	with	without	with	without	with	without	with	without	with	without
Albizia	DL	5.6 (0.1) dB	5.9 (0.0) dA	3.9 (0.0) d	3.7 (0.0) d	0.4 (0.0) d	0.3 (0.0) c	10.1 (0.1) a	11.4 (0.2) a	27.3 (0.4) dB	28.6 (0.6) dA	0.02 (0.0) d	0.01 (0.0) c
		ns	ns	*	*	*	*	ns	*	ns	ns		
	AG	7 (0.1) aB	7.4 (0.0) aA	12.1 (0.2) a	11.9 (0.7) a	1 (0.0) a	1 (0.1) a	11.6 (0.1) a	11.6 (0.1) a	74.1 (2.0) bB	86.9 (3.9) bA	0.05 (0.0) aA	0.04 (0.0) aB
		*	*	ns	ns	ns	ns	*	*	*	ns		
	CM	6.5 (0.0) bA	6.6 (0.1) bA	8.3 (0.1) b	8.6 (0.4) b	0.8 (0.0) b	0.8 (0.0) b	10.7 (0.1) b	10.8 (0.1) b	34.5 (0.8) c	47.3 (1.1) c	0.04 (0.0) b	0.04 (0.0) a
		ns	ns	ns	ns	ns	ns	ns	ns	ns	ns		
	JP	6 (0.0) cB	6.3 (0.0) cA	7.3 (0.1) c	7.7 (0.1) c	0.7 (0.0) c	0.7 (0.0) b	10.6 (0.1) b	10.7 (0.0) b	74.2 (3.7) aB	94.8 (1.5) aA	0.03 (0.0) c	0.03 (0.0) b
		*	*	*	ns	ns	ns	ns	ns	ns	*		

Different superscript lowercase letters indicate significant differences among soil origins for each species; different superscript uppercase letters indicate significant differences between soil treatments (Tukey HSD, *p < 0.05*, n = 8) and asterisks indicate statistically significant differences between before (Table 1) and after GH experiments in the same soil origin (* *p ≤ 0.05*; and ns, not significant). Values are mean (±standard error).

For *Albizia* seedlings, soil pH was significantly affected by soil treatments and soil origins and their interactions, whereas for *Olea* seedlings, soil origins and its interaction with soil treatments were significant (Table S1). Soil pH for *Olea* seedlings with non-sterilized JP soil was significantly higher than for seedlings in sterilized soil, whereas with AG soil, the opposite trend was observed (Table 2). Except for CM soil, there was a significant difference in pH between non-sterilized and sterilized soil for *Albizia* seedlings, i.e., in DL, AG, and JP soils, pH was significantly higher for seedlings in sterilized soil (Table 2). For both *Olea* and *Albizia* seedlings, regardless of soil treatments, pH of DL was significantly lower than AG and CM but was comparable with JP (Table 2).

Generally, SOC, TN, C/N ratio, and available P for *Olea* seedlings were not significantly affected by soil treatments, soil origins, and their interaction. SOC, TN, and C/N ratio for *Olea* seedlings did not significantly vary between soil origins (Table S1). Sterilized AG soil had a higher SOC (10.10%) and TN (0.92%) contents, whereas the lowest SOC (4.48%) and TN (0.40%) contents were found in non-sterilized DL soil (Table 2). The highest (11.40) and the lowest (10.90) C/N ratios for *Olea* were measured in CM sterilized and non-sterilized soils, respectively (Table 2).

Soil available P significantly differed among soil origins for *Olea* seedlings (Table S1). The non-sterilized AG soil had a significantly higher available P (88.00 mg kg⁻¹) than the others. The lowest content of available P (28.40 mg kg⁻¹) was found in DL non-sterilized soil for *Olea*. Available P in the forest soils was highest in AG followed by JP then CM (Table 2). In contrast to *Olea* seedlings, SOC, TN, C/N, and available P for *Albizia* seedlings significantly varied among soil origins, with SOC and TN in particular being significantly influenced by soil origins.

However, except for available P content and C/N ratio, other soil characteristics did not significantly differ between sterilized and non-sterilized soil (Table 2). SAS was numerically higher in non-sterilized than in sterilized soil for both *Olea* and *Albizia* seedlings (Table 2), significantly so for AG and CM soil with *Olea* seedlings, and for AG soil with *Albizia*. Regardless of soil treatments, for both *Olea* and *Albizia* seedlings, SAS for all of the forest soils (AG, CM, and JP) was significantly higher than for DL soil (Table 2).

4. Discussion

The soil from the four origins in this study caused significant variation in plant characteristics (Figure 7). *Olea* seedlings with non-sterilized soil of *Albizia gummifera*, *Croton macrostachyus*, and *Juniperus procera* showed higher plant growth compared to seedlings with sterilized soil. Plant characteristics of *Albizia* seedlings were consistently affected by non-sterilized soil from *Croton macrostachyus*, whereas for other soil origins the effect was not consistent. *Olea* and *Albizia* are species that co-exist in dry Afromontane forest of Ethiopia [29] and have arbuscular mycorrhizal associations [53]. Our results agree with other studies that have shown that inoculation with the appropriate microbes can significantly modify or improve seedling growth [53], pathogen resistance [12], and biomass production [54]. Conversely, seedlings of both species in non-sterilized *Croton macrostachyus* soil consistently varied in plant growth (Figure 7). This may be linked to the abundance of *Actinobacteria* and *Nitrospirae* for seedlings in non-sterilized soil, which is strongly related to the bacterial community of forest soils in particular with *Croton macrostachyus* soil (Figures 3 and 6). Several reports have indicated that members of the phylum *Actinobacteria* are involved in organic matter decomposition, plant growth promotion, and soil pathogen control [54]. Also, *Nitrospirae* is one of the phyla whose members are involved in soil nitrification [55]; therefore, their abundance in soil may influence the availability of soil nitrogen [56]. However, the phylum *Proteobacteria* was highly correlated with seedlings in sterilized soil. Furthermore, relatively high soil pH was observed for seedlings in non-sterilized soils (Table 2 and Figure 5). Studies have reported that the abundance of *Proteobacteria* is positively correlated to soil pH [56].

Likewise, *Ascomycota* and *Basidiomycota* are found to be the most abundant fungi phyla in forest soils. *Ascomycota* is saprophytic in the soil, which had high resistance and better environmental adaptability [57]. It can also decompose most plant and animal residue into nutrients that can be available for plants [58]. Similarly, *Basidiomycota* plays a key role in the decomposition of organic matter in the soil, such as lignins, resins, tannin, and other compounds which might affect soil properties [59]. However, in this study, the abundance of *Ascomycota* and *Basidiomycota* was slightly influenced by species type (Figure 4). Several studies have reported that plant species type can influence the structure of the soil microbial community by producing a different amount of organic matter, altering the soil moisture and nutrition status [60–62]. Moreover, the quality and quantity of organic compounds released by plant roots such as carbohydrates, amino acids, organic acids, and enzymes can also influence the soil microbial community by exerting stimulatory and/or inhibiting effects [63,64]. In this study, lower forest soil TN (0.07%) and P (39.84 mg kg^{-1}) contents were found in *Croton macrostachyus* soil. According to Wubet et al. [53], native tree species of the dry Afromontane forest in Ethiopia have mycorrhizal associations, which are effective tools when the soil nutrients (i.e., N and P) are limited [53]. However, fungal abundance in the experimental soil was low, and fungal communities did not correlate with the soil properties, and had lower diversity, evenness, and number of operational taxonomic units than bacterial communities. This result is in line with other studies [65,66]. Thus, when a difference in pH range preference for optimum growth pertains, soil pH is often a factor exerting more control over the abundance of the bacterial community than the fungal community. Additionally, the beneficial effect of soil from beneath *Croton macrostachyus* on the regeneration of *Olea* seedlings [67] has been reported in the highlands of Ethiopia.

As expected, *Olea* seedlings in degraded land soil did not vary in size among soil treatments (Figure 7). This could be due to the low level of soil microbial diversity and abundance, which are common soil characteristics of degraded land [68]. Soil microbes are widely known to enhance plant growth, increase efficiency of nutrient uptake, and facilitate establishment and competitive ability of seedlings. Moreover, in the present study, degraded land soil had lower fungal diversity and number of operational taxonomic units than forest soil. Correspondingly, SOC, TN, available P, and moisture content were found at lower levels in degraded land soil than in forest soil (Table 1).

Plant biomass allocation strategy is species-dependent and varies with environmental factors [68]. Studies have shown that resource availability controls biomass allocation patterns in plants [69,70], especially for the root to shoot ratio. In the present study, root to shoot ratios for *Olea* seedlings in *Albizia gummifera*, *Croton macrostachyus* and *Juniperus procera* sterilized soil were higher than in non-sterilized soil. The root to shoot ratio was highest in *Olea* seedlings, reaching 3.44 in degraded land soil, a value not influenced by soil sterilization. This could be because plants under conditions of low soil nutrients and limited water are obliged to allocate high biomass to their roots to exploit the soil resources more effectively [11,71]. In contrast, low root to shoot ratios were found in seedlings in non-sterilized forest soils, which could be because seedlings in non-sterilized soils have greater access to water and nutrients, provided by the microbial association, meaning that seedlings were able to allocate more biomass to the shoot. A similar finding was also reported by Zandavalli et al. [72].

Soil aggregate stability and the process of structure formation are complex, influenced by soil properties (e.g., clay content, organic matter), plant root development, and soil microbial activity [10]. Soil aggregate stability is an indicator of soil aeration and nutrient availability, soil erosion resistance, root penetration, and water regime of the soil [10]. In this study, soil aggregate stability was significantly higher for forest soils than degraded land soil. This result is in agreement with the findings by Delelegn et al. [73], who reported higher soil aggregate stability in natural forest soil than degraded croplands in the highlands of northern Ethiopia. A similar result was also reported by Caravaca et al. [74]. SOC is the main element in soil aggregate formation and is directly related to soil microbial diversity for Caravaca, Lax, and Albaladejo [74]. Loss of SOC results in significant deterioration in soil structure [75], which is a key indicator of soil degradation [75]. Moreover, loss of beneficial soil microbes (mainly fungi and bacteria) significantly affects soil aggregate stability [76]. Furthermore, fungi play a significant role in endorsing the formation of macro-aggregates through their hyphae, which "glue" the micro-aggregates together [77]. However, as mentioned above, the degraded land soil had lower fungal diversity than forest soil. Thus, greater soil aggregate stability under seedlings with non-sterilized Albizia gummifera and Croton macrostachyus soils (Table 2) can be attributed to the higher abundance of beneficial microbes (Figure 4) that facilitate the formation of micro and macro-aggregates [78].

5. Conclusions

Higher growth in non-sterilized, than sterilized forest (*Albizia gummifera, Croton macrostachyus and Juniperus procera*) soil indicates a microbial benefit to seedling growth from forest soil. We also observed higher plant growth in forest soils than in degraded soils mainly due to a higher relative abundance of beneficial bacterial phyla (*Acidobacteria, Actinobacteria,* and *Nitrospirae*). Soil pH showed a strong correlation with the abundance of the bacterial community, but no relationship was found between soil properties and fungal communities. Moreover, the effect of soil microbiome was more noticeable on the performance of *Olea* seedlings grown in the soil from *Croton macrostachyus*. This suggests that soils from *Croton macrostachyus* can promote growth and survival of *Olea* and *Albizia* seedlings in degraded lands. Overall, the results of this study imply that soils from the remnant church forests could serve as a potential source of soil microbiome for the restoration of degraded lands using native tree species.

Supplementary Materials: The following are available online at http://www.mdpi.com/2071-1050/12/12/4976/s1; Figure S1: Rarefication curve of soil bacteria (a) and fungi (b) communities. Table S1: ANOVA of *Olea, Albizia* and both seedlings showing results for plant height (H), root collar diameter (RCD), survival rate (SR), shoot mass (SB), root mass (RB), the root to shoot ratio (R/S), total mass (TB), soil pH (pH), soil organic carbon (SOC), total nitrogen (TN), carbon to nitrogen ratio (C/N), and available phosphorus (P). Figure S2: Effect of soil origins on shoot and root growth of *Olea* seedling at end of the experiment. With treatment (non-sterilized soil) and without treatment (sterilized soil) of DL: Degraded land, AG: *Albizia* gummifera, CM: Croton macrostachyus and JP: Juniperus procera. Figure S3: Effect of soil origins on shoot and root growth of *Albizia* seedling at end of the experiment. With (non-sterilized soil) and without (sterilized soil) of DL: Degraded land, AG: *Albizia* gummifera, CM: Croton macrostachyus and JP: Juniperus procera. Figure S4: Effects of soil from different origins on root collar diameter (RCD) (a, b), shoot mass (c, d), and root mass (e, f), in *Olea* and *Albizia* seedlings, respectively, with treatment (non-sterilized soil) and without treatment (sterilized soil). DL, AG, CM, and JP stand for soil origins from degraded land, or from beneath *Albizia* gummifera, Croton macrostachyus, and Juniperus procera, respectively.

Author Contributions: Conceptualization, G.A. and T.T.; data curation, G.A. and K.E.; validation, A.T. and N.H.; Methodology, G.A., M.W., T.M., A.M. and T.T.; formal analysis, G.A., T.T. and A.M.; investigation, G.A.; resources A.T. and N.H.; writing—original draft preparation, G.A.; writing—review and editing, A.T., E.A., N.H., T.T., M.W., K.E. and D.T.M.; supervision, A.T., N.H., T.T. and M.T.; project administration, A.T., E.A., N.H. and D.T.M.; funding acquisition, A.T., E.A. and N.H. All authors have read and agreed to the published version of the manuscript.

Funding: This research was funded by the Science and Technology Research Partnership for Sustainable Development (grant no. JPMJSA1601), Japan Science and Technology Agency/Japan International Cooperation Agency.

Acknowledgments: We are grateful to Anteneh Wubet and Agerselam Gualie for the facilitation of our field and laboratory work. We thank Ayele Almaw Fenta and Brihanu Kebede for the watering of seedlings during our absence. We also thank the Arid Land Research Center of Tottori University for providing a convenient research environment and facilities throughout our work.

Conflicts of Interest: The authors declare no conflict of interest.

References

1. Nkonya, E.; Johnson, T.; Kwon, H.Y.; Kato, E. Economics of land degradation in sub-Saharan Africa. In *Economics of Land Degradation and Improvement–A Global Assessment for Sustainable Development*; Springer: Cham, Switzerland, 2016; pp. 215–259.

2. Mekuria, W.; Wondie, M.; Amare, T.; Wubet, A.; Feyisa, T.; Yitaferu, B. Restoration of degraded landscapes for ecosystem services in North-Western Ethiopia. *Heliyon* **2018**, *4*, e00764. [CrossRef] [PubMed]

3. Nunes, J.S.; Araujo, A.S.F.; Nunes, L.; Lima, L.M.; Carneiro, R.F.V.; Salviano, A.A.C.; Tsai, S.M. Impact of Land Degradation on Soil Microbial Biomass and Activity in Northeast Brazil. *Pedosphere* **2012**, *22*, 88–95. [CrossRef]

4. Meshesha, D.T.; Tsunekawa, A.; Tsubo, M.; Ali, S.A.; Haregeweyn, N. Land-use change and its socio-environmental impact in Eastern Ethiopia's highland. *Reg. Environ. Chang.* **2014**, *14*, 757–768. [CrossRef]

5. Araujo, A.S.; Borges, C.D.; Tsai, S.M.; Cesarz, S.; Eisenhauer, N. Soil bacterial diversity in degraded and restored lands of Northeast Brazil. *Antonie Van Leeuwenhoek* **2014**, *106*, 891–899. [CrossRef] [PubMed]

6. Singh, J.S. Microbes: The chief ecological engineers in reinstating equilibrium in degraded ecosystems. *Agric. Ecosyst. Environ.* **2015**, *203*, 80–82. [CrossRef]

7. Zhang, P.; Li, L.; Pan, G.; Ren, J. Soil quality changes in land degradation as indicated by soil chemical, biochemical and microbiological properties in a karst area of southwest Guizhou, China. *Environ. Geol.* **2006**, *51*, 609–619. [CrossRef]

8. Neuenkamp, L.; Prober, S.M.; Price, J.N.; Zobel, M.; Standish, R.J. Benefits of mycorrhizal inoculation to ecological restoration depend on plant functional type, restoration context and time. *Fungal Ecol.* **2019**, *40*, 140–149. [CrossRef]

9. Van der Heijden, M.G.A.; Horton, T.R. Socialism in soil? The importance of mycorrhizal fungal networks for facilitation in natural ecosystems. *J. Ecol.* **2009**, *97*, 1139–1150. [CrossRef]

10. Rashid, M.I.; Mujawar, L.H.; Shahzad, T.; Almeelbi, T.; Ismail, I.M.; Oves, M. Bacteria and fungi can contribute to nutrients bioavailability and aggregate formation in degraded soils. *Microbiol. Res.* **2016**, *183*, 26–41. [CrossRef]

11. Classen, A.T.; Sundqvist, M.K.; Henning, J.A.; Newman, G.S.; Moore, J.A.M.; Cregger, M.A.; Moorhead, L.C.; Patterson, C.M. Direct and indirect effects of climate change on soil microbial and soil microbial-plant interactions: What lies ahead? *Ecosphere* **2015**, *6*, 21. [CrossRef]

12. Wehner, J.; Antunes, P.M.; Powell, J.R.; Mazukatow, J.; Rillig, M.C. Plant pathogen protection by arbuscular mycorrhizas: A role for fungal diversity? *Pedobiologia* **2010**, *53*, 197–201. [CrossRef]

13. Rillig, M.C.; Aguilar-Trigueros, C.A.; Bergmann, J.; Verbruggen, E.; Veresoglou, S.D.; Lehmann, A. Plant root and mycorrhizal fungal traits for understanding soil aggregation. *New Phytol.* **2015**, *205*, 1385–1388. [CrossRef] [PubMed]

14. Calderon, K.; Spor, A.; Breuil, M.C.; Bru, D.; Bizouard, F.; Violle, C.; Barnard, R.L.; Philippot, L. Effectiveness of ecological rescue for altered soil microbial communities and functions. *ISME J.* **2017**, *11*, 272–283. [CrossRef] [PubMed]

15. Ambrosini, A.; de Souza, R.; Passaglia, L.M.P. Ecological role of bacterial inoculants and their potential impact on soil microbial diversity. *Plant Soil* **2015**, *400*, 193–207. [CrossRef]

16. SER. The SER international primer on ecological restoration. *Ecol. Restor* **2004**, *2*, 206–207.

17. Sultan, D.; Tsunekawa, A.; Haregeweyn, N.; Adgo, E.; Tsubo, M.; Meshesha, D.T.; Masunaga, T.; Aklog, D.; Fenta, A.A.; Ebabu, K. Impact of Soil and Water Conservation Interventions on Watershed Runoff Response in a Tropical Humid Highland of Ethiopia. *Environ. Manag.* **2018**, *61*, 860–874. [CrossRef]

18. Damene, S.; Tamene, L.; Vlek, P.L.G. Performance of exclosure in restoring soil fertility: A case of Gubalafto district in North Wello Zone, northern highlands of Ethiopia. *CATENA* **2013**, *101*, 136–142. [CrossRef]

19. Aerts, R.; Negussie, A.; Maes, W.; November, E.; Hermy, M.; Muys, B. Restoration of dry Afromontane forest using pioneer shrubs as nurse-plants for *Olea europaea* ssp. cuspidata. *Restor. Ecol.* **2007**, *15*, 129–138. [CrossRef]

20. Delelegn, Y.T.; Purahong, W.; Sanden, H.; Yitaferu, B.; Godbold, D.L.; Wubet, T. Transition of Ethiopian highland forests to agriculture-dominated landscapes shifts the soil microbial community composition. *BMC Ecol.* **2018**, *18*, 58. [CrossRef]

21. Wassie, A.; Sterck, F.J.; Bongers, F. Species and structural diversity of church forests in a fragmented Ethiopian Highland landscape. *J. Veg. Sci.* **2010**, *21*, 938–948. [CrossRef]

22. Abiyu, A.; Teketay, D.; Glatzel, G.; Gratzer, G. Seed production, seed dispersal and seedling establishment of two afromontane tree species in and around a church forest: Implications for forest restoration. *For. Ecosyst.* **2016**, *3*, 16. [CrossRef]

23. Berhane, A.; Totland, Ø.; Moe, S.R. Woody plant assemblages in isolated forest patches in a semiarid agricultural matrix. *Biodivers. Conserv.* **2013**, *22*, 2519–2535. [CrossRef]

24. Aerts, R.; Overtveld, K.; November, E.; Wassie, A.; Abiyu, A.; Demissew, S.; Daye, D.D.; Giday, K.; Haile, M.; TewoldeBerhan, S.; et al. Conservation of the Ethiopian church forests: Threats, opportunities and implications for their management. *Sci. Total Environ.* **2016**, *551*. [CrossRef]

25. Wassie, A.; Sterck, F.J.; Teketay, D.; Bongers, F. Tree Regeneration in Church Forests of Ethiopia: Effects of Microsites and Management. *Biotropica* **2009**, *41*, 110–119. [CrossRef]

26. Wassie, A.; Sterck, F.J.; Teketay, D.; Bongers, F. Effects of livestock exclusion on tree regeneration in church forests of Ethiopia. *For. Ecol. Manag.* **2009**, *257*, 765–772. [CrossRef]

27. Friis, I. *Forests and Forest Trees of Northeast Tropical Africa: Their Natural Habitats and Distribution Patterns in Ethiopia, Djibouti and Somalia*; Her Majesty's Stationery Office: London, UK, 1992.

28. Tesemma, A.B. *Useful Trees and Shrubs for Ethiopia: Identification, Propagation and Management for 17 Agro-Climatic Zones*; Regional Land Management Unit.: Nairobi, Kenya, 2007.

29. Aynekulu, E.; Denich, M.; Tsegaye, D.; Aerts, R.; Neuwirth, B.; Boehmer, H.J. Dieback affects forest structure in a dry Afromontane forest in northern Ethiopia. *J. Arid Environ.* **2011**, *75*, 499–503. [CrossRef]

30. Yibeltal, M.; Tsunekawa, A.; Haregeweyn, N.; Adgo, E.; Meshesha, D.T.; Aklog, D.; Masunaga, T.; Tsubo, M.; Billi, P.; Vanmaercke, M.; et al. Analysis of long-term gully dynamics in different agro-ecology settings. *Catena* **2019**, *179*, 160–174. [CrossRef]

31. Cardelús, C.; Scull, P.; Hair, J.; Baimas-George, M.; Lowman, M.; Eshete, A. A Preliminary Assessment of Ethiopian Sacred Grove Status at the Landscape and Ecosystem Scales. *Diversity* **2013**, *5*, 320–334. [CrossRef]

32. Mekonnen, G. *Soil Characterization Classification and Mapping of Three Twin Watersheds in the Upper Blue Nile basin, Ethiopia*; Final Project Report; Amhara Design and Supervision Works Enterprise: Bahir Dar, Ethiopia, 2016.

33. McNamara, N.P.; Black, H.I.J.; Beresford, N.A.; Parekh, N.R. Effects of acute gamma irradiation on chemical, physical and biological properties of soils. *Appl. Soil Ecol.* **2003**, *24*, 117–132. [CrossRef]

34. Yibeltal, M.; Tsunekawa, A.; Haregeweyn, N.; Adgo, E.; Meshesha, D.T.; Masunaga, T.; Tsubo, M.; Billi, P.; Ebabu, K.; Fenta, A.A.; et al. Morphological characteristics and topographic thresholds of gullies in different agro-ecological environments. *Geomorphology* **2019**, *341*, 15–27. [CrossRef]

35. Orlowsky, B. iki. dataclim: Consistency, Homogeneity and SummaryStatistics of Climatological Data. R Package Version 1.0. 2014. Available online: https://rdrr.io/cran/iki.dataclim/ (accessed on 4 September 2019).

36. Chandra, N.; Kumar, S. Antibiotics Producing Soil Microorganisms. In *Antibiotics and Antibiotics Resistance Genes in Soils: Monitoring, Toxicity, Risk Assessment and Management*; Hashmi, M.Z., Strezov, V., Varma, A., Eds.; Springer International Publishing: Cham, Switzerland, 2017; pp. 1–18. [CrossRef]

37. Bray, R.H.; Kurtz, L.T. Determination of total, organic, and available forms of phosphorus in soils. *Soil Sci.* **1945**, *59*, 39–46. [CrossRef]

38. Levy, G.J.; Mamedov, A.I. High-Energy-Moisture-Characteristic Aggregate Stability as a Predictor for Seal Formation. *Soil Sci. Soc. Am. J.* **2002**, *66*, 1603–1609. [CrossRef]

39. Bouyoucos, G.j. Directions For Making Mechanical Analyses Of Soils By The Hydrometer Method. *Soil Sci.* **1936**, *42*, 225–230. [CrossRef]

40. Klindworth, A.; Pruesse, E.; Schweer, T.; Peplies, J.; Quast, C.; Horn, M.; Glockner, F.O. Evaluation of general 16S ribosomal RNA gene PCR primers for classical and next-generation sequencing-based diversity studies. *Nucleic Acids Res.* **2013**, *41*, e1. [CrossRef] [PubMed]

41. Toju, H.; Tanabe, A.S.; Yamamoto, S.; Sato, H. High-coverage ITS primers for the DNA-based identification of ascomycetes and basidiomycetes in environmental samples. *PLoS ONE* **2012**, *7*, e40863. [CrossRef]

42. Tian, Q.; Taniguchi, T.; Shi, W.Y.; Li, G.; Yamanaka, N.; Du, S. Land-use types and soil chemical properties influence soil microbial communities in the semiarid Loess Plateau region in China. *Sci. Rep.* **2017**, *7*, 45289. [CrossRef]

43. Tedersoo, L.; Sánchez-Ramírez, S.; Kõljalg, U.; Bahram, M.; Döring, M.; Schigel, D.; May, T.; Ryberg, M.; Abarenkov, K. High-level classification of the Fungi and a tool for evolutionary ecological analyses. *Fungal Divers.* **2018**, *90*, 135–159. [CrossRef]

44. Caporaso, J.G.; Kuczynski, J.; Stombaugh, J.; Bittinger, K.; Bushman, F.D.; Costello, E.K.; Fierer, N.; Peña, A.G.; Goodrich, J.K.; Gordon, J.I.; et al. QIIME allows analysis of high-throughput community sequencing data. *Nat. Methods* **2010**, *7*, 335–336. [CrossRef]

45. Bengtsson-Palme, J.; Ryberg, M.; Hartmann, M.; Branco, S.; Wang, Z.; Godhe, A.; De Wit, P.; Sánchez-García, M.; Ebersberger, I.; de Sousa, F.; et al. Improved software detection and extraction of ITS1 and ITS2 from ribosomal ITS sequences of fungi and other eukaryotes for analysis of environmental sequencing data. *Methods Ecol. Evol.* **2013**. [CrossRef]

46. Edgar, R.C. Search and clustering orders of magnitude faster than BLAST. *Bioinformatics* **2010**, *26*, 2460–2461. [CrossRef] [PubMed]

47. Edgar, R.C.; Haas, B.J.; Clemente, J.C.; Quince, C.; Knight, R. UCHIME improves sensitivity and speed of chimera detection. *Bioinformatics* **2011**, *27*, 2194–2200. [CrossRef] [PubMed]

48. Team, R.C. *R: A Language and Environment for Statistical Computing*; R Foundation for Statistical Computing: Vienna, Austria, 2015.

49. De Mendiburu, M.F. Package 'agricolae'. R Package, Version 2019, 1.2-1. Available online: https://tarwi.lamolina.edu.pe/~{}fmendiburu/ (accessed on 4 September 2019).

50. Allignol, A.; Latouche, A. CRAN Task View: Survival Analysis. Version 2017-04-25. Available online: https://CRAN.R-project.org/view=Survival.2017 (accessed on 4 September 2019).

51. Oksanen, J.; Blanchet, F.; Friendly, M.; Kindt, R.; Legendre, P.; McGlinn, D.; Minchin, P.; O'Hara, R.; Simpson, G.; Solymos, P. vegan: Community Ecology Package. R Package Version 2.5-2. 2018. Available online: https://rdrr.io/cran/vegan/ (accessed on 4 September 2019).

52. Brady, N.; Ray, W. *The Nature and Properties of Soils*, 15th ed.; Pearson Education: London, UK, 2017.

53. Wubet, T.; Kottke, I.; Teketay, D.; Oberwinkler, F. Mycorrhizal status of indigenous trees in dry Afromontane forests of Ethiopia. *For. Ecol. Manag.* **2003**, *179*, 387–399. [CrossRef]

54. Zhang, Q.; Zhang, L.; Weiner, J.; Tang, J.; Chen, X. Arbuscular mycorrhizal fungi alter plant allometry and biomass-density relationships. *Ann. Bot.* **2011**, *107*, 407–413. [CrossRef] [PubMed]

55. Altmann, D.; Stief, P.; Amann, R.; De Beer, D.; Schramm, A. In situ distribution and activity of nitrifying bacteria in freshwater sediment. *Environ. Microbiol.* **2003**, *5*, 798–803. [CrossRef] [PubMed]

56. Li, Y.; Chapman, S.J.; Nicol, G.W.; Yao, H. Nitrification and nitrifiers in acidic soils. *Soil Biol. Biochem.* **2018**, *116*, 290–301. [CrossRef]

57. Deacon, J. Fungal Biology. *Fungal Biol.* **2005**. [CrossRef]

58. Jia, T.; Wang, R.; Fan, X.; Chai, B. A Comparative Study of Fungal Community Structure, Diversity and Richness between the Soil and the Phyllosphere of Native Grass Species in a Copper Tailings Dam in Shanxi Province, China. *Appl. Sci.* **2018**, *8*, 1297. [CrossRef]

59. Hewelke, E.; Górska, E.B.; Gozdowski, D.; Korc, M.; Olejniczak, I.; Prędecka, A. Soil Functional Responses to Natural Ecosystem Restoration of a Pine Forest Peucedano-Pinetum after a Fire. *Forests* **2020**, *11*, 286. [CrossRef]

60. Garau, G.; Morillas, L.; Roales, J.; Castaldi, P.; Mangia, N.P.; Spano, D.; Mereu, S. Effect of monospecific and mixed Mediterranean tree plantations on soil microbial community and biochemical functioning. *Appl. Soil Ecol.* **2019**, *140*, 78–88. [CrossRef]

61. Ushio, M.; Kitayama, K.; Balser, T.C. Tree species-mediated spatial patchiness of the composition of microbial community and physicochemical properties in the topsoils of a tropical montane forest. *Soil Biol. Biochem.* **2010**, *42*, 1588–1595. [CrossRef]

62. Prescott, C.E.; Grayston, S.J. Tree species influence on microbial communities in litter and soil: Current knowledge and research needs. *For. Ecol. Manag.* **2013**, *309*, 19–27. [CrossRef]

63. Haichar, F.E.Z.; Santaella, C.; Heulin, T.; Achouak, W. Root exudates mediated interactions belowground. *Soil Biol. Biochem.* **2014**, *77*, 69–80. [CrossRef]

64. Philippot, L.; Raaijmakers, J.M.; Lemanceau, P.; van der Putten, W.H. Going back to the roots: The microbial ecology of the rhizosphere. *Nat. Rev. Microbiol.* **2013**, *11*, 789–799. [CrossRef] [PubMed]

65. Smith, S.E.; Jakobsen, I.; Gronlund, M.; Smith, F.A. Roles of arbuscular mycorrhizas in plant phosphorus nutrition: Interactions between pathways of phosphorus uptake in arbuscular mycorrhizal roots have important implications for understanding and manipulating plant phosphorus acquisition. *Plant Physiol.* **2011**, *156*, 1050–1057. [CrossRef]

66. St-Denis, A.; Kneeshaw, D.; Bélanger, N.; Simard, S.; Laforest-Lapointe, I.; Messier, C. Species-specific responses to forest soil inoculum in planted trees in an abandoned agricultural field. *Appl. Soil Ecol.* **2017**, *112*, 1–10. [CrossRef]

67. Abiyu, A.; Teketay, D.; Glatzel, G.; Aerts, R.; Gratzer, G. Restoration of degraded ecosystems in the Afromontane highlands of Ethiopia: Comparison of plantations and natural regeneration. *South. For.* **2017**, *79*, 103–108. [CrossRef]

68. Asmelash, F.; Bekele, T.; Birhane, E. The Potential Role of Arbuscular Mycorrhizal Fungi in the Restoration of Degraded Lands. *Front. Microbiol.* **2016**, *7*, 1095. [CrossRef]

69. Atkin, O.K.; Loveys, B.R.; Atkinson, L.J.; Pons, T.L. Phenotypic plasticity and growth temperature: Understanding interspecific variability. *J. Exp. Bot.* **2006**, *57*, 267–281. [CrossRef]

70. Jiang, Y.; Wang, L. Pattern and control of biomass allocation across global forest ecosystems. *Ecol. Evol.* **2017**, *7*, 5493–5501. [CrossRef]

71. Eziz, A.; Yan, Z.; Tian, D.; Han, W.; Tang, Z.; Fang, J. Drought effect on plant biomass allocation: A meta-analysis. *Ecol. Evol.* **2017**, *7*, 11002–11010. [CrossRef] [PubMed]

72. Zandavalli, R.B.; Dillenburg, L.R.; de Souza, P.V.D. Growth responses of Araucaria angustifolia (Araucariaceae) to inoculation with the mycorrhizal fungus Glomus clarum. *Appl. Soil Ecol.* **2004**, *25*, 245–255. [CrossRef]

73. Delelegn, Y.T.; Purahong, W.; Blazevic, A.; Yitaferu, B.; Wubet, T.; Göransson, H.; Godbold, D.L. Changes in land use alter soil quality and aggregate stability in the highlands of northern Ethiopia. *Sci. Rep.* **2017**, *7*, 1–12. [CrossRef] [PubMed]

74. Caravaca, F.; Lax, A.; Albaladejo, J. Aggregate stability and carbon characteristics of particle-size fractions in cultivated and forested soils of semiarid Spain. *Soil Tillage Res.* **2004**, *78*, 83–90. [CrossRef]

75. Daynes, C.N.; Field, D.J.; Saleeba, J.A.; Cole, M.A.; McGee, P.A. Development and stabilisation of soil structure via interactions between organic matter, arbuscular mycorrhizal fungi and plant roots. *Soil Biol. Biochem.* **2013**, *57*, 683–694. [CrossRef]

76. Obalum, S.E.; Chibuike, G.U.; Peth, S.; Ouyang, Y. Soil organic matter as sole indicator of soil degradation. *Environ. Monit. Assess.* **2017**, *189*, 176. [CrossRef]

77. Blankinship, J.C.; Fonte, S.J.; Six, J.; Schimel, J.P. Plant versus microbial controls on soil aggregate stability in a seasonally dry ecosystem. *Geoderma* **2016**, *272*, 39–50. [CrossRef]

78. Bearden, B. Influence of arbuscular mycorrhizal fungi on soil structure and soil water characteristics of vertisols. *Plant Soil* **2001**, *229*, 245–258. [CrossRef]

Article

Constructability Criteria for Farmland Reclamation and Vegetable Cultivation Using Micro-Dam Sediments in Tigray, Ethiopia

Kazuhisa Koda [1,*], **Gebreyohannes Girmay** [2] **and Tesfay Berihu** [2]

[1] Rural Development Division, Japan International Research Center for Agricultural Sciences, Ibaraki 3058686, Japan
[2] College of Dryland Agriculture and Natural Resources, Mekelle University, Mekelle 231, Ethiopia; gebreyohannes.girmay@gmail.com (G.G.); tesra@yahoo.com (T.B.)
* Correspondence: kodakazu@affrc.go.jp; Tel.: +81-29-8386676

Received: 29 June 2020; Accepted: 5 August 2020; Published: 7 August 2020

Abstract: The Ethiopian agriculture sector is characterized by rain-fed smallholder systems. The Ethiopian Government has promoted micro-dam construction in micro-watershed in Tigray for the past two decades. The lack of proper conservation measures to control severe soil erosion at the micro-watershed level, however, has often filled downstream micro-dams in with sediments. Sedimentation has affected the irrigation performance of micro-dams due to their bottom pipes becoming clogged with nutrient-rich soils eroded from upstream farmlands. While there is a growing need for adequate resource management to mitigate severe soil erosion at the watershed-level, it is urgent that methods to make use of the sediments deposited in micro-dam reservoirs to facilitate rural agricultural development are discovered. One practical solution is to use sediments to rehabilitate the bare land excavated for micro-dam embankment construction and turn it into reclaimed farmland. The purpose of this paper is to relate the constructability criteria to the farmland reclamation to solve sedimentation problems. This case study reports the yield of vegetable cultivation on farmland reclaimed using sediments from a micro-dam reservoir in Tigray. This case study highlights the practical potential of such a method to contribute to the livelihoods of farmers through the production of vegetable cash crops. The future research needs cost reduction factors on durability, safety or other related aspects to improve our "Constructability Criteria" approach.

Keywords: micro-dam; sedimentation; reclaimed farmland; constructability; Ethiopian highlands

---

1. Introduction

Ethiopia is a Sub-Saharan African country that covers an area of 109.7 million km^2. Its climate is classified as tropical monsoon due to the low latitude location, and weather varies depending on the topographic elevation. During intense rainfall events, surface runoff flows down slopes and carves deep V-shaped valleys. About 45% of the land in Ethiopia is located at elevations greater than 1500 m above sea level. [1]. The Ethiopian agriculture sector has continuously suffered from poor soil nutrients levels caused by long-lasting cultivation and accompanying soil losses in the form of sheet, rill, and gully erosion [2]. Soil erosion in the Ethiopian highlands is particularly severe due to the sparse vegetation cover. Woldearegay et al. [3] found that the region loses nutrient-rich topsoil at the rate of over 130 tons ha^{-1} year^{-1}. Although farmers in the Ethiopian highlands cultivate wheat as a staple food, the wheat growth period and the wet season coincide. Especially in the northern Ethiopian highlands, soil loss can cause sudden great damage, and effective control of this is difficult [4–7]. Sedimentary, igneous, and metamorphic rocks are found in Tigray [8]. The micro-watershed with steep slopes is subject to soil erosion, which is driven by conventional tillage practices.

Ethiopia's main industry is agriculture with cereals (wheat, barley, and indigenous teff, etc.), pulses (bean and chickpea, etc.), and coffee. The agriculture, forestry, and fishing sector accounts for 31% of GDP, the agriculture sector accounts for 65% of employment in 2018, and the agricultural land area accounts for 36% of the total land [9]. Agricultural development is needed in Ethiopia. While there is a great demand for food for the increasing population, Ethiopia's food security is severely undermined by soil nutrient loss. While sediments are deposited in micro-dam reservoirs, watershed lose a considerable quantity of soil nutrients by losing sediments to micro-dams. Sediment-fixed nutrient export due to soil erosion is a severe nutrient loss process that contributes to soil degradation [10]. The sediments in micro-dams with a high nutrient accumulation are left unused, while degraded areas, which used to be farmlands, have been abandoned, despite the increasing demand for higher-yield farmland to produce food for the growing population [10]. The storage volume and lifetime of the micro-dams are reduced because the soil erosion dumps sediments in downstream micro-dams [3,5]. The effects of fast population growth, dry weather, and the small area of arable farmland have extended desertification and land deterioration. These problems need to be mitigated because more than 60% of people in the Ethiopian highlands work in the agriculture sector.

In the Ethiopian highlands, poor water resources significantly affect the yield of agricultural crops. Most farming areas in Ethiopia are rain-fed and are therefore vulnerable to the highly variable rainfall distribution. Growth of the agriculture sector is affected by droughts, which occur once every 2.4 years. The effect has been very severe, especially in northern Ethiopia [11]. To mitigate water shortages in efforts to cultivate more crops for the increasing population, the Ethiopian government has, for more than two decades, constructed water-harvesting facilities, including micro-dams [5,8,10]. The construction of micro-dams was planned to achieve various economic, hydrologic, and ecological goals, including increased food production, easy access to available drinking water for people as well as livestock, a rise in the groundwater level, and the emergence of new springs [5]. A micro-dam is a small dam or reservoir to store water for domestic, livestock, and irrigation purposes. It has been estimated that 50% of the micro-dams in northern Ethiopia have seen their life expectancy decline from 26 to 13 years due to sedimentation [10–12]. Sedimentation has resulted in lower levels of soil nutrients, such as organic carbon (OC), total nitrogen (N), available phosphorus (P), and exchangeable cations, being found in the soils of the watershed than in micro-dam sediments [10].

The Ethiopian government, in collaboration with international organizations, embarked on a large-scale soil and water conservation strategy through the construction of micro-dams in Tigray; indeed, more than 50 micro-dams were built in the region from 1996 to 2001 [5,13]. Due to a lack of good planning, however, including the selection of appropriate micro-dam sites and technologies, these micro-dams suffered from serious sedimentation and water leakage, resulting in the failure of expected functions [1,12,14,15]. Based on a study conducted in Tigray on micro-dam sedimentation in relation to soil erosion in the watershed, Tamene et al. [5] reported that most of the micro-dams constructed to harvest rainwater lost 50% of their storage capacity less than five years after becoming operational. Haregeweyn et al. [12] showed that 50% of the 13 studied micro-dams had lost half of their life expectancy, while only three micro-dams were estimated to operate for their total expected lifespan [16]. Furthermore, based on an analysis of 92 micro-dams, Berhane et al. [8] found that 61% had sedimentation/siltation problems, 53% suffered from leakage, 22% experienced insufficient inflow, 25% had structural damage, and 21% had spillway erosion problems. Rapid sedimentation is mainly caused by poor planning of the micro-dams [12].

The purpose of this study was to demonstrate the utility of constructability criteria for reclaimed farmland to mitigate the sediment accumulation in micro-dams. This study showed the vegetable cultivation on reclaimed farmland using sediments from the micro-dam reservoir. It was found that micro-dam sediments consisted of fine clay and some were suitable for farmland reclamation on bare land [10,16]. However, the erosion process and the sedimentation usage method in the micro-dam has not been well studied, so onion was cultivated on reclaimed farmland by using the sediments and its yield was compared against the Ethiopian average.

2. Materials and Methods

2.1. Study Site

The Tigray region is situated in the northern Ethiopian highlands. The climate is characterized as tropical alpine and semi-arid. The topography of the region mainly consists of highland plateaus up to 3900 m a.s.l. (above sea level), which are divided by gorges [12]. In terms of geohydrology, the region is dominated by different types of rocks and soils [3]. The rough topography with rocky geohydrology is very sensitive to erosion, making effective utilization and management measures important. The highland climate has sustained a high population density with a long cultivation history, which is estimated to date back to 3000 BCE [12]. The long-term use of farmlands for crop cultivation, combined with the steep topography, water erosion, and insufficient vegetation cover caused by cutting almost all of the residues of wheat, teff, and barley, etc., has caused serious land degradation. Consequently, the Tigray region is considered to be one of the most degraded (and still degrading) regions in Ethiopia [3,5]. The land degradation, coupled with abnormal rainfall distribution, has caused recurring drought and famine, which was historically demonstrated during 1888–1892, 1973–1974, and 1984–1985 [13].

This research was conducted from 2017 to 2020 in the Adizaboy micro-watershed, which covers an area of about 8.5 km^2 and is located between latitudes of 13.64° N and 13.68° N, and between longitudes of 39.56° E and 39.6° E (Figure 1). Adizaboy micro-watershed ranges in altitude from 2050 to 2275 m a.s.l. The annual rainfall in Wukuro city near this area varies from about 300 to 1000 mm. Slope survey results along the representative survey line (Figure 1c) in the Adizaboy micro-watershed show that the average slope of the survey line was 8.8%, which is classified as a steep slope.

The Adizaboy micro-dam is situated at the exit of the Adizaboy micro-watershed (Figure 2). The Adizaboy micro-dam was built in 2009 with local soils and rocks taken from an upland field near the dam site. It is labeled as a zoned rockfill type, which has a crest height of 9.06 m from the bottom pipe and is somewhat smaller than the average of other micro-dams in Tigray [8]. The water surface of the Adizaboy micro-dam seasonally changes. It sometimes becomes dry in March–April. The upstream slope of the Adizaboy micro-dam embankment (H3: V1) is gentler than the downstream slope (H2.55: V1). The trail for farmers in the micro-watershed traverses in a north–south direction over the embankment. The reclaimed farmland that used the micro-dam sediments is situated near the Adizaboy micro-dam and is above the micro-dam crest.

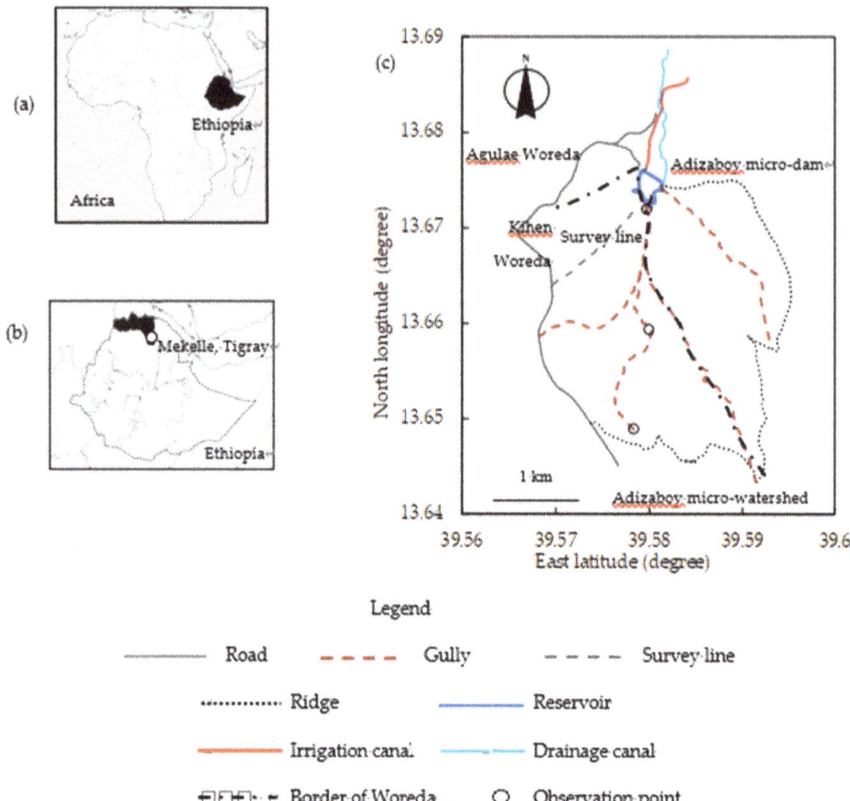

Figure 1. Location map of the Adizaboy micro-watershed in Tigray: (**a**) Location map of Ethiopia in Africa (black area displays Ethiopia); (**b**) location map of Mekelle, Tigray in Ethiopia (bold dotted line shows the national boundary of Ethiopia, and the black area is Tigray); and (**c**) overview of Adizaboy micro-watershed.

Figure 2. Adizaboy micro-dam (photo aspect, NW–SE).

2.2. Gullies in the Micro-Watershed

Some gullies developed in the upper reaches of the micro-watershed, which caused sediment accumulation in the Adizaboy micro-dam. The uncovered soil mainly included Cambisols, which is suitable for agriculture [17]. In the upstream and middle reaches of upland areas of the Adizaboy micro-watershed wheat is mainly cultivated by rainwater. There is about 1.82 km^2 of cropland in the area of the Adizaboy micro-watershed, accounting for 21.4% of the total; 87.1% of the cropland is situated on comparatively flat land with a slope of less than 10° [18].

The geological layers in the Adizaboy micro-watershed consist of weathered marl, shale, and limestone [15]. The rainfall and water flow in the micro-watershed caused fast outflow of fine particles and slow outflow of coarse particles. The location of the three observation points is shown in Figure 1c.

1). Upstream

Four major gullies formed in the upper reaches of the Adizaboy micro-watershed as a result of surface outflow and underground infiltration caused by rainfall. At the starting point of the gully, the depth of the surface collapses and a valley-shaped topography formed of about 2 to 2.5 m height (Figure 3). At this point, water entered through rill erosion due to intense rainfall. Even after rainfall stopped, permeated water continued to enter the gully erosion channel. The starting point of gully erosion then gradually extended upwards.

Figure 3. The starting point of gully erosion in the upper reaches of the Adizaboy micro-watershed (photo aspect, NE–SW).

Figure 4 shows gully erosion in the upper reaches at the observation point. The shape of the cross-sectional views of the gully was deeper around the outside part because the water velocity was relatively faster there. Sand, stone, and gravel collapsed at the slopes of the gullies and this material was transported downstream. Inflows of sand and stone were greater than the outflows in the gully from the starting point to 60 m, although outflows were larger than inflows in other parts. The soil erosion volume depended on the distance from the starting point of the gully because of the slope collapse and water velocity effect.

Figure 4. Gully erosion upstream of the Adizaboy micro-watershed (photo aspect, NE–SW).

2). Middle reaches (Figure 5)

The water velocity was higher in the middle reaches of the Adizaboy micro-watershed and the depth of gully erosion was about 1.0–1.5 m. Gullies in the middle reaches were slightly wider than in the upper reaches. The slope in the middle reaches was often steeper than the one in the upstream, and many boulders existed as well. The widths were sometimes very narrow and, at depths of more than 2 m, the soft geological layers were eroded.

Figure 5. Gully erosion in the middle reaches of the Adizaboy micro-watershed (photo aspect, S–N).

3). Near micro-dam

The gully near the micro-dam is gently sloped (Figure 6) but wider, with a width from 8 to 10 m. The depth is under 0.5 m. The shape of gravel and stone transported is rounder and smaller but there are still some big pieces of stone with a diameter over 0.5 m. As sand, silt and clay are smaller and lighter, they are transported to the micro-dam. Some of them go down-stream in the irrigation canal through the bottom pipe, and some of them stay in the micro-dam and accumulate at the bottom of the micro-dam.

Figure 6. Gully erosion near Adizaboy micro-dam (photo aspect, S–N).

2.3. Sedimentation in Micro-Dam and Reclaimed Farmland

The new bathymetric survey method (topographic survey) was applied to estimate the sediment volume for a short period [19]. An echo-sounder was used to measure the sediment surface depth from a boat. The coordinates around the perimeter of the micro-dam and the water depth were recorded (Figure 7). The sediment volume between the measured sediment surface and the estimated bottom was calculated to be 6400 m^3. Table 1 shows the sediment characteristics versus the bare land.

Micro-dam sediments were transported to the reclaimed farmland in March and April 2017 when the surface water of the Adizaboy micro-dam disappeared in the dry season. The construction period was about 2 weeks. The micro-dam sediment was excavated manually and transported by donkey to the farmland. The reclaimed farmland was constructed in the following sequence: 1) stone bunds were installed around the perimeter of the reclaimed farmland, with a height of 0.5 m and width of 0.6 m; 2) sediments were transported and leveled in the farmland; and 3) a diversion canal was constructed on the ground around the reclaimed farmland. The farmland reclamation was conducted on bare land about 100 m from the Adizaboy micro-dam. The dimensions of the farmland were 23 × 14 m (322 m^2).

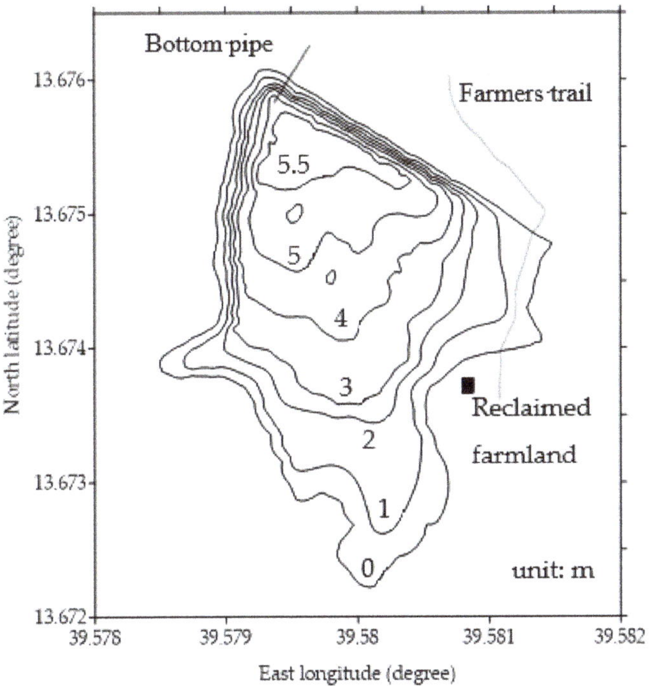

Figure 7. Sediment survey results for Adizaboy micro-dam. The micro-dam depth contour map was measured with an echo-sounder; numerical characters show the water depth (m) compared to the full water level.

Table 1. Micro-dam sediment and bare land soil sample characteristics [19].

Items	Sediments	Bare Land
Sand (g/kg)	70	102
Silt (g/kg)	536	501
Clay (g/kg)	394	397
Bulk density (g/cm^3)	1.1	1.2
Field capacity (v/v%)	28.8	15
Permanent wilting point (v/v%)	13.3	9.2
Available water capacity (v/v%)	15.6	5.9
Available water capacity (mm/15cm depth)	23.3	8.8
pH (H$_2$O)	7.3	8.1
Organic carbon (g/kg)	24.7	17.6
Total nitrogen (g/kg)	3.3	3.4
Available phosphorus (mg/kg)	9.2	8.8
Exchangeable potassium (cmol(+)/kg)	25.0	14.1

2.4. Constructability Concepts

The Construction Industry Institute (CII) Constructability Task Force defines constructability as the "optimum use of construction knowledge and experience in planning, design, procurement, and field operations to achieve overall project objectives" [20].

A general framework to be covered with constructability research was worked out by Vanegas [21]. Research from the CII established 17 constructability concepts, and the CII built up a Constructability Concepts File [22] that describes useful examples regarding the application of each concept. These constructability concepts were classified into three major project delivery phases: 1) conceptual planning, 2) design and procurement, and 3) field operations. The CII also published a Constructability Implementation Guide [23], which outlines a system of methods through which to apply constructability. Table 2 provides the constructability concepts.

Table 2. Constructability concepts.

Conceptual Planning Phase
Concept 1-A: The constructability program should be made an integral part of the project execution plan.
Concept 1-B: Special emphasis should be placed on maintaining an effective project team.
Concept 1-C: Early project planning should actively involve individuals with current construction knowledge and experience.
Concept 1-D: This early construction involvement should be a consideration in developing the contracting strategy.
Concept 1-E: The master project schedule should be start-up and construction-sensitive.
Concept 1-F: Major construction methods should be analyzed in-depth early on and should be facilitated through proper facility design.
Concept 1-G: Site layouts should promote efficient construction, operation, and maintenance.
Design and Procurement Phase
Concept 2-A: Design and procurement schedules should be construction-driven.
Concept 2-B: The capabilities and benefits of advanced information technology should be exploited.
Concept 2-C: Designs should be configured to enable efficient construction.
Concept 2-D: Design elements should be standardized.
Concept 2-E: Technical specifications should promote construction efficiency.
Concept 2-F: Detailed designs of modules and preassemblies should be prepared to facilitate efficient fabrication, transport, and installation.
Concept 2-G: Project designs should promote accessibility to materials and equipment by construction personnel.
Concept 2-H: Designs should allow for and enable construction under adverse weather conditions.
Field Operation Phase
Concept 3-A: Special effort should be applied toward developing innovative construction methods.

3. Results

3.1. Vegetable Cultivation on Reclaimed Farmland

Onion (*Allium cepa*) cultivation began in January 2019 and harvest took place in June 2019. We believed onions should grow with a sediment depth of about 20 cm, which included soil loss of 5 cm, in terms of cost/benefit ratio [10]. The shallow-rooted onion was selected as it does not require a deep cover of soil. Furrow and drip irrigation methods were adopted. Irrigation water was taken from the Adizaboy micro-dam and was stored in a concrete farm pond located at the highest point in the reclaimed farmland site (capacity of 85 m^3). The two irrigation methods for onion planting had 12 replications each and were grown at 25 cm intervals in rows, with 10 cm between plants. The irrigation requirement was 17.5 mm/day.

Figure 8 shows the onion yield on reclaimed farmland. Micro-dam sediments are classified into cohesive soil, which holds nutrients, fertilizer, and water, but does not have good air permeability. Abundant irrigation water naturally moved out through the stone bund, which surrounded the perimeter of the reclaimed farmland. The drainage problem did not occur on reclamation farmland. According to the Tukey–Kramer method (multiple comparison procedure for statistical analysis), the onion yield on reclaimed farmland with drip irrigation was significantly higher than that on reclaimed farmland with furrow irrigation and that of the national average (2006 to 2017).

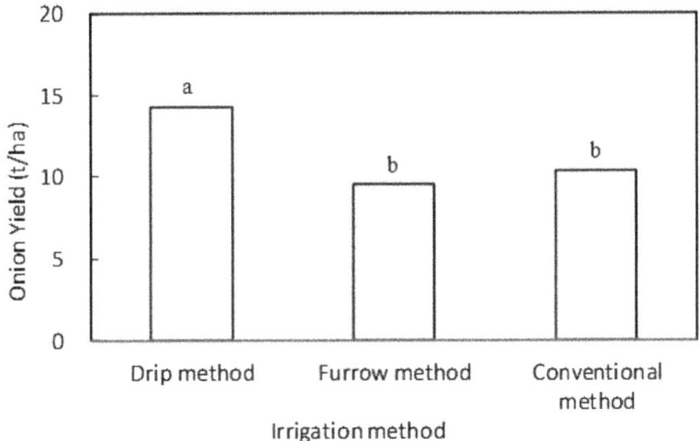

Figure 8. Onion yield on reclaimed farmland for different methods. Letters above onion yields show significant differences ($p < 0.01$) among treatments.

3.2. Constructability Criteria

Farmland reclamation processes included the construction of 1) a stone bund; 2) sediment layers; 3) a drainage canal; 4) a weather observation device; 5) irrigation facilities such as a pump, a hose, a farm pond, and a water tank; 6) a warehouse to store equipment, as well as to accommodate a guard; and 7) a barbed wire fence with prickly timber attached to metal columns to protect agricultural products against attacks by wild animals when deemed necessary, depending on local conditions.

After the farmland site was selected, weeding, shrub-clearing, and the removal of large stones began. Stones were utilized to construct the bunds along the boundaries of the farmland. Micro-dam sediments were collected by shovels and transported by donkeys to the farmland. Transferred sediments were layered and leveled on the reclaimed farmland, from which small stones and weeds were removed, so that the farmland surface was made flat and conducive to farming. Before seeding on the farmland, fences with barbed wire were established to keep away domestic animals such as goats, sheep, and cattle, which might have grazed the farmland, destroying planted crops. A warehouse was built to accommodate a guard to watch the crops and prevent theft or damage by wild animals, as well as to store equipment such as water pumps and drip irrigation. After water flows from upstream catchments to the reclaimed farmland were observed during rains, a drainage system was established so that the water eventually flowed into the micro-dam without causing erosion problems for the reclaimed farmland.

The constructability criteria and their attributes based on the authors' experience were used to guide the implementation of the farmland reclamation practice for each phase. Each criterion is examined here and the key attributes are summarized. We participated in the stakeholder meeting held at the public hall in Agulae Woreda on June 1, 2019. The meeting was attended by the Ministry of Agriculture and Rural Development in Tigray, Mekelle University, and about 50 farmers. After the meeting was over, all participants moved to the reclaimed farmland site and we explained our activities to them. Our guidance for the technical feasibility and acceptance by the community and other stakeholders was validated. These factors had a large effect on the up-scalability of the approach.

3.2.1. Conceptual Planning Phase

(1) Building an effective project team

The success of a farmland reclamation project depends on the capacity of project team members in designing, procurement, and field operations through relevant training, incentives, and communication. Labor productivity could be improved by concrete mixing training. This criterion is related to the constructability concepts 1-B, 1-C, and 1-D. The attributes required are described as follows: 1) a training program for specific crafts, 2) daily allowances for on-site jobs, 3) on-site communication with persons who have construction expertise, 4) on-site teamwork under the leadership of those with construction expertise, 5) availability of delivery systems such as a horse cart and a car, and 6) availability of special craftsmen and equipment for metal welding.

(2) Facilitating proper designs and layouts

Decisions on appropriate construction methods, facility designs, and site-layouts must be based on in-depth analyses to promote efficient construction, operation, and maintenance, utilizing information and survey data. This criterion is related to constructability concepts 1-E, 1-F, and 1-G. The attributes required are as follows: 1) the amount of storage water in the micro-dam required for irrigation, and 2) the availability of standard designs of farm pond, tank, and drip irrigation.

(3) Choosing suitable sites for reclaimed farmland considering the whole implementation planning

The selection of suitable sites for reclaimed farmland comes at the beginning of the implementation planning process (Figure 9). It is especially important for the sustainability of farmland reclamation works to consider engineering factors from a farmer's perspective. For example, the dimensions of reclaimed farmland should be consistent with the farmer's capacity and demand. Distance, as well as height difference, between micro-dams and reclaimed farmland, should be kept to a minimum because it is difficult to transport sediment and to pump irrigation water from micro-dams to reclaimed farmlands if the distance becomes too great. Farmland should not have steep slopes, so as to prevent soil loss and retain the sediment thickness.

Figure 9. Drip and furrow irrigation were conducted on the reclaimed farmland (photo aspect, S–N).

It is also important during site selection to consider the logistics, as farmland reclamation works are often implemented in rural remote areas. The site conditions must be exploited to maximize efficiency. For example, the availability of construction materials (e.g., stone, gravel, and sand) and sediment to be reclaimed in surrounding areas facilitates mobilizing locally available resources for operation. In turn, procuring external construction materials and tools (cement for foundation work, nails, iron bars, galvanized sheets, barbed wire, and wood poles) as well as inputs, such as extra chemical fertilizer to complement the fertility of sediments, requires good accessibility to reclaimed farmlands for easy delivery of supplies. The work area should be large enough to allow on-site activities such as assembly and concrete mixing for the construction of fences, along with space for the construction of a warehouse to store equipment, a guardhouse for protection against wild animals, and a farm pond with tanks to store irrigation water. After the farmland has been reclaimed, accessibility is still critical for the farmers who will be commuting to the farmland to grow crops and deliver crops to market. There should also be sufficient space to expand reclaimed farmland in the future. This criterion is related to constructability concepts 1-A and 1-G. The following attributes should be taken into consideration: 1) compatibility of reclaimed farmland dimensions with farmer's operability and maintainability; 2) the distance and the height difference between micro-dam and reclaimed farmland; 3) slope of farmland and thickness of sediments; 4) clearing and leveling of reclaimed farmland; 5) fencing the reclaimed farmland; 6) site accessibility for material delivery, as well as for farmers; 7) adequate laydown area availability for requisite working space, as well as a warehouse in which to keep necessary heavy equipment and tools; and 8) extra space for future site expansion.

3.2.2. Designing and Procurement Phase

(1) Designing efficient construction elements

When designing farmland, costs should be minimized, including procurement costs and costs for labor, materials, equipment, and guards. This criterion is related to constructability concepts 2-B, 2-C, 2-D, and 2-E. In doing so, the following points must be considered for designing and procuring schedules: 1) to minimize the complexity of design details and reduce the need for overly detailed specifications, 2) to make use of past survey results and water balance analysis results, and 3) to use standard dimensions and sizes for the reclaimed farmland system.

(2) Preparing for preassemblies and logistics

In order to facilitate efficient field operations, detailed designs of modules, including fabrication, transport, and installation of materials and equipment, should be prepared by construction personnel in advance. Procurement schedules must be planned and designed to minimize potential factors that could delay field operations, such as delays in equipment delivery, customs clearance, and permission processes. An inventory of construction and delivery components can help minimize the costs and time involved in on-site and off-site field operations, thus maximizing efficiency. This criterion is related to constructability concepts 2-F and 2-G. The key attributes are described as follows: 1) construction processes involving maximum use of on-site equipment and minimum labor, 2) off-site preassembly of some materials requiring prefabrication (i.e., weather observation devices) and cutting/welding (e.g., construction components, such as L-type metal columns) by skilled labor, 3) plan to maximize the use of the same transportation system for material and equipment delivery, and 4) facilitation of customs inspection for equipment made abroad.

(3) Preparing for adverse weather conditions

Negative effects due to bad weather must be minimized. This criterion is related to constructability concept 2-H. The key attributes are described as follows: 1) reclamation work, such as sediment transportation in micro-dams and concrete work for a farm pond and foundation work to fix poles, should be limited under rainy conditions; 2) site access through submerged farming roads under rainy conditions should be restricted; and 3) temporary storage for weather-sensitive equipment and materials should be provided.

(4) Planning, design, and procurement schedules and flexibility

The interaction and interface of activities must be well managed in the planning, design, and procurement schedules. Applications for farmland reclamation permits should be made at the earliest available opportunity. Schedules should include flexibility to deal with potential factors that could delay field operations and procurement processes. This criterion is related to constructability concept 2-A. The key elements related to this concept are as follows: 1) land permit processes to obtain the reclaimed land, 2) adaptability to withstand unexpected field conditions, such as extremely high run-off volume, dropping or rising groundwater level, and water consumption by people and livestock in the vicinity, and 3) potential delays due to the unavailability of specialized equipment, material, and labor.

3.2.3. Field Operation Phase

It is necessary to maximize the use of advanced and innovative technology and construction techniques. This criterion is related to constructability concepts 2-B and 3-A. The following points must be considered in the operation works: 1) to maximize the use of advanced materials (solar light), and 2) to maximize innovative survey equipment (GPS, note PCs, cameras, weather observation devices, and echo-sounders).

4. Discussion and Conclusions

Micro-dams have been constructed to deal with water shortage problems related to producing more crops in Tigray, Ethiopia. However, problems associated with sedimentation have occurred in many micro-dams. This issue precludes their ability to meet their intended performance levels. This paper showed one possible solution to solve the micro-dam sedimentation problem in the Adizaboy micro-dam, by constructing reclaimed farmland using the micro-dam sediments. The constructability concepts could facilitate farmland reclamation.

Constructability criteria, which generally facilitate the quality control, safety management, and schedule management of a project, were evaluated by the cost reduction of the farmland reclamation. The cost of labor (including guardianship), material, equipment, and groundwork was included in the project. The land use permit issued from Agulae Woreda administrative office could justify and formalize the financing of meaningful farmland reclamation for rural development (relating to Section 3.2.1. (1)). If there was no water in the micro-dam, the water stored in the farm pond was used. If there was no water in the farm pond, then the expensive option of arranging for a water tank truck to transport groundwater in Wukro city to the farm pond was required in our research. Drip irrigation was conducted to save water and avoid the expensive option (relating to Section 3.2.1. (2)). It is important that the effect of sedimentation is mitigated by conducting manual excavation and watershed management. Sediment excavation and transportation to the reclaimed farmland was carried out by preparing and organizing donkeys and local labor (relating to Section 3.2.1. (3)). Watershed management was carried out using conservation agriculture on wheat farmland by wheat residue to protect the farmland from water erosion. Machine excavation and the lost storage replacement were not conducted because they were prohibitively expensive for farmers, although the farming road could have been used for machine transportation.

Table 3 shows the cost reductions made by design change and construction material recycling. Land rent fees for the reclaimed farmland and the use of water and sediments in the micro-dam were waived after we obtained the land-use permit (relating to Section 3.2.2. (4)). Once the reclaimed farmland was handed over to the farmers from the project, they did not have to pay for the guardianship costs as long as they stayed near the reclaimed farmland (relating to Section 3.2.3.). The future research needs cost reduction factors on durability, safety or other related aspects to improve our "Constructability Criteria" approach.

Table 3. Cost reduction by design change and construction material recycling.

Item	Original Design	Modified Design	Rough Cost Estimation	Relation
Farm pond	Concrete stairs	Wooden ladder	12,000 JPY	3.2.2 (1)
Warehouse	New corrugated metal plate	Used corrugated metal plate	8000 JPY	3.2.2 (2)
Farm pond cover	Use of eucalyptus wood	Reduced use of eucalyptus wood	1000 JPY	3.2.2 (3)

We aimed to mitigate soil erosion problems in a micro-watershed. JIRCAS supported the research activities alongside Mekelle University in terms of the international joint research project. There would have been further room for cost-savings associated with farmland reclamation if the work was not restricted for a short period during the dry season. There are many young landless farmers as well as some conflict victims in the project site. It will be necessary to construct more reclaimed farmland to meet demand in the future. Farmers will conduct rotational vegetable cultivation of crops such as garlic, onion, potato, and carrot.

The authors have mapped the constructability criteria to solve the micro-dam sedimentation problem through the farmland reclamation by making use of micro-dam sediments, which administrative officers, farmers, and researchers have faced in Tigray. The good or bad performance of planning, design, and construction of the reclaimed farmland is decided by participants' experience, knowledge, teamwork, communication, and leadership. The constructability criteria will produce an optimum reclaimed farmland model to make the most of benefits and reduce costs to the minimum, to build the sustainable food production practices, and to increase the agricultural productivity and incomes of small-scale farmers.

Author Contributions: Conceptualization, K.K. and G.G.; methodology, K.K. and G.G.; software, K.K.; validation, K.K. and G.G.; formal analysis, K.K.; investigation, K.K., G.G., and T.B.; resources, K.K., G.G., and T.B.; data curation, K.K. and G.G.; writing—original draft preparation, K.K.; writing—review and editing, K.K.; visualization, K.K.; supervision, G.G. and T.B.; project administration, K.K. and G.G. All authors have read and agreed to the published version of the manuscript.

Funding: This research received no external funding.

Acknowledgments: Japan International Research Center for Agricultural Sciences and Mekelle University implemented the work, which was conducted as part of the African Watershed Management Project in Ethiopia. The authors thank JIRCAS research coordinator Miyuki Iiyama, and JIRCAS project members Keiichi Hayashi and Tomohiro Nishigaki, for commenting on the manuscript. Thanks also to Mekelle University, Ethiopia for providing a base for this joint project.

References

1. Haregeweyn, N.; Berhe, A.; Tsunekawa, A.; Tsubo, M.; Meshesha, D. Integrated watershed management as an effective approach to curb land degradation: A case study of the Enabered Watershed in Northern Ethiopi. *Environ. Manag.* **2012**, *50*, 1219–1233. [CrossRef] [PubMed]

2. Hurni, H.; Berhe, W.A.; Chadhokar, P.; Daniel, D.; Gete, Z.; Grunder, M.; Kassaye, G. *Soil and Water Conservation in Ethiopia: Guidelines for Development Agents*; Centre for Development and Environment (CDE), University of Bern, with Bern Open Publishing (BOP): Bern, Switzerland, 2016; ISBN1 978-3-906813-13-4. ISBN2 978-3-906813-14-1. [CrossRef]

3. Woldearegay, K.; Tamene, L.; Mekonnen, K.; Kizito, F.; Bossio, D. Fostering food security and climate resilience through integrated landscape restoration practices and rainwater harvesting/management in arid and semi-arid areas of Ethiopia. In *Rainwater-Smart Agric. Arid Semi-Arid Areas*; Springer: Berlin, Germany, 2018. [CrossRef]

4. Haregeweyn, N.; Melesse, B.; Tsunekawa, A.; Tsubo, M.; Meshesha, D.; Balana, B. Reservoir sedimentation and its mitigating strategies: A case study of Angereb reservoir (NW Ethiopia). *J. Soils Sediments* **2012**, *12*, 291–305. [CrossRef]

5. Tamene, L.; Park, S.; Dikau, R.; Vlek, P. Analysis of factors determining sediment yield variability in the highlands of northern Ethiopia. *Geomorphology* **2006**, *76*, 76–91. [CrossRef]

6. Tamene, L.; Park, S.; Dikau, R.; Vlek, P. Reservoir siltation in the semi-arid highlands of northern Ethiopia: Sediment yield-catchment area relationship and a semi-quantitative approach for predicting sediment yield. *Earth Surf. Process. Landf.* **2006**, *31*, 1364–1383. [CrossRef]

7. Tamene, L.; Abegaz, A.; Aynekulu, E.; Woldearegay, K.; Vlek, P. Estimating sediment yield risk of reservoirs in northern Ethiopia using expert knowledge and semi-quantitative approaches. *Lakes Reserv: Res. Manag.* **2011**, *16*, 293–305. [CrossRef]

8. Berhane, G.; Gebreyohannes, T.; Martens, K.; Walraevens, K. Overview of micro-dam reservoirs (MDR) in Tigray (Northern Ethiopia): Challenges and benefits. *J. Afr. Earth Sci.* **2016**, *123*, 210–222. [CrossRef]

9. The World Bank. Data Bank, World Development Indicators. Available online: http://databank.worldbank.org/data/reports.aspx?source=2&country=ETH (accessed on 3 January 2020).

10. Girmay, G.; Nyssen, J.; Poesen, J.; Bauer, H.; Merckx, R.; Haile, M.; Deckers, J. Land reclamation using reservoir sediments in Tigray, northern Ethiopia. *Soil Use Manag.* **2012**, *28*, 113–119. [CrossRef]

11. Yagi, M.; Madoo, T.; Shiratopi, K. *Agriculture and Forestry in Ethiopia, Overseas Agricultural Development Study*; Japan Association for International Collaboration of Agriculture and Forestry: Tokyo, Japan, 2006; p. 29.

12. Haregeweyn, N.; Poesen, J.; Nyssen, J.; Govers, G.; Verstraeten, G.; De Vente, J.; Deckers, J.; Moeyersons, J.; Haile, M. Sediment yield variability in Northern Ethiopia: A quantitative analysis of its controlling factors. *Catena* **2008**, *75*, 65–76. [CrossRef]

13. Gebremeskel, G.; Gebremicael, T.G.; Girmay, A. Economic and environmental rehabilitation through soil and water conservation, the case of Tigray in northern Ethiopia. *J. Arid Environ.* **2018**, *151*, 113–124. [CrossRef]

14. Haregeweyn, N.; Poesen, J.; Nyssen, J.; De Wit, J.; Haile, N.; Govers, G.; Deckers, S. Reservoirs (MDR) in Tigray (Northern Ethiopia): Characteristics and sediment deposition problems. *Land Degrad. Dev.* **2006**, *17*, 211–230. [CrossRef]

15. Berhane, G.; Martens, K.; Nawal, F.; Walraevens, K. Water leakage investigation of micro-dam reservoirs in Mesozoic sedimentary sequences in Northern Ethiopia. *J. Afr. Earth Sci.* **2013**, *79*, 98–110. [CrossRef]

16. Girmay, G.; Mitiku, H.; Singh, B. Agronomic and economic performance of reservoir sediment for rehabilitating degraded soils in Northern Ethiopia. *Nutr. Cycl. Agroecosystems* **2009**, *84*, 23–38. [CrossRef]

17. Rabia, A.; Afifi, R.; Gelaw, A.; Bianchi, S.; Figueredo, H.; Huong, T.; Lopez, A.; Mandala, S.; Matta, E.; Ronchi, M.; et al. Soil mapping and classification: A case study in the Tigray Region, Ethiopia. *J. Agric. Environ. Int. Dev.* **2013**, *107*, 73–99.

18. Ogawa, O.; Hirata, M.; Gebremedhin, G.B.; Uchida, S.; Sakai, T.; Koda, K.; Takenaka, K. Impact of differences in land management on natural vegetation in semi-dry areas: The case study of the Adi Zaboy watershed in the Kilite Awlaelo district, eastern Tigray region, Ethiopia. *Environments* **2019**, *6*, 2. [CrossRef]

19. Koda, K.; Girmay, G.; Berihu, T.; Nagumo, F. Reservoir Conservation in a Micro-Watershed in Tigray, Ethiopian Highlands. *Sustainability* **2019**, *11*, 2038. [CrossRef]

20. Construction Industry Institute (CII). *Constructability: A Primer. Publication 3–1*; Construction Industry Institute, University of Texas at Austin: Austin, TX, USA, 1986.

21. Vanegas, J.A. *A General Framework for Constructability Research. Independent Study*; Stanford University: Stanford, CA, USA, 1987.

22. Construction Industry Institute (CII). *Constructability Concept File. Publication 3–3*; Construction Industry Institute, the University of Texas at Austin: Austin, TX, USA, 1987.

23. Construction Industry Institute (CII). *Constructability Implementation Guide. Publication 34–1*; Construction Industry Institute, University of Texas at Austin: Austin, TX, USA, 1993.

 *sustainability*

Article

Communities' Livelihood Vulnerability to Climate Variability in Ethiopia

Misganaw Teshager Abeje [1,2,*], Atsushi Tsunekawa [3], Nigussie Haregeweyn [4], Zerihun Nigussie [3,5], Enyew Adgo [5], Zemen Ayalew [5], Mitsuru Tsubo [3], Asres Elias [6], Daregot Berihun [7], Amy Quandt [8], Mulatu Liyew Berihun [1,9] and Tsugiyuki Masunaga [10]

1   The United Graduate School of Agricultural Sciences, Tottori University, 4-101 Koyama-Minami, Tottori 680-8553, Japan; mulatuliyew@yahoo.com
2   Institute of Disaster Risk Management and Food Security Studies, Bahir Dar University, P.O. Box 79, Bahir Dar 6000, Ethiopia
3   Arid Land Research Center, Tottori University, 1390 Hamasaka, Tottori 680-0001, Japan; tsunekawa@tottori-u.ac.jp (A.T.); zeriye@gmail.com (Z.N.); tsubo@tottori-u.ac.jp (M.T.)
4   International Platform for Dryland Research and Education, Tottori University, 1390 Hamasaka, Tottori 680-0001, Japan; nigussie_haregeweyn@yahoo.com
5   College of Agriculture and Environmental Sciences, Bahir Dar University, Bahir Dar 5501, Ethiopia; enyewadgo@gmail.com (E.A.); zayalew@gmail.com (Z.A.)
6   Faculty of Agriculture, Tottori University, 4-101 Koyama-Minami, Tottori 680-8550, Japan; asres97@yahoo.com
7   College of Business and Economics, Bahir Dar University, P.O. Box 79, Bahir Dar 6000, Ethiopia; daregot21@gmail.com
8   Department of Geography, San Diego State University, San Diego, CA 92182, USA; Amy.Quandt@colorado.edu
9   Faculty of Civil and Water Resource Engineering, Bahir Dar Institute of Technology, Bahir Dar University, P.O. Box 26, Bahir Dar 6000, Ethiopia
10 Faculty of Life and Environmental Science, Shimane University, Shimane Matsue 690-0823, Japan; masunaga@life.shimane-u.ac.jp
*   Correspondence: tmisganaw16@gmail.com

Received: 26 August 2019; Accepted: 7 November 2019; Published: 9 November 2019

Abstract: Ethiopia has experienced more than 10 major drought episodes since the 1970s. Evidence has shown that climate change exacerbates the situation and presents a daunting challenge to predominantly rain-fed agricultural livelihoods. The aim of this study was to analyze the extent and sources of smallholder famers' livelihood vulnerability to climate change/variability in the Upper Blue Nile basin. We conducted a household survey ($n = 391$) across three distinct agroecological communities and a formative composite index of livelihood vulnerability (LVI) was constructed. The Mann–Kendall test and the standard precipitation index (SPI) were employed to analyze trends of rainfall, temperature, and drought prevalence for the period from 1982 to 2016. The communities across watersheds showed a relative difference in the overall livelihood vulnerability index. Aba Gerima (midland) was found to be more vulnerable, with a score of 0.37, while Guder (highland) had a relatively lower LVI with a 0.34 index score. Given similar exposure to climate variability and drought episodes, communities' livelihood vulnerability was mainly attributed to their low adaptive capacity and higher sensitivity indicators. Adaptive capacity was largely constrained by a lack of participation in community-based organizations and a lack of income diversification. This study will have practical implications for policy development in heterogeneous agroecological regions for sustainable livelihood development and climate change adaptation programs.

Keywords: climate change; drought; livelihood vulnerability; Shannon-entropy index

---

1. Introduction

1.1. Livelihood Vulnerability to Climate Change

A livelihood is a means by which individuals or households make a living [1]. Household livelihood outcomes are a function of a range of components, including livelihood assets, activities, processes, and structures. Smallholder farmers, accounting for 75% of the world's agricultural area [2] and 60% of employment [3], produce over 80% of the food consumed in the developing world [4] and are one of the most vulnerable groups of people to climate change [5,6]. The vulnerability has been attributed to their high dependence on ecosystem services and goods, exposure and sensitivity to climate variability, low adaptive capacity, reliance on rain-fed livelihood activities, and often marginal locations in the landscapes [7–9]. Moreover, adverse consequences of environmental challenges (e.g., climate change) on crop sustainability and productivity could affect farmers' livelihood activities and lower their adaptive capacity [10]. According to the Intergovernmental Panel for Climate Change (IPCC), climate change is projected to increase climate-related shocks (e.g., drought) and disproportionately manifest its adverse consequences through human health, food security, and water resources, specifically for rural poor households [11]. The IPCC report emphasized that efforts should focus on enhancing adaptation, reducing exposure, and decreasing the vulnerability of small-scale farmers while enhancing their resilience to shock impacts. These interventions in turn should be informed by evidence of livelihood vulnerabilities [12], which are shaped by physical, economic, social, and ecological factors and processes [13].

Livelihood vulnerability to climate variability and change is a function of exposure, sensitivity, and adaptive capacity [14]. Exposure refers to changes in climate variability explained by seasonal variations, and it is essentially associated with precipitation and temperature [15]. Sensitivity refers to the degree to which a system could be adversely affected, and it is explained by the potential impact's net effect and people's potential to cope with any adverse consequences. It essentially captures a system's susceptibility to harm associated with environmental and social changes. Adaptive capacity entails the capacity of the system to withstand variability and changes in order to minimize potential damages, to cope with negative consequences, and possibly even benefit from these changes [11,13,16–18]. Several researchers have attempted to explore the blend between livelihood approaches and vulnerability dimensions as part of a broader study of sustainable livelihood development. Many studies in Africa and elsewhere have found varied results in terms of which factors contribute to overall livelihood vulnerability [19–25]. Other studies in this continent (e.g., [7–9]) have indicated that lower adaptive capacity and higher exposure to climate-related hazards (e.g., drought) are the major contributors to livelihood vulnerability and consequently undermine the sustainability of small-scale farmers' livelihood bases.

To empirically understand livelihood vulnerability, the fundamental work by Hahn et al. [26] to assess community livelihood vulnerability in Mozambique is of great importance. Employing the Sustainable Livelihoods Approach [27], Hahn et al. used the IPCC-LVI (livelihood vulnerability index) to investigate communities in terms of their endowment of human capital, financial capital, physical capital, social capital, and natural capital. Environmental shocks and stresses (e.g., drought) related to climate change were viewed from the perspective of each of these types of capital. The methodology has since been applied to study communities and regions elsewhere in the developing world and Ethiopia and provides the foundation for this research as well [20,22,25,28–33].

1.2. The Ethiopian Context

Climate variability and change is repeatedly cited as the source of vulnerability in Ethiopia [34–36]. The country has experienced more than 10 drought events since the 1970s [9]; hence, sensitivities are also a product of large inter-seasonal climate variability and the reliance of the economy on rain-fed agriculture [37]. Vulnerability to climate variability is evident in terms of social and economic institutional sensitivity to variability in rainfall and the occurrence of extreme climate related shock events (e.g., drought and flood) [31,34,38]. The climate projections for 2040 to 2059 show a 1.8 °C increase in temperature and, with a higher inter-annual variability in the northern part, rainfall is anticipated to decline [39]. Therefore, as noted by Simane, Zaitchik, and Foltz [31], amalgamation of sensitivity to past climate variability and limited adaptive capacity in terms of socioeconomic and institutional aspects, coupled with the projections of anticipated future climate change, suggests that the country will be adversely affected by climate patterns in the years to come [39]. Ethiopia's efforts to address the negative impact of climate variability are in the process of shifting from a technocratic perspective of climate and disaster science to long-term efforts at reducing livelihood vulnerability and attaining sustainable livelihoods [34]. The country's effort to develop and operationalize the Climate Resilient Green Economy as a guiding strategy for climate change adaptation and mitigation efforts is a huge step forward in this regard [40].

Some research has been conducted to support policies and strategies to reduce the exposure and vulnerability of rural livelihoods to the effects of climate change/variability. In the Ethiopian highlands, Simane, Zaitchik, and Foltz [31] revealed that climate change vulnerability is context specific and agroecosystems should be at the centre of future studies. The authors found that midland areas are better off in terms of climate change vulnerability as compared to both high and lowland areas. A similar study in Tigray revealed that climate change exposure and low adaptive capacity were substantially associated, and moreover, they were the major causes of vulnerability among farming communities [29]. Similarly, in their comparative study of Ethiopian highlands, Siraw, Adnew Degefu, and Bewket [32] found that watersheds that received soil and water conservation works were less vulnerable to the effects of climate change than those that did not receive any. Owing to the above facts, there are still gaps in our study area where variations exist in agroecological locations, which indicates smaller scale studies are essential to better inform planners. Moreover, there are methodological gaps in the weightage procedure of indicators to measure vulnerability [13,41,42] and livelihood vulnerability studies need to emphasize one of the most important concerns of Ethiopian rural smallholder farmers, particularly drought [9].

1.3. Study Objectives

By applying the IPCC-LVI [26] at agroecologically contrasting environments, this study aimed at analyzing the livelihood vulnerability to climate change/variability for small-scale farmers in north western Ethiopia, Upper Blue Nile basin. First, the major objective of this manuscript is to provide empirical evidence at smaller scales of how livelihood vulnerability may vary across diverse agroecological ecosystems [43]. A second, and much more minor, objective is to address methodological gaps related to weightage of indicators so that robust conclusions can be drawn [13,41,42]. Lastly, we aim to shed light on the community level obstruction (constraints) indicators to limiting climate change/variability adaptive response mechanisms [44].

1.4. Significance of the Study

Despite being few in numbers, recently, livelihood vulnerability to climate variability/change studies have received growing attention in Ethiopia. Most of these studies were broader in scale and used political administrations (e.g., districts, regions, and national) as unit of analysis, while others focused on vulnerability to food insecurity and poverty. Almost all of them have adopted the LVI-IPCC livelihood measure, using subjectively evaluated indicators to construct the indices. However, studies at the national level are not believed to show the full picture of socioeconomic livelihood and the variability of other adaptive capacities at lower scales, and findings may not precisely indicate the necessary information for practical implications. Furthermore, these studies have underestimated the importance of studying past drought episodes on the current livelihoods of communities. The present study provides a sounder quantitative analysis of livelihood vulnerability using the Shannon entropy weighting procedure at a lower scale (watersheds in this study), whereby the locations represent contrasting agroecological environments. Therefore, in this study, we argue that small-scale farmer level studies would help to better understand the adaptive capacities of communities, and hence would help decision-makers tailor policies to the local conditions. Moreover, to further help fine-scale decision making at the community level, we adopted an obstruction degree analysis, which enabled us to bring up the specific constraint indicators of adaptive capacity that varied by study locations. Hence, the current study can be adopted for similar agroecological environments, watersheds, and communities in Ethiopia and other developing countries. More broadly, we also contribute to the limited existing literature of rural livelihood vulnerability analysis studies in Ethiopia and other developing countries. For instance, the methodology could be adopted for national level objective climate vulnerability studies and promotes the inclusion of climate related shocks as part of the analysis.

2. Materials and Methods

2.1. Study Area

This study was carried out in the following three different agroecological environments of the northwestern highlands of Ethiopia in the Upper Blue Nile basin: Guder (highland), Aba Gerima (midland), and Dibatie (lowland) (Figure 1). Area selection was determined by their differences in elevation, cropping system, and precipitation [45], and these three watersheds were selected because they represent a range of different agroecological and socioeconomic characteristics. Households in the study areas primarily make their livelihoods from a mixed crop–livestock production system. The major crops grown are barley (*Hordeum vulgare* L.), teff (*Eragrostis tef* Zucc.), wheat (*Triticum aestivum* L.), and potato (*Solanum tuberosum* L.) [46]. Cattle, sheep, goats, donkeys, and horses are the dominant livestock raised.

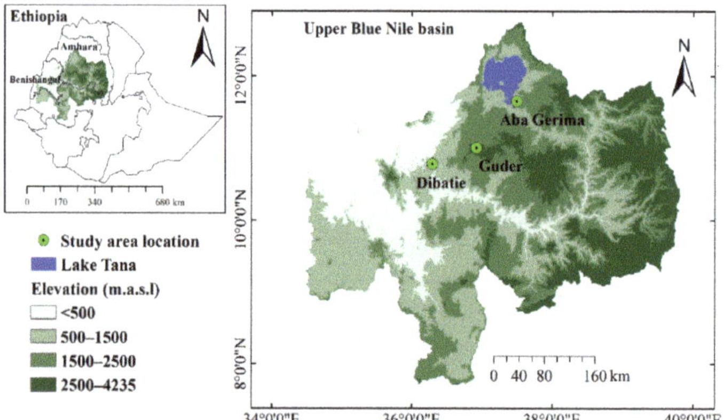

Figure 1. Location map of the study areas.

Aba Gerima watershed is categorized as humid sub-tropical, with an annual rainfall that ranges from 895 to 2037 mm. Guder is moist tropical, with an annual rainfall of 1951–3424 mm, and Dibatie is tropical hot humid, with an annual rainfall of 850–1200 mm [46]. Data from nearby meteorological stations shows that there has been notable climate variability since 1982 [47]. Despite being in different agroecological environments, the watersheds are characterized by similar rainy (June to October) and dry (November to May) seasons, and more than 86% of the rainfall is concentrated during the rainy season.

Most farmers in these areas have subsistence-based livelihoods supplemented by additional sources of non-and off-farm income. In the past decade, most farmers in Guder and Aba Gerima have shifted from growing food crops to growing green wattle (Acacia decurrens) (*Wendl. f.) Willd.* and khat (*Catha edulis*), respectively [48,49]. Notable droughts have occurred in these areas, for example, in 1984/1985, 1992/1993, 2000/2001, and 2015/2016 [9].

2.2. Data and Sampling Procedures

We used a mix of data collection methods and sources. First, meteorological data (temperature and rainfall) were gathered from the Ethiopian National Meteorology Station for the period from 1982 to 2016. Second, we conducted a participatory rural appraisal [27] to select the specific local indicators of livelihood vulnerability, as defined by the communities themselves. Twenty individuals from each community participated in this appraisal (for a total of 60). Third, we considered the newly identified livelihood vulnerability indicators and developed a draft questionnaire to be administered in a pilot test administered to five households in each watershed.

Finally, primary data were obtained from sampled respondents using a structured questionnaire administered in face-to-face interviews conducted in October and December 2018. The topics included sociodemographic profiles, food, water, social networks, livelihood strategies, health, and climatic shocks. To select sample households for the study, we used a two-stage sampling procedure. We first selected watersheds according to their agroecological and socioeconomic differences and household respondents from each watershed were selected randomly by using a probability proportional to size sampling procedure. Due to its simplicity, the purposive selection of the watersheds, and the proportional to size sampling, we followed and adopted Cochran's representative size to proportions formula. Therefore, based on the sampling procedures of Cochran [50], we first calculated the required sample size to be 391 from the full household lists of all three watersheds. We then randomly selected 130 households from Aba Gerima, 132 households from Guder, and 129 households from Dibatie. The questionnaire was first developed in English and then translated into the Amharic language. Enumerators were trained on each question and practiced doing mock interviews. Under supervision, the enumerators interviewed each head of household (male or female). In the absence of the household head, elders who were willing to participate in the interview were interviewed. The first author did all the supervisory work and quality checks throughout the data collection period. On average, each interview took 60 min.

2.3. Data Analysis

This study generally followed the IPCC-LVI construction methodology, which is essentially based on a sustainable livelihood framework (SLF) [26]. Data management and analyses were performed with Stata ver. 15 (StataCorp LLC, College Station, TX, USA), MS Excel, and XLSTAT.

2.3.1. Meteorological Data Analysis

To understand the long-term trends of rainfall and temperature, we analyzed the climate variability/change and the significance of monotonic trends by using the Mann–Kendall (MK) test for long-term metrological records [51]. The MK test and Sen's slope estimator (with Pettitt's homogeneity test) were applied to the time-series data from 1982 to 2016 for the three watersheds.

The standardized precipitation index (SPI) was used to characterize historical drought patterns and periods in the study watersheds. The SPI uses a Z-score to explore unusual weather events that happened in the past. SPI is a normalized index in time and space and was computed following Thomas B. McKee [52]. To characterize drought intensity, the author suggested that SPI can be calculated on a time scale from 1 month to 72 months. For the purpose of this study, we chose a 3-month time scale to assess the main rainy season. We used statistical software developed by Tigkas et al. [53], called the Drought indices Calculator (DrinC), in the drought analysis.

2.3.2. Measures of Livelihood Vulnerability

We followed an inductive approach for the construction of an overall formative composite index where we could explore the community (watershed)-scale livelihood vulnerability based on SLF [27,54] and the pragmatic approach of Hahn, Riederer, and Foster [26]. Presently, many different composite index constructions are usually criticized for their subjective weighting procedures because they may result in misleading information [55,56]. Including the pioneering work by Hahn et al., scholars in Ethiopia and elsewhere in the world followed the subjective weighting of components/indicators to construct the composite index. This is based on the number of questions included under the indicated component, which is inappropriate whereby the information is not quantitative, exclusive, and partial or incomplete. Unlike the subjective methods, in this study, we applied an objective weighting procedure by which it is more appropriate to give precise evidences for prioritizing planning areas (based on the value of sub-components), which seeks attention and remedies to reducing the livelihood vulnerability of communities to the adverse effects of climate variability/change. The Shannon entropy method, an objective weighting method that has been recommended as robust [20,22,57], was used to generate an evaluation score for each indicator. The following computation procedures were used (for a more detailed description of the procedures, see the Supplementary Materials): (1) We standardized all 33 indicators (as provided in Table 1 in the results section) by using a dimensionless processing technique that helps facilitate easy comparison of score values. In this case, the variables have a positive functional relationship with vulnerability, and a higher value is generally understood to indicate greater vulnerability. (2) The proportion of indicators in an evaluation matrix was computed. (3) An individual entropy value for each indicator was calculated. (4) The entropy weight for each indicator was computed. (5) A comprehensive index value was constructed for each community. The minimum value was scaled to 0 (least vulnerable) and the maximum was scaled to 1 (most vulnerable) (see the Supplementary Materials for more details).

In the results, we found that adaptive capacity was the most salient factor influencing overall IPCC-based livelihood vulnerability in the combined study area. As a result, we additionally followed a weighting and aggregation procedure for this dimension. Williams et al. (Williams, Crespo, Abu, and Simpson) [42] suggested that, in Africa, the adaptive capacity dimension of smallholder farmers' livelihood vulnerability assessments should be emphasized to better inform decision-making. Therefore, we adopted a degree of obstruction model [44] to discover which factors limit adaptive capacity (see the Supplementary Materials for more details). A higher value (percentage points) indicates that the indicators could have a higher hindering capacity in terms of limiting households' capacity to respond to the effects of climate change. This model is widely applicable in urban land-use management as a mathematical decision approach [58].

3. Results

3.1. Vulnerability Indicators

Profiling of livelihood vulnerability indicators was documented and analyzed for the three contrasting agroecological environments. Table 1 presents descriptions and summary statistics of vulnerability indicators used for the development of IPCC-LVI. It highlights the major differences between the three research areas in regards to climate, demographics, livelihoods, and other metrics. These differences allow for analysis and discussion of how livelihood vulnerability may vary in different agroecosystems. Guder had the highest deviation from the rainfall trend (87.1 mm), as compared to Aba Gerima (44.8 mm) and Dibatie (43.4 mm). The highest values for both average minimum (14.6 °C) and maximum temperatures (28.1 °C) were recorded in Dibatie ($p < 0.001$), which also had the lowest SPI and the most drought episodes. Aba Gerima (79.2%) had the highest proportion of households who did not attend school and the highest proportion of households (96.2%) with members who work outside their community. A larger percentage of sampled households in Dibatie (65%) did not participate in natural resource management works ($p < 0.001$) as compared to the other two communities. More households reported chronic illness in Dibatie (49.6%) as compared to the other sites ($p < 0.001$), and Dibatie households reported less contact with the local government offices for any kind of service ($p < 0.001$). Households in Aba Gerima reported more water-related conflicts as compared to Dibatie and Guder, and they also had the highest percentage (37%) of households who did not save seed for the next growing season. Most households in Aba Gerima (95.4%) and Dibatie (91.5%) reported that their own farm was their main source of food, whereas the proportion was lower in Guder (74.2%) ($p < 0.001$). From the participatory rural appraisal, communities in all three watersheds were able to bring two new indicators to be used as part of the overall IPCC-LVI. These indicators were environment related indicators; namely, the level of household participation in natural resource management works and the soil erosion status in their farms. These indicators showed significant difference amongst the study watersheds ($p < 0.001$).

Table 1. Description of vulnerability indicators and summary statistics.

Dimensions	Components	Vulnerability Indicators	Summary Statistics			p-Values	References
			Aba Gerima	Guder	Dibatie		
Exposure (Exp)	Climate	Mean standard deviation of monthly average rainfall (mm)	44.8	87.1	43.4		[31,57]
		Mean standard deviation of monthly average of average minimum daily temperature (°C)	12.7	9.5	14.6	a **	[32]
		Mean standard deviation of monthly average of average maximum daily temperature (°C)	27.5	25.2	28.1	a **	[32]
		Frequency of climate-related hazards (no.)	2.9	2.3	3.6	a **	[32]
		Access to warning information (%)	55.4	44	91.5	b ***	[57]
		SPI for the wet season	0.016	0.02	0.01		[57,58]
Adaptive capacity (AdapCap)	Sociodemographic	Age of household head (years)	48	53	45	a **	[32]
		Dependency ratio (%)	0.91	0.72	0.94	a **	[31]
		Households with female heads (%)	13	20	10	b **	[31]
		Household heads who have not attended school (%)	79.2	72.7	62.0	b ***	[29,31]
	Livelihood strategies	Households without members working outside the community (%)	3.9	38.6	27.9	b ***	[29,31]
		Households with no other source of income	25.8	34.2	40.1	b ***	[29,31]
		Agricultural Livelihood Diversification Index	0.29	0.29	0.26	a ***	[29,31]
		Livestock Diversification Index	0.6	0.24	0.61	a ***	[29,31]
		Households who have not participated in natural resource management activities (%)	30	16	65	b ***	[31,58]
		Average receive-give ratio (%)	1.06	1.27	1.07	a ***	[29,58]
		Borrow-lend ratio (%)	0.32	0.18	0.16	a **	[29,58]
	Social networks	Households are not members of community-based organizations (%)	4	2	11	b **	[29,58]
		Households are not members of farmer-based organizations (%)	6.15	57.6	24.8	b ***	[29,58]
		Households who have not gone to local government (%)	33.33	13.6	82.2	b ***	[58]
		Households who have no communication devices (%)	52.31	39.4	20.2	b ***	[58]
Sensitivity (Sen)	Health	Walking distance to health service (')	56.77	68.4	34.2	a ***	[32]
		Households who reported chronic illness (%)	25.38	14.4	49.6	b ***	[32]
		Households where a member missed work/school due to illness (%)	35.38	30.3	53.5	b ***	[32]
		Number of times where households were exposed to epidemics	1.4	0.5	1.5	a ***	[32]
		Households mainly dependent on family farm for food (%)	95.4	74.2	91.5	a ***	[31,58]
	Food	Average number of months households struggle to get food (no.)	0.2	1.3	0.3	a ***	[32]
		Households who did not save seed (%)	37	34	25	b *	[32]
		Average crop diversification (index)	0.24	0.39	0.37	a ***	[29,58]
		Households with plots with high soil erosion status (%)	8	14	6	b ***	[32]
	Water	Households who reported water conflicts (%)	12.3	3.0	9.3	b **	[31]
		Households who reported lack of consistent access to water (%)	68.5	47	83.7	b ***	[29,31]
		Average walking time to water source (')	0.09	0.15	0.1	a ***	[31]

Source: Field survey. * $p < 0.10$. ** $p < 0.05$.*** $p < 0.001$. [a] F-statistic and [b] χ2 test for mean differences "Average Agricultural Livelihood Diversification Index (range: 0.20–1): The inverse of (the number of agricultural livelihood activities +1) reported by a household, e.g., a household that farms, raises animals, and collects natural resources will have a Livelihood Diversification Index = 1/(3 + 1) = 0.25; Average Receive:Give ratio (range: 0–15): Ratio of (the number of types of help received by a household in the past month + 1) to (the number of types of help given by a household to someone else in the past month + 1); Average Borrow:Lend Money ratio (range: 0.5–2): Ratio of a household borrowing money in the past month to a household lending money in the past month, e.g., if a household borrowed money but did not lend money, the ratio = 2:1 or 2 and if they lent money but did not borrow any, the ratio = 1:2 or 0.5; Average Crop Diversity Index (range: >0–1)a: The inverse of (the number of crops grown by a household +1), e.g., a household that grows wheat, maize, beans, and barley will have a Crop Diversity Index = 1/(4 + 1) = 0.20" [26]. [a] and [b] are referring to the P values in the table.

3.2. IPCC-Based Livelihood Vulnerability

In terms of overall IPCC-LVI, Aba Gerima was found to be more vulnerable, with an aggregate score of 0.37 (on a scale of 0 to 1), followed by Dibatie and Guder at 0.35 and 0.34, respectively (Figure 2a). The adaptive capacity of smallholder farmers made the greater contribution relative to the other dimensions in terms of explaining livelihood vulnerability (0.15, 0.14, and 0.13 in Aba Gerima, Guder, and Dibatie, respectively), followed by sensitivity (0.14, 0.12, and 0.13) and exposure (0.08, 0.08, and 0.09) (Figure 2b).

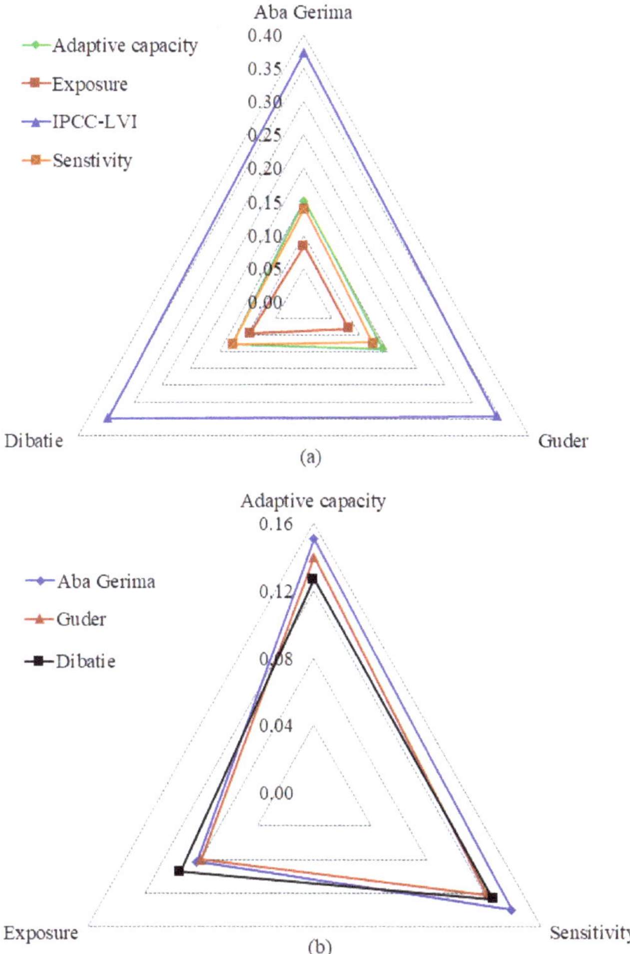

Figure 2. IPCC-based livelihood vulnerability (IPCC-LVI) (**a**) and its dimensions (**b**) in the three watersheds.

3.2.1. Exposure to Climate Shocks

Figure 3 shows the climate exposure trends (rainfall and temperature) in the three agroecological areas. In the IPCC-LVI, the exposure score at Dibatie was slightly higher (0.09) than that of both Aba Gerima and Guder (0.08). This dimension was aggregated from rainfall and temperature data, the number of climate-related shocks, household access to warning information about these shocks, and

SPI. Although there was no significant SPI trend in the watersheds, there has been recurrent drought episodes at the Dibatie and Aba Gerima sites (Figure 4). Climate variability anomalies, as measured by SPI, made a substantial contribution to the exposure to livelihood vulnerability in Dibatie and Aba Gerima. Conversely, although Guder experienced relatively severe drought episodes in 1984/1985, this watershed has been less vulnerable to rainfall deficits since then. A higher temperature and lack of access to warning information made a notable contribution to the overall exposure in Aba Gerima (Table 1).

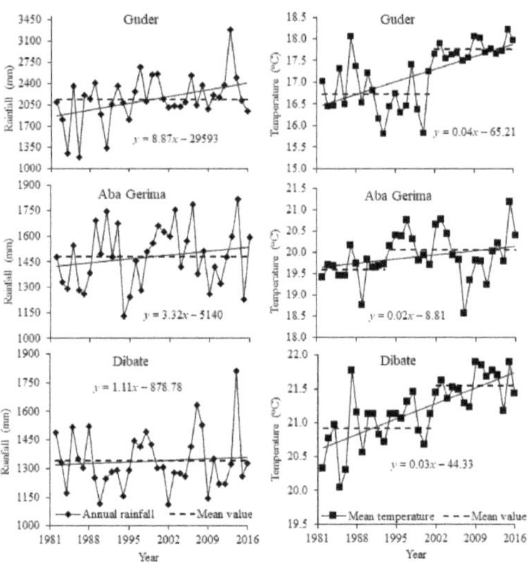

Figure 3. Climate trends (rainfall and temperature) in the three watersheds.

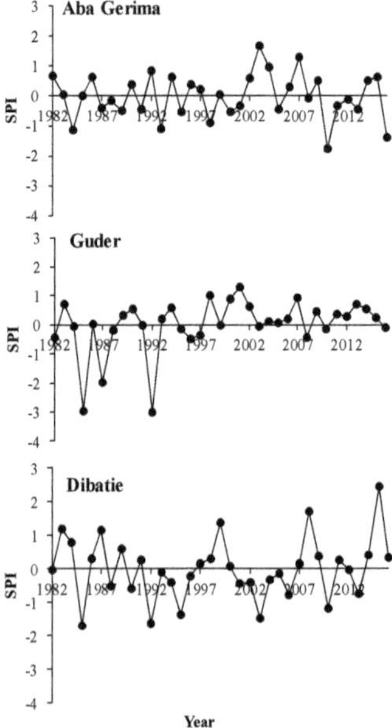

Figure 4. Standard Precipitation Index (SPI) for the three study sites.

The mean annual rainfall of Aba Gerima was 1482.5 mm, with a SD of 178.6 mm and a coefficient of variation (CV) of 12.1%. The value was much higher in Guder (2145.3 ± 395.3 mm, CV = 18.4%) and lower in Dibatie (1339.5 ± 150.1 mm, CV = 11.20%). The average monthly temperature decreased from lowland to highland; that is, in the order Dibatie > Aba Gerima > Guder (Figure 3). Dibatie had more hot years, with the average maximum and minimum ever-recorded temperature of 21.9 °C in 2015 and 20.0 °C in 1985. The average maximum ever-recorded annual temperature in Guder was 18.2 °C in 2015 and the minimum was 15.8 °C in 1993. The Mann–Kendall test showed no significant long-term monotonic trend in the rainfall amount, with Z_c of 1.3, 0.9, and 0.4, for Guder, Aba Gerima, and Debatie watersheds, respectively (Table 2, Figure 3) in the study watersheds. Similarly, Pettitt's test showed strong homogeneity in annual rainfall in the three watersheds, indicating that annual rainfall did not change significantly over the study period (Table 2, Figure 3). Therefore, the null hypotheses $H_o a$ and $H_o b$ for the two tests for annual rainfall in the three watersheds were accepted. In contrast, the watersheds showed a significant increasing trend in temperature during the study period ($p < 0.05$). The Z_c values of 3.92 in Guder, 2.07 in Aba Gerima, and 4.55 in Dibatie confirmed that there were significant changes in annual temperature over the study period (Table 2). The mean annual temperature increased by 0.04 °C per year in Guder watershed, 0.02 °C per year in Aba Gerima, and 0.03 °C per year in Dibatie from 1982 to 2016.

Table 2. Monotonic trend (Mann–Kendall) test and significant change (Pettitt's homogeneity) test for two climate variables (annual rainfall and mean annual temperature time series) for 1982–2016 in three watersheds.

Climate Variable	Watershed	Mann–Kendall Test			Pettitt's Test		
		Zc	p	H_0^a	K	p	H_0^b
Rainfall	Guder	1.30	0.20	A	134.00	0.20	A
	Aba Gerima	0.90	0.30	A	108.00	0.60	A
	Dibatie	0.40	0.90	A	68.00	0.40	A
Temperature	Guder	3.92	<0.0001	R	272.00	<0.0001	R
	Aba Gerima	2.07	0.04	R	172.00	0.03	R
	Dibatie	4.55	<0.0001	R	258.00	<0.0001	R

H_0^a is the null hypothesis that there is no monotonic trend in the time series for annual rainfall or mean temperature; H_0^b is the null hypothesis that there is no significant change in the time series data for annual rainfall or mean temperature (the data are homogeneous). The null hypotheses are accepted (A) or rejected (R) at significance level $\alpha = 0.05$.

The three watersheds experienced drought episodes with varied intensities (Figure 4). For example, 1984/1985, 1992/1993, and 2015/2016 were recorded as drought years of moderate intensity for Aba Gerima watershed, but 2003/2004 and 2007/2008 were very wet and moderately wet years, respectively. Guder and Dibatie also experienced severe drought in 1984/1985 and1992/1993, but 2001/2002 was very wet. Dibatie also experienced drought in 1994/1995, 2003/2004, and 2010/2011. Except Guder, the MK trend test showed a decrease in SPI values across two other watersheds, suggesting a frequent drought incidence at a 3-month (Ethiopian summer) time scale, but there was no statistical evidence of any positive or negative trend. However, we can still justify that there have been recurrent drought episodes in Dibatie and Aba Gerima sites.

3.2.2. Sensitivity

Aba Gerima had the highest score for the sensitivity dimension of the IPCC-LVI (0.14), followed by Dibatie (0.13) and Guder (0.12) (Figure 2). Combined across all three watersheds, the food component had a higher score than the water and health components (Figure 5), particularly in Guder (0.07). Crop diversification (0.03) and on-farm food source (0.02) were the main contributors among the five indicators that made up the food component. In Aba Gerima, reliance on on-farm agriculture (0.03) and a low tradition of saving seed (0.01) substantially contributed to the overall sensitivity score. Compared to Aba Gerima (0.006) and Dibatie (0.007), Guder had a lower level of lower crop diversification (0.028). Guder also had a relatively low contribution from the health component (0.025) as compared to Dibatie (0.042) and Aba Gerima (0.048). In addition, inconsistent access to water (part of the water component) played a relatively higher role in Dibatie (0.025) and Aba Gerima (0.021) relative to Guder (0.014).

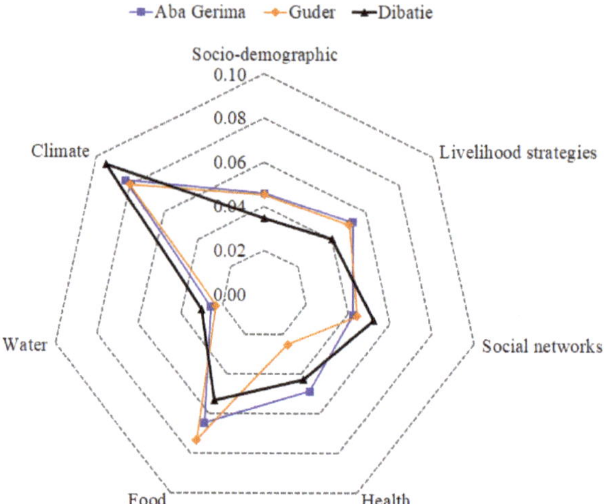

Figure 5. Contribution of major components to the overall IPCC-LVI in all watersheds.

3.2.3. Adaptive Capacity

A slightly higher vulnerability due to lower adaptive capacity was revealed in Aba Gerima (0.15) as compared to Guder (0.14) and Dibatie (0.13). The livelihood strategies component had the highest contribution (0.052) in Aba Gerima, followed by the sociodemographic (0.046) and social networking (0.042) components. Overall, literacy status (0.024) contributed the most to aggregate adaptive capacity, followed by age (0.0173) and livestock ownership (0.0171). In Guder watershed, low adaptive capacity was attributable to households' restricted options for livelihood strategies (0.050) and low sociodemographic characteristics (0.040). Social networking (0.052) had the greatest contribution in Dibatie, but a lack of contact with the local government office (0.024) contributed to their lower adaptive capacity.

3.3. Factors Obstructing Adaptive Capacity

Overall, availability of a higher number of dependents, low participation in community-based organizations (CBOs), a higher borrowing-lending ratio, and being a female-headed household were the most important limiting factors for adapting to climate change. In Aba Gerima, the top three indicators obstructing a household's adaptive capacity were the dependency ratio (9.8%), a low degree of participation in CBOs (9.0%), and fewer household members working outside the community (9.0%). In Guder, the main factors were the dependency ratio (9.8%), low participation in CBOs (9.1%), and a higher borrowing-lending ratio (8.9%), and in Dibatie, they were dependency ratio (9.4%), lack of other sources of income (9.0%), and a higher borrowing-lending ratio (8.6%) (Figure 6).

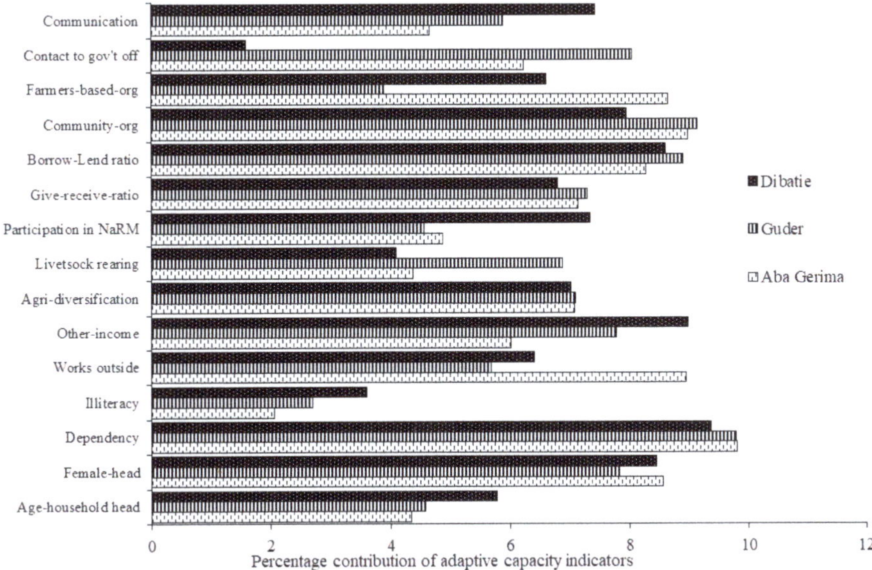

Figure 6. Obstacles and the degree of obstruction in the adaptive capacity of the three study areas.

4. Discussion

4.1. Livelihood Vulnerability Index

Communities in all three watersheds were vulnerable to the adverse effects of climate change variability because of high exposure, high sensitivity, and low adaptive capacity. In relative terms, Aba Gerima (midland) watershed was found to be the most vulnerable and Guder (highland) the least. The total score for Aba Gerima was 0.37, with 0.15, 0.14, and 0.08 contributed by adaptive capacity, sensitivity, and exposure, respectively. The corresponding scores for Dibatie were 0.35, 0.13, 0.13, and 0.09 and for Guder, they were 0.34, 0.14, 0.13, and 0.08. Interestingly, despite the very different contexts on each of the three communities, they showed relatively similar scores in the dimensions of the IPCC-LVI. Conversely, in their national scale study Ferede, Ayenew, Hanjra, and Hanjra [43] noted that the highland part of Ethiopia is more vulnerable compared to other agroecological zones. A possible reason for the two differences could be emanated from the scale of unit of analysis, whereby we used a watershed scale which helped us to gain the necessary details about livelihood vulnerability. Moreover, we also showed that adaptive capacity had relatively higher contribution for the overall livelihood vulnerability compared to its counterparts.

Aba Gerima was found to be the most vulnerable area, mainly because of its limited adaptive capacity and higher sensitivity. The sensitivity may be a result of the area's severe soil erosion status and the fact that households are not sufficiently participating in sustainable land management activities. In the same study area, References [46,59] found that the community's lower economic adaptive capacity affected its adoption of natural resource management practices. Ecological indicators such as soil depletion have also been shown to contribute to the vulnerability of village communities in Tanzania and South Africa [60]. In contrast, Guder was relatively less exposed to climate-related shocks. We found few drought episodes and an increasing rainfall trend in this area, so it is possible that communities in this watershed were less exposed to drought episodes and did not experience water shortages. An agroecological-based climate change vulnerability study in the Ethiopian highlands also found that highland agroecosystems are relatively less vulnerable to climate change shocks [31].

4.2. Exposure

Although the study area has had increasing rainfall, the trend was not significant at any of the sites. The watersheds studied have a unimodal rainfall pattern, and the rainy months occur mostly during the summer season (usually from mid-June to mid-September). Temperature did show a significant increasing trend at the study sites. Berihun, Tsunekawa, Haregeweyn, Meshesha, Adgo, Tsubo, Masunaga, Fenta, Sultan, and Yibeltal [47] also found that station data showed a significant upward trend in temperature but not rainfall in north-western Ethiopia. Moreover, Fenta et al. [61] also did not observe a monotonic trend in rainfall time series data in Ethiopia, and Teshome [62] found an increasing temperature trend in the Dembia District in the Upper Blue Nile basin. Furthermore, our findings are in line with other similar studies, which revealed observed trend changes in the Ethiopian highlands and elsewhere, with a mean annual temperature increase range between 0.028 °C and 1.65 °C from 1955 to 2016 [63,64]. In contrast, Samy et al. [65] reported a significant decreasing trend in rainfall in the southwestern part of the Upper Blue Nile basin. SPI results indicated that all the three watersheds were affected by drought episodes in 1984/1985, 1987/1988, 1992/1993, 2000/2001, 2010/2011, and 2015/2016, which is in line with other research results of drought experiences in Ethiopia [9,66]. Similarly, in their work in the northern part of Ethiopia, Kasie et al. [67] indicated that household livelihood systems were very much connected to the increment in their income, and this has been hurdled by the recurrent drought episodes (e.g., the 2015 El Niño).

Communities in Guder were found to be less exposed to climate-related shocks in five indicators (mean annual rainfall, mean annual maximum temperature, mean annual minimum temperature, number of shocks, and access to warning information). This could be partly attributed to the fact that the area has received more rainfall than the others (Figure 3). Moreover, despite experiencing drought episodes in 1984/1985 and 1992/1993, the watershed had a positive SPI and an increasing rainfall tendency in the study period. In Ethiopia, rainfall is crucial for predominantly rain-fed agriculture; hence, having more years of normal rainfall and more rainy days means better production and productivity for small-scale farmers [38]. In turn, having better agricultural productivity may enhance the adaptive capacity of the people. This result is in line that of with Simane, Zaitchik, and Foltz [31] who reported that highland agroecological systems were relatively less exposed to the effects of climate variability. Despite the higher amount of rainfall, Guder has had less exposure to soil erosion problems, which could be attributable to the unprecedented expansion of *A. decurrens* plantations across the watershed in the past decade [49]. The plantations might have helped the watershed by restoring the ecosystem of degraded hillsides and intercepting rainfall [68]. In addition, the contribution of agroforestry to rural livelihood resilience was noted in Quandt et al. [69], who showed the contribution of agroforestry in response to the impacts of climate-related hazards like drought and flooding. For example, during drought, many tree species still produced fruit for household consumption and sale, while staple food crops, such as maize, did not survive.

As compared to Guder and Aba Gerima watersheds, Dibatie watershed has been more exposed to climate-induced shocks and experienced more anomalies in the last 35 years. Despite that this watershed is located in a regional state known for its forest resources, it has been subject to overgrazing, deforestation, and poor farming practices, which has advanced desertification in the Nile basin (UNDP, 2017) [70]. Our findings are not in agreement with those of Deressa, Hassan, and Ringler [8], who revealed that households in Benishangul Gumuz had experienced fewer numbers of droughts and floods. A possible reason for this difference could be that they did not use station-level data to study vulnerability, whereas our SPI analysis did, and it showed that the area has experienced some types of meteorological and agricultural drought in the past 35 years. The SPI for Dibatie indicated relatively high climate variability, particularly dry spells and a significant increasing temperature trend ($p < 0.001$), which might have had negative impacts on crop production and livestock rearing. Late and untimely rainfall arrival in lowland agroecosystems and/or high temperatures during the crop development stage could cause a decline in yield and increase the community's exposure and ultimately contribute to increased vulnerability [14]. Rurinda et al. [71] noted that increased rainfall variability together with rising temperatures reduces soil moisture availability and increases the risk of crop failure.

4.3. Sensitivity

The overall sensitivity score, which included the health, food, and water components, was lower than the overall adaptive capacity score for all three watersheds. The food component was the primary contributor in all three cases. A study carried out in Myanmar similarly revealed the food component substantially contributed to the sensitivity dimension of farm households' climate change vulnerability [72]. A slightly higher sensitivity score was estimated for the Aba Gerima watershed as compared to Guder and Dibatie, which was mainly attributable to its sensitivity to the food and health components. Aba Gerima farmers generally had a lower level of food source diversification, which could trigger enhanced sensitivity. In addition, in Aba Gerima, land is being utilized to expand production of khat, which reduces the amount of land available for food production [48]. Moreover, no access to agricultural technology, a high degree of water abstraction from ground water aquifers for khat fields, and higher soil erosion severity also could have played roles in the sensitivity of households in this study site. Similarly, a higher livelihood sensitivity associated with landscape greenness, soil fertility, soil erosion, water availability, pasture availability, and plot condition was reported by Siraw, Adnew Degefu, and Bewket [32] in the Ethiopian highlands.

Dibatie scored better on crop diversification as compared to its counterparts. Crop diversification should make household livelihoods less sensitive to climate-related adverse effects. The Benishnagul Gumuz region, particularly Dibatie, besides other crops, has a notable tradition of ground nut cultivation [73], which is well known as a drought-resistant crop [74]. Dibatie, however, also had a relatively higher contribution from the soil erosion and seed-saving indicators. Ebabu et al. [75] and Abeje, Tsunekawa, Adgo, Haregeweyn, Nigussie, Ayalew, Elias, Molla, and Berihun [48] also reported that this watershed had higher land degradation problems as compared to Aba Gerima and Guder. Guder, representing the highland area, had lower crop diversification and, consequently, is relatively more sensitive to the effects of climate-related shocks. Saving seed for the next growing season is also not often practiced in Aba Gerima and Guder watersheds. This could possibly be related to their shift to the more remunerative cash-based khat plantation and A. decurrens monocropping [48].

Health-related problems made a greater contribution to the sensitivity dimension in Dibatie and Aba Gerima as compared to Guder. Community discussion participants at both of these sites reported that government health services in their proximity do not work properly, and they usually use private health services in the nearest town, which usually costs more than the government services. More household members were reported to miss school due to health problems in Dibatie, which is most likely related to their greater experience with chronic illness. Health problems could result in a shortage of family labour for operating agricultural lands. A similar study in Ghana revealed that the sensitivity of farmers to the impacts of climate variability was partly contributed from their higher exposure, especially in households who do not own enough livelihood capital to support agricultural labour [24].

The water component comprised consistent water access, water resource conflict, and average time to fetch water. Guder was relatively better off than the other two watersheds, and its lower sensitivity score might be associated with recent improvements in its broader ecosystem, and the presence of A. decurrens plantations and other sustainable land management activities [48,49,68]. Excessive extraction of water for irrigating khat farmlands could serve as a potential point of conflict among downstream and upstream farmers. Farmers in Aba Gerima reported that they had no consistent access to water sources, and the time it took them to fetch water was longer as compared farmers in the other watersheds. Moreover, the distance from the river to their farms was somewhat longer, which largely limited their ability to irrigate their farms. Whereas irrigation directly minimizes the impacts of climatic stresses such as droughts, farmers at all three sites are at increased risk of water availability due their dependence on natural water sources. A similar study noted a more pronounced livelihood vulnerability to drought in rural Iran, and it was mainly associated with access to water sources [76].

4.4. Adaptive Capacity

Adaptive capacity plays an essential role in responding to the adverse impacts of climate change/variability, reducing livelihood vulnerability, and helping people to achieve sustainable livelihoods [77]. Communities in all three watersheds were significantly less able to adapt to the effects of climate change. A household's endowment of essential livelihood assets contributes to its adaptive capacity, which ultimately determines its livelihood vulnerability to certain negative consequences. Overall, communities in Aba Gerima were found to be more vulnerable, in a large part because of their low adaptive capacity, which was mainly attributable to limitations in the livelihood strategies and sociodemographic components. These components were, in turn, made up of elements, such as a relatively higher rate of illiteracy, higher age, and lower rate of engagement in livestock rearing. Possible reasons for lower livestock production may be a reduced availability of feed and restrictions on grazing in some parts of the watershed [78]. Their lower engagement in livestock rearing may also be associated with their larger family size and their relative affluence due to their proximity to Bahir Dar city and their greater engagement in the production of cash crops (i.e., khat). In a study conducted in Kenya, livestock diversification was shown to be an important indicator in terms of lowering sensitivity to multiple stresses related to climate change [23]. Even though the Dibatie (lowland) watershed exhibits a slightly better adaptive capacity as compared to the other two sites, in part because of its better social networking component score, households had little contact with local government offices, which ultimately constrained their institutional adaptive capacity to climate change vulnerability. Government offices are responsible for services related to response to climate-related shocks, such as issuing warnings, training residents about climate-smart technologies, providing information about markets, and delivering inputs. Hence, weaker contact with government offices may negatively affect the provision of these services. In addition, as we learned from the community discussions, most of the interviewed households are not entitled to own land, and thus cultivate land through informal renting arrangements. Land tenure insecurity has been cited as one of the constraints of production and productivity for households in Benishangul Gumuz [79,80]. In addition, effective, timely, and appropriate delivery of climate related warning information is of greater advantage at lower levels of administration (e.g., district level) planning and decision making for adaptation planning in the Ethiopian case [81]. In a similar study done along the Nile basin, access to climate related information was indicated as a significant driver for household adaptation to the adverse effects of climate variability/change [82].

A lower extent of livelihood diversification contributed to lower adaptive capacity in the Guder and Aba Gerima watersheds, possibly because the wider coverage of *A. decurrens* and khat plantations in these areas reduced the amount of land that could have been used for the production of food crops [48]. Climate variability in the form of drought could have more impact on communities with less diversified livelihood strategies [83]. A study of weather shocks in Ethiopia indicated that off-farm livelihood diversification enhanced the capacity households to cope with climate-related shocks [38]. A similar study in Kenya revealed that, as part of adaptive capacity, short- and long-term climate change adaptation should be supported by CBOs and enhanced social networking [19]. Communities in Guder showed a relatively high level of vulnerability in terms of participating in farmer-based organizations and a lack of communication devices to receive climate-related information. Rural community awareness on the causes and impacts of climate calamities on their lives and livelihoods can be improved by education. In addition, the more educated community members are, it is highly likely that they will adopt climate smart technologies [11]. Guder had a relatively higher education level, which indicates communities in this area should be more likely to adapt to the effects of climate change. A similar study carried out by Deressa et al. [84] indicated that level of education plays an essential role in terms of choice of climate adaptation strategies and, hence, affects livelihoods. Likewise, low literacy was shown to contribute to higher vulnerability of households in Ghana [24].

4.5. Obstruction Factors of Communities' Adaptive Capacity

A larger number of dependents, low degree of participation in community based organizations (CBOs) and farmer-based organizations (FBOs), and lack of contact with the local government were major obstacles limiting community adaptive capacity. Moreover, being a female-headed household, not having other sources of income, and having a shortage of family labour were also major obstacles in all studied watersheds. As climate change/variability puts households under extra pressure, households with a shortage of labour could be more vulnerable by limiting their ability to diversify their income sources or go outside the community for employment. The availability and quality of human capital, including active working labour, has been shown to affect household adaptive capacity to climate change [85]. A lower degree of participation in community- and farmer-based organizations was a substantial obstacle, primarily because these organizations provide important services to the communities. For example, CBOs include informal social networking schemes where households help each other during periods of social, as well as economic, problems. Social networking essentially reduces a community's vulnerability to the adverse effects of climate change [86]. Female-headed households generally have a marginal role in all walks of life and are denied many opportunities, which makes them more vulnerable to the effects of climate change [87]. Female-headed households may not have access to social services and information because of socially constructed problems related to inequality, and this could limit their capacity to mobilize available resources to adapt to the negative effects of climate-related shocks them [66,84]. Similarly, a gender specific study in Ghana revealed that female headed households, due to their low sociodemographic profile, low social network, and lack of access to water and food, were indicated as vulnerable to the impacts of climate change/variability compared to their counterparts[88]. Moreover, in Central Nepal, limited access to communication and reliable information on climate related hazards and low participation in local based organizations made female head households vulnerable [89].

5. Conclusions and Implications

As a consequence of the great heterogeneity in socioeconomic capacity and livelihoods across different communities, similar exposure to climate variability and climate change poses differential impacts on different groups of people at different scales. Rural households in the Upper Blue Nile basin rely on rain-fed agriculture for their livelihoods. This sector is vulnerable and sensitive to the risks and impacts of climate-related shocks/hazards, particularly considering the accumulated negative effects of past droughts. This study aimed to analyze the vulnerability of smallholder farmer livelihoods to climate change/variability in three watersheds of the Upper Blue Nile basin in Ethiopia. We developed and applied a multi-indicator and quantitative methodology that helped show the relative livelihood vulnerability differences among agroecologically different watersheds as a function of exposure, sensitivity, and adaptive capacity. To compute more precise indices, we utilized Shannon's entropy evaluation computation to assign objective weights to the proxy indicators that made up the composite index.

Temperature showed a significant increasing trend over the 35-year study period. Extreme drought episodes were more pronounced in Guder (the highland). In contrast, Aba Gerima (midland) and Dibatie (lowland) experienced more frequent drought episodes. Drought episodes were observed in 1984/1985, 1987/1988, 1992/1993, 2000/2001, 2010/2011, and 2015/2016 in all study watersheds, which is consistent with the national historical record.

In terms of the overall IPCC-LVI score, Aba Gerima was found to be relatively more vulnerable, with a score of 0.37. With similar exposure to climate variability, communities' livelihood vulnerability was mainly attributed to their low adaptive capacity and higher sensitivity to proxy indicators. Guder had the lowest IPCC-LVI score (0.34). Smallholder farmers' adaptive capacity contributed the most, relative to other dimensions, in terms of explaining livelihood vulnerability, with scores of 0.15, 0.14, and 0.13 in Aba Gerima, Guder, and Dibatie, respectively. Dibatie watershed, representing the lowland agroecological setting, had the greatest contribution from the exposure dimension (climate-related indicators) as compared to the other two sites. Results indicated that communities with more diversified livelihood strategies are less vulnerable to the impacts of climate change. The obstruction degree analysis showed that some indicators were turned out to have constrained the adaptive capacity of communities to climate variability effects. These indicators include, but are not limited to the availability of a higher number of dependents and being a female-headed household, low participation in CBOs, lack of alternative income sources, and less engagement in community borrowing and lending cultural practices. In this study, by adopting the Shannon entropy weightage procedure, we tried to address the challenges of subjective measurement of livelihood vulnerability to better inform interventions actions.

Given the predominant rain-fed agricultural system in the Upper Blue Nile basin, enhancing adaptive capacity and mitigating sensitivity should be prioritized to help communities adapt to the adverse impacts of climate change. For example, empowerment of female-headed households (e.g., by improving their livelihood bases, increasing their social role, etc.) will enhance their social position, increase their opportunities, and help reduce their vulnerability. Diversifying livelihood options and food sources will also help reduce sensitivity.

This study has practical implications for agroecological heterogeneous policy development and program design in sustainable livelihood development and climate change adaptation programs. More specifically, it will have practical implications by filling the gap between the broader theoretical aspect of livelihood vulnerability to climate change to the day to day decision making at lower administration and planning scales. In addition, providing sound scientific evidence employing objective weighting and appropriate aggregation procedures will not only further improve livelihood vulnerability analysis methods, but also has practical implications for providing better information in the prioritization of countermeasures to address factors that contribute to vulnerability.

Supplementary Materials: The following are available online at http://www.mdpi.com/2071-1050/11/22/6302/s1.

Author Contributions: Conceptualization, M.T.A.; Data curation, M.T.A.; Formal analysis, M.T.A.; Funding acquisition, A.T., E.A., and N.H.; Investigation, M.T.A.; Methodology, M.T.A. and A.Q; Project administration, A.T., E.A., N.H. and Z.N.; Resources, A.T. and N.H.; Supervision, A.T., E.A., N.H., Z.N., Z.A., A.E., and D.B.; Validation, A.T.; Writing—original draft, M.T.A.; Writing—Review & editing, A.T., E.A., N.H., Z.N., A.Q., M.L.B., M.T.,T.M., and Z.A.

Funding: This research was funded by the Science and Technology Research Partnership for Sustainable Development (SATREPS)—Development of a Next-Generation Sustainable Land Management (SLM) Framework to Combat Desertification project, Grant Number JPMJSA1601, Japan Science and Technology Agency (JST)/Japan International Cooperation Agency (JICA).

Acknowledgments: The authors are grateful to all respondent farmers for their willingness to provide data and to Yoshimi Katsumata, Nigus Tadesse, Bilew Getenet, and Anteneh Wubet for their field assistance.

Conflicts of Interest: The authors declare no potential conflict of interest.

References

1. Ellis, F. Household strategies and rural livelihood diversification. *J. Dev. Stud.* **1998**, *35*, 1–38. [CrossRef]
2. Lowder, S.K.; Skoet, J.; Raney, T. The number, size, and distribution of farms, smallholder farms, and family farms worldwide. *World Dev.* **2016**, *87*, 16–29. [CrossRef]
3. Griek, L.; Penikett, J.; Hougee, E. Bitter Harvest: Child Labour in the Cocoa Supply Chain. Sustainalytics. 2010. Available online: http://www.cocoainitiative.org/wp-content/uploads/2017/09/Bitter-Harvest-Child-Labour-in-the-Cocoa-Supply-Chain (accessed on 21 May 2019).
4. UNEP. *Smallholders, Food Security, and the Environment*; International Fund for Agricultural Development (IFAD) and the United Nations Environment Programme (UNEP): Rome, Italy, 2013; pp. 1–54.
5. IPCC. *Climate Change 2007: Impacts, Adaptation and Vulnerability*; IPCC: Genebra, Switzerland, 2007; pp. 1–23.
6. Morton, J.F. The impact of climate change on smallholder and subsistence agriculture. *Proc. Natl. Acad. Sci. USA* **2007**, *104*, 19680–19685. [CrossRef] [PubMed]
7. Cooper, S.J.; Wheeler, T. Rural household vulnerability to climate risk in Uganda. *Reg. Environ. Chang.* **2017**, *17*, 649–663. [CrossRef]

8. Deressa, T.; Hassan, R.M.; Ringler, C. *Measuring Ethiopian Farmers' Vulnerability to Climate Change across Regional States*; International Food Policy Research Institute: Washington, DC, USA, 2008.

9. Mohammed, Y.; Yimer, F.; Tadesse, M.; Tesfaye, K. Meteorological drought assessment in north east highlands of Ethiopia. *Int. J. Clim. Chang. Strateg. Manag.* **2018**, *10*, 142–160. [CrossRef]

10. Lobell, D.B.; Burke, M.B.; Tebaldi, C.; Mastrandrea, M.D.; Falcon, W.P.; Naylor, R.L. Prioritizing climate change adaptation needs for food security in 2030. *Science* **2008**, *319*, 607–610. [CrossRef]

11. IPCC. *Climate Change 2014Synthesis Report Summary for Policymakers*; IPCC: Geneva, Switzerland, 2014; pp. 1–32.

12. De Haan, L.; Zoomers, A. Exploring the frontier of livelihoods research. *Dev. Chang.* **2005**, *36*, 27–47. [CrossRef]

13. Adger, W.N. Vulnerability. *Glob. Environ. Chang.* **2006**, *16*, 268–281. [CrossRef]

14. Parry, M.; Parry, M.L.; Canziani, O.; Palutikof, J.; Van der Linden, P.; Hanson, C. Climate Change 2007-Impacts, Adaptation and Vulnerability: Working Group II Contribution to the Fourth Assessment Report of the IPCC. Cambridge University Press: Cambridge, UK, 2007; Volume 4.

15. Gornall, J.; Betts, R.; Burke, E.; Clark, R.; Camp, J.; Willett, K.; Wiltshire, A. Implications of climate change for agricultural productivity in the early twenty-first century. *Philos. Trans. R. Soc. B* **2010**, *365*, 2973–2989. [CrossRef]

16. Heltberg, R.; Siegel, P.B.; Jorgensen, S.L. Addressing human vulnerability to climate change: Toward a 'no-regrets' approach. *Glob. Environ. Chang.* **2009**, *19*, 89–99. [CrossRef]

17. Hertel, T.W.; Rosch, S.D. *Climate Change, Agriculture and Poverty*; The World Bank: Washington, DC, USA, 2010.

18. Snover, A.K.; Whitely Binder, L.C.; Lopez, J.; Willmott, E.; Kay, J.; Howell, D.; Simmonds, J. *Preparing for Climate Change: A Guidebook for Local, Regional, and State Governments*; The Climate Impacts Group, King County: Washington, DC, USA, 2007.

19. Dzoga, M.; Simatele, D.; Munga, C. Assessment of ecological vulnerability to climate variability on coastal fishing communities: A study of Ungwana Bay and Lower Tana Estuary, Kenya. *Ocean Coast. Manag.* **2018**, *163*, 437–444. [CrossRef]

20. Peng, L.; Xu, D.; Wang, X. Vulnerability of rural household livelihood to climate variability and adaptive strategies in landslide-threatened western mountainous regions of the Three Gorges Reservoir Area, China. *Clim. Dev.* **2018**, 1–16. [CrossRef]

21. Quandt, A. Measuring livelihood resilience: The household livelihood resilience approach (HLRA). *World Dev.* **2018**, *107*, 253–263. [CrossRef]

22. Tjoe, Y. Measuring the livelihood vulnerability index of a dry region in Indonesia: A case study of three subsistence communities in West Timor. *World J. Sci. Technol. Sustain. Dev.* **2016**, *13*, 250–274. [CrossRef]

23. Unks, R.R.; King, E.G.; Nelson, D.R.; Wachira, N.P.; German, L.A. Constraints, multiple stressors, and stratified adaptation: Pastoralist livelihood vulnerability in a semi-arid wildlife conservation context in Central Kenya. *Glob. Environ. Chang.* **2019**, *54*, 124–134. [CrossRef]

24. Williams, P.A.; Crespo, O.; Abu, M. Assessing vulnerability of horticultural smallholders' to climate variability in Ghana: Applying the livelihood vulnerability approach. *Environ. Dev. Sustain.* **2018**, *19*, 1–22. [CrossRef]

25. Zhang, Q.; Zhao, X.; Tang, H. Vulnerability of communities to climate change: Application of the livelihood vulnerability index to an environmentally sensitive region of China. *Clim. Dev.* **2018**, *11*, 1–18. [CrossRef]

26. Hahn, M.B.; Riederer, A.M.; Foster, S.O. The Livelihood Vulnerability Index: A pragmatic approach to assessing risks from climate variability and change—A case study in Mozambique. *Glob. Environ. Chang.* **2009**, *19*, 74–88. [CrossRef]

27. Chambers, R.; Conway, G. *Sustainable Rural Livelihoods: Practical Concepts for the 21st Century*; Institute of Development Studies (UK): Brighton, UK, 1992.

28. Adu, D.T.; Kuwornu, J.K.; Anim-Somuah, H.; Sasaki, N. Application of livelihood vulnerability index in assessing smallholder maize farming households' vulnerability to climate change in Brong-Ahafo region of Ghana. *Kasetsart J. Soc. Sci.* **2018**, *39*, 22–32. [CrossRef]

29. Gebrehiwot, T.; van der Veen, A. Climate change vulnerability in Ethiopia: Disaggregation of Tigray Region. *J. East. Afr. Stud.* **2013**, *7*, 607–629. [CrossRef]

30. Huong, N.T.L.; Yao, S.; Fahad, S. Assessing household livelihood vulnerability to climate change: The case of Northwest Vietnam. *Hum. Ecol. Risk Assess.* **2018**, *25*, 1–19. [CrossRef]

31. Simane, B.; Zaitchik, B.F.; Foltz, J.D. Agroecosystem specific climate vulnerability analysis: Application of the livelihood vulnerability index to a tropical highland region. *Mitig. Adapt. Strateg. Glob. Chang.* **2016**, *21*, 39–65. [CrossRef] [PubMed]

32. Siraw, Z.; Adnew Degefu, M.; Bewket, W. The role of community-based watershed development in reducing farmers' vulnerability to climate change and variability in the northwestern highlands of Ethiopia. *Local Environ.* **2018**, *23*, 1190–1206. [CrossRef]

33. Sujakhu, N.M.; Ranjitkar, S.; Niraula, R.R.; Salim, M.A.; Nizami, A.; Schmidt-Vogt, D.; Xu, J. Determinants of livelihood vulnerability in farming communities in two sites in the Asian Highlands. *Water Int.* **2018**, *43*, 165–182. [CrossRef]

34. Conway, D.; Schipper, E.L.F. Adaptation to climate change in Africa: Challenges and opportunities identified from Ethiopia. *Glob. Environ. Chang.* **2011**, *21*, 227–237. [CrossRef]

35. USAID, Ethiopia. Climate change risk profile: Climate links. 2016. Available online: https://www.climatelinks.org/resources/climate-change-risk-profile-ethiopia (accessed on 21 May 2019).

36. Thornton, P.K.; Jones, P.G.; Owiyo, T.; Kruska, R.L.; Herrero, M.; Orindi, V.; Bhadwal, S.; Kristjanson, P.; Notenbaert, A.; Bekele, N. Climate change and poverty in Africa: Mapping hotspots of vulnerability. *Afr. J. Agric. Resour. Econ.* **2008**, *2*, 24–s44.

37. Spielman, D.J.; Byerlee, D.; Alemu, D.; Kelemework, D. Policies to promote cereal intensification in Ethiopia: The search for appropriate public and private roles. *Food Policy* **2010**, *35*, 185–194. [CrossRef]

38. Gao, J.; Mills, B.F. Weather shocks, coping strategies, and consumption dynamics in rural Ethiopia. *World Dev.* **2018**, *101*, 268–283. [CrossRef]

39. McSweeney, C.; New, M.; Lizcano, G.; Lu, X. The UNDP Climate Change Country Profiles: Improving the accessibility of observed and projected climate information for studies of climate change in developing countries. *Bull. Am. Meteorol. Soc.* **2010**, *91*, 157–166. [CrossRef]

40. FDRE. *Ethiopia's Climate Resilient Green Economy: Green Economy Strategy*; FDRE: Addis Ababa, Ethiopia, 2011.

41. Salvati, L.; Carlucci, M. A composite index of sustainable development at the local scale: Italy as a case study. *Ecol. Indic.* **2014**, *43*, 162–171. [CrossRef]

42. Williams, P.A.; Crespo, O.; Abu, M.; Simpson, N.P. A systematic review of how vulnerability of smallholder agricultural systems to changing climate is assessed in Africa. *Environ. Res. Lett.* **2018**, *13*, 103004. [CrossRef]

43. Ferede, T.; Ayenew, A.B.; Hanjra, M.A.; Hanjra, M. Agroecology matters: Impacts of climate change on agriculture and its implications for food security in Ethiopia. *Glob. Food Secur.* **2013**, *3*, 71–112.

44. Huang, X.; Huang, X.; He, Y.; Yang, X. Assessment of livelihood vulnerability of land-lost farmers in urban fringes: A case study of Xi'an, China. *Habitat Int.* **2017**, *59*, 1–9. [CrossRef]

45. Hurni, H.; Berhe, W.; Chadhokar, P.; Daniel, D.; Gete, Z.; Grunder, M.; Kassaye, G. Soil and water conservation in Ethiopia: Guidelines for development agents. In *Centre for Development and Environment (CDE)*; Bern Open Publishing (BOP): Bern, Switzerland, 2016.

46. Nigussie, Z.; Tsunekawa, A.; Haregeweyn, N.; Adgo, E.; Nohmi, M.; Tsubo, M.; Aklog, D.; Meshesha, D.T.; Abele, S. Farmers' perception about soil erosion in Ethiopia. *Land Degrad. Dev.* **2017**, *28*, 401–411. [CrossRef]

47. Berihun, M.L.; Tsunekawa, A.; Haregeweyn, N.; Meshesha, D.T.; Adgo, E.; Tsubo, M.; Masunaga, T.; Fenta, A.A.; Sultan, D.; Yibeltal, M. Hydrological responses to land use/land cover change and climate variability in contrasting agro-ecological environments of the Upper Blue Nile basin, Ethiopia. *Sci. Total Environ.* **2019**, *689*, 347–365. [CrossRef]

48. Abeje, M.T.; Tsunekawa, A.; Adgo, E.; Haregeweyn, N.; Nigussie, Z.; Ayalew, Z.; Elias, A.; Molla, D.; Berihun, D. Exploring Drivers of Livelihood Diversification and Its Effect on Adoption of Sustainable Land Management Practices in the Upper Blue Nile Basin, Ethiopia. *Sustainability* **2019**, *11*, 1–23.

49. Nigussie, Z.; Tsunekawa, A.; Haregeweyn, N.; Adgo, E.; Nohmi, M.; Tsubo, M.; Aklog, D.; Meshesha, D.T.; Abele, S. Factors affecting small-scale farmers' land allocation and tree density decisions in an acacia decurrens-based taungya system in Fagita Lekoma District, North-Western Ethiopia. *Small-Scale For.* **2017**, *16*, 219–233. [CrossRef]

50. Cochran, W.G. *Sampling Techniques*; John Wiley & Sons: New York, NY, USA, 1977.

51. Abdi, H. The Kendall rank correlation coefficient. In *Encyclopedia of Measurement and Statistics*; Sage: Thousand Oaks, CA, USA, 2007; pp. 508–510.

52. Thomas, B.; McKee, N.J.D. The Relationship of Drought Frequency and Duration to Time Scale. In Proceedings of the 8th conference on Applied Climatology, Boston, MA, USA, January 1993; pp. 179–183.

53. Tigkas, D.; Vangelis, H.; Tsakiris, G. DrinC: A software for drought analysis based on drought indices. *Earth Sci. Inf.* **2015**, *8*, 697–709. [CrossRef]

54. DFID, U. *Sustainable Livelihoods Guidance Sheets*; DFID: London, UK, 1999.

55. Beccari, B. A comparative analysis of disaster risk, vulnerability and resilience composite indicators. *PLoS Curr.* **2016**, *8*. [CrossRef]

56. Miller, H.J.; Witlox, F.; Tribby, C.P. Developing context-sensitive livability indicators for transportation planning: A measurement framework. *J. Transp. Geogr.* **2013**, *26*, 51–64. [CrossRef]

57. Yang, W.; Xu, K.; Lian, J.; Ma, C.; Bin, L. Integrated flood vulnerability assessment approach based on TOPSIS and Shannon entropy methods. *Ecol. Indic.* **2018**, *89*, 269–280. [CrossRef]

58. Shen, P.; You, H.; Wu, C. A study on obstacle diagnosis and support system of sustainable urban land use of Huangshi city in Hubei Province. In Proceedings of the 18th international symposium on advancement of construction management and real estate, Berlin, Germany, May 2014; pp. 139–146.

59. Nigussie, Z.; Tsunekawa, A.; Haregeweyn, N.; Adgo, E.; Nohmi, M.; Tsubo, M.; Aklog, D.; Meshesha, D.T.; Abele, S. Factors influencing small-scale farmers' adoption of sustainable land management technologies in north-western Ethiopia. *Land Use Policy* **2017**, *67*, 57–64. [CrossRef]

60. Grothmann, T.; Petzold, M.; Ndaki, P.; Kakembo, V.; Siebenhüner, B.; Kleyer, M.; Yanda, P.; Ndou, N. Vulnerability assessment in african villages under conditions of land use and climate change: Case studies from Mkomazi and Keiskamma. *Sustainability* **2017**, *9*, 976. [CrossRef]

61. Fenta, A.A.; Yasuda, H.; Shimizu, K.; Haregeweyn, N.; Kawai, T.; Sultan, D.; Ebabu, K.; Belay, A.S. Spatial distribution and temporal trends of rainfall and erosivity in the Eastern Africa region. *Hydrol. Process.* **2017**, *31*, 4555–4567. [CrossRef]

62. Teshome, M. Rural households' agricultural land vulnerability to climate change in Dembia woreda, Northwest Ethiopia. *Environ. Syst. Res.* **2016**, *5*, 14. [CrossRef]

63. Abebe, G. Long-term climate data description in Ethiopia. *Data Brief* **2017**, *14*, 371–392. [CrossRef]

64. Alemayehu, A.; Bewket, W. Local spatiotemporal variability and trends in rainfall and temperature in the central highlands of Ethiopia. *Geogr. Ann.* **2017**, *99*, 85–101. [CrossRef]

65. Samy, A.; Ibrahim, M.G.; Mahmod, W.E.; Fujii, M.; Eltawil, A.; Daoud, W. Statistical Assessment of Rainfall Characteristics in Upper Blue Nile Basin over the Period from 1953 to 2014. *Water* **2019**, *11*, 468. [CrossRef]

66. Deressa, T.T.; Hassan, R.M.; Ringler, C. Perception of and adaptation to climate change by farmers in the Nile basin of Ethiopia. *J. Agric. Sci.* **2011**, *149*, 23–31. [CrossRef]

67. Kasie, T.A.; Adgo, E.; Botella, A.G.; García, I.G. Measuring resilience properties of household livelihoods and food security outcomes in the risky environments of Ethiopia. *Rev. Iberoam. Estud. Desarro.* **2018**, *7*, 52–80. [CrossRef]

68. Berihun, M.L.; Tsunekawa, A.; Haregeweyn, N.; Meshesha, D.T.; Adgo, E.; Tsubo, M.; Masunaga, T.; Fenta, A.A.; Sultan, D.; Yibeltal, M. Exploring land use/land cover changes, drivers and their implications in contrasting agro-ecological environments of Ethiopia. *Land Use Policy* **2019**, *87*, 104052. [CrossRef]

69. Quandt, A.K.; Neufeldt, H.; McCabe, J.T. The role of agroforestry in building livelihood resilience to floods and drought in semiarid Kenya. *Ecol. Soc.* **2017**, *22*. [CrossRef]

70. UNDP. *Benishangul-Gumuz Regional State REDD+Design: A Regional Model for REDD+ under the UNFCCC Warsaw Framework*; Center for International Forestry Research (CIFOR): Bogor, Indonesia, 2017; pp. 1–12.

71. Rurinda, J.; Mapfumo, P.; Van Wijk, M.; Mtambanengwe, F.; Rufino, M.C.; Chikowo, R.; Giller, K.E. Sources of vulnerability to a variable and changing climate among smallholder households in Zimbabwe: A participatory analysis. *Clim. Risk Manag.* **2014**, *3*, 65–78. [CrossRef]

72. Oo, A.T.; Van Huylenbroeck, G.; Speelman, S. Assessment of climate change vulnerability of farm households in Pyapon District, a delta region in Myanmar. *Int. J. Disaster Risk Reduct.* **2018**, *28*, 10–21. [CrossRef]

73. Nega, F.; Mausch, K.; Rao, K.; Legesse, G. *Scoping Study on Current Situation and Future Market Outlook of Groundnut in Ethiopia, Socioeconomics*; Discussion Paper Series 38; International Crops Research Institute for the Semi-Arid Tropics (ICRISAT): Andhra Pradesh, India, 2015.

74. Reddy, T.; Reddy, V.; Anbumozhi, V. Physiological responses of groundnut (Arachis hypogea L.) to drought stress and its amelioration: A critical review. *Plant Growth Regul.* **2003**, *41*, 75–88. [CrossRef]

75. Ebabu, K.; Tsunekawa, A.; Haregeweyn, N.; Adgo, E.; Meshesha, D.T.; Aklog, D.; Masunaga, T.; Tsubo, M.; Sultan, D.; Fenta, A.A. Analyzing the variability of sediment yield: A case study from paired watersheds in the Upper Blue Nile basin, Ethiopia. *Geomorphology* **2018**, *303*, 446–455. [CrossRef]

76. Keshavarz, M.; Maleksaeidi, H.; Karami, E. Livelihood vulnerability to drought: A case of rural Iran. *Int. J. Disaster Risk Reduct.* **2017**, *21*, 223–230. [CrossRef]

77. Jamshed, A.; Rana, I.A.; Mirza, U.M.; Birkmann, J. Assessing relationship between vulnerability and capacity: An empirical study on rural flooding in Pakistan. *Int. J. Disaster Risk Reduct.* **2019**, *36*, 101109. [CrossRef]

78. Nigussie, Z.; Tsunekawa, A.; Haregeweyn, N.; Adgo, E.; Cochrane, L.; Floquet, A.; Abele, S. Applying Ostrom's institutional analysis and development framework to soil and water conservation activities in north-western Ethiopia. *Land Use Policy* **2018**, *71*, 1–10. [CrossRef]

79. Lavers, T. Responding to land-based conflict in Ethiopia: The land rights of ethnic minorities under federalism. *Afr. Aff.* **2018**, *117*, 462–484. [CrossRef]

80. Teklemariam, D.; Azadi, H.; Nyssen, J.; Haile, M.; Witlox, F. How sustainable is transnational farmland acquisition in Ethiopia? Lessons learned from the Benishangul-Gumuz Region. *Sustainability* **2016**, *8*, 213. [CrossRef]

81. Kassie, B.T.; Hengsdijk, H.; Rötter, R.; Kahiluoto, H.; Asseng, S.; Van Ittersum, M. Adapting to climate variability and change: Experiences from cereal-based farming in the Central Rift and Kobo Valleys, Ethiopia. *Environ. Manag.* **2013**, *52*, 1115–1131. [CrossRef] [PubMed]

82. Di Falco, S.; Veronesi, M.; Yesuf, M. Does adaptation to climate change provide food security? A micro-perspective from Ethiopia. *Am. J. Agric. Econ.* **2011**, *93*, 829–846. [CrossRef]

83. Call, M.; Gray, C.; Jagger, P. Smallholder responses to climate anomalies in rural Uganda. *World Dev.* **2019**, *115*, 132–144. [CrossRef] [PubMed]

84. Deressa, T.T.; Hassan, R.M.; Ringler, C.; Alemu, T.; Yesuf, M. Determinants of farmers' choice of adaptation methods to climate change in the Nile Basin of Ethiopia. *Glob. Environ. Chang.* **2009**, *19*, 248–255. [CrossRef]

85. Burton, I.; Soussan, J.; Hammill, A. *Livelihoods and Climate Change: Combining Disaster Risk Reduction, Natural Resource Management and Climate Change Adaptation in a New Approach to the Reduction of Vulnerability and Poverty*; International Inst. for Sustainable Development: Winnipeg, Canada, 2003.

86. Adger, W.N. Social vulnerability to climate change and extremes in coastal Vietnam. *World Dev.* **1999**, *27*, 249–269. [CrossRef]

87. Nabikolo, D.; Bashaasha, B.; Mangheni, M.; Majaliwa, J. Determinants of climate change adaptation among male and female headed farm households in eastern Uganda. *Afr. Crop Sci. J.* **2012**, *20*, 203–212.

88. Alhassan, S.I.; Kuwornu, J.K.; Osei-Asare, Y.B. Gender dimension of vulnerability to climate change and variability: Empirical evidence of smallholder farming households in Ghana. *Int. J. Clim. Chang. Strateg. Manag.* **2019**, *11*, 195–214. [CrossRef]

89. Sujakhu, N.M.; Ranjitkar, S.; He, J.; Schmidt-Vogt, D.; Su, Y.; Xu, J. Assessing the Livelihood Vulnerability of Rural Indigenous Households to Climate Changes in Central Nepal, Himalaya. *Sustainability* **2019**, *11*, 2977. [CrossRef]

Article

Exploring Drivers of Livelihood Diversification and Its Effect on Adoption of Sustainable Land Management Practices in the Upper Blue Nile Basin, Ethiopia

Misganaw Teshager Abeje [1,2,*], Atsushi Tsunekawa [3], Enyew Adgo [4], Nigussie Haregeweyn [5], Zerihun Nigussie [3,4], Zemen Ayalew [4], Asres Elias [6], Dessalegn Molla [7] and Daregot Berihun [8]

[1]   The United Graduate School of Agricultural Sciences, Tottori University, 4-101 Koyama-Minami, Tottori 680-8553, Japan

[2]   Institute of Disaster Risk Management and Food Security Studies, Bahir Dar University, P.O. Box 79, Bahir Dar 6000, Ethiopia

[3]   Arid Land Research Center, Tottori University, 1390 Hamasaka, Tottori 680-0001, Japan; tsunekawa@tottori-u.ac.jp (A.T.); zeriye@gmail.com (Z.N.)

[4]   College of Agriculture and Environmental Sciences, Bahir Dar University, P.O. Box 79, Bahir Dar 6000, Ethiopia; enyewadgo@gmail.com (E.A.); zayalew@gmail.com (Z.A.)

[5]   International Platform for Dryland Research and Education, Tottori University, 1390 Hamasaka, Tottori 680-0001, Japan; nigussie_haregeweyn@yahoo.com

[6]   Faculty of Agriculture, Tottori University, 4-101 Koyama-Minami, Tottori 680-8550, Japan; asres97@yahoo.com

[7]   College of Agriculture, Woldia University, P.O. Box 400, Woldia 7220, Ethiopia; dessmoll@gmail.com

[8]   College of Business and Economics, Bahir Dar University, P.O. Box 79, Bahir Dar 6000, Ethiopia; daregot21@gmail.com

*   Correspondence: D17A4005M@edu.tottori-u.ac.jp or tmisganaw16@gmail.com

Received: 25 April 2019; Accepted: 16 May 2019; Published: 27 May 2019

Abstract: Land degradation poses a major threat to agricultural production and food security in Ethiopia, and sustainable land management (SLM) is key in dealing with its adverse impacts. This paper examines the covariates that shape rural livelihood diversification and examines their effects on the intensity of adoption of SLM practices. Household-level data were collected in 2017 from 270 households in three drought-prone watersheds located in northwestern Ethiopia. We used the Herfindahl–Simpson diversity index to explore the extent of livelihood diversification. A stochastic dominance ordering was also employed to identify remunerative livelihood activities. A multivariate probit model was employed to estimate the probability of choosing simultaneous livelihood strategies, and an ordered probit model was estimated to examine the effect of livelihood diversification on the adoption intensity of SLM practices. In addition to mixed cropping and livestock production, the production of emerging cash crops (e.g., *Acacia decurrens* for charcoal, and khat) dominated the overall income generation of the majority of farmers. Stress/shock experience, extent of agricultural intensification, and agro-ecology significantly affected the probability of choosing certain livelihood strategies. Livelihood diversification at the household level was significantly associated with the dependency ratio, market distance, credit access, extension services, membership in community organizations, level of income, and livestock ownership. A greater extent of livelihood diversification had a significant negative effect on adopting a greater number of SLM practices, whereas it had a positive effect on lower SLM adoption intensity. Overall, we found evidence that having greater livelihood diversification could prompt households not to adopt more SLM practices. Livelihood initiatives that focus on increasing shock resilience, access to financial support mechanisms, improving livestock production, and providing quality extension services, while also considering agro-ecological differences, are needed. In addition, development planners should take into account the livelihood portfolios of rural households when trying to implement SLM policies and programs.

Keywords: Herfindahl–Simpson diversity index; multivariate probit; drought prone; ordered probit; livelihood diversification; sustainable land management

---

1. Introduction

Globally, agriculture accounts for 67% of employment, 39.4% of national gross domestic product, and 43% of export goods [1]. This sector can continue to be a major source of the world's food and fiber if ecosystem balance is maintained. The world's population continues to grow and is forecast to reach 9.7 billion by 2050 [2]; the demand for food and livelihood security is therefore a pressing concern of development planners and researchers. This is especially true in Sub-Saharan African countries, where more than three-fourths of the population is essentially dependent on rain-fed agriculture and land degradation is the principal cause of the reduction in production and productivity [3].

In rural areas of developing countries, households combine diverse portfolios of activities in their pursuit of alleviating poverty and improving living standards [4,5]. As defined by Ellis [6], livelihood diversification refers to the process by which households pool a wide range of activities and social support systems to deal with shocks and improve their welfare. It is recognized as a way to confront the various idiosyncratic risks and shocks that people face [4]. For most smallholder farmers in developing countries, diversification away from agriculture accounts for 30–40% of their overall incomes [4,6]. A more specific study carried out in Ghana pointed out that livelihood diversification by smallholder farmers prioritize less viable livelihoods attributable to their current food demand, availability of livelihood alternatives and level of entry barriers [7]. Diversifying livelihood activities may affect the environment and the natural resource base [8–10]. The use of natural resource-based livelihood sources to increase income and reduce poverty has a two-way cause–effect association with environmental depletion [10–14]. Rural households' engagement in livelihood activities that increase income, and thus reduce poverty, could have varied effects on the environment [15]. A positive effect could be through the decision to allocate labor away from livelihood activities that exploit natural resources to other, less exploitative, livelihood options. Rural households' engagement in these latter activities may induce less environmental degradation [5,15]. On the other hand, although such livelihood diversification could increase the income of the rural poor, it is possible that some related activities could have a degrading effect on the environment, and poor households could end up living in worse conditions [16,17].

Previous studies related to livelihood diversification focused mainly on identifying its extent and determinants, and examining dominant income sources among sets of livelihood activities [4,18–24]. Some of these studies have identified two main types of livelihood diversification: distressed diversification, in which poor households are motivated to address the shocks they are facing [4,6,21,25], and progressive diversification, which is mostly regarded as an ex-ante strategy implemented by relatively well-off households [7,21]. According to these studies, livelihood diversification could be motivated by asset ownership, market accessibility, credit accessibility, education, and income, among other factors [4,6,21,22].

In Ethiopia, rural households combine a broad array of livelihood activities, most of which are depend mainly on the exploitation of natural resources and subsistence farming systems [4,26,27]. In studies conducted in different parts of the country, high population growth, land scarcity among youth, and lack of agricultural inputs and the associated low productivity have all been reported to drive diversification away from agriculture [28–32]. On the other hand, evidence has shown that land degradation in Ethiopia has been profoundly associated with an insufficient regulatory environment, weak institutions, population increase and high population density, the land tenure system (land right to the state), and lack of participation from the local community. Moreover, the proximate causes are believed to be unsustainable agricultural practices, uneven topography, and high fuel-wood consumption [33–37]. Therefore, in the context of land degradation and food security problems, the essential role of livelihood diversification in Ethiopia has focused on addressing shocks and enhancing household coping strategies [38–41].

Despite the inseparable and practical links between livelihood diversification and sustainability of the farming system in Ethiopia [42,43], few studies have sought to understand the linkage between these interdependent goals. As in other developing countries, rural households in Ethiopia have been undergoing considerable socioeconomic and environmental transitions in recent years, and this has brought both opportunities and challenges in terms of livelihoods [27,41,44–46]. As a result, rural households are trying to diversify their household economies to either survive or generate additional income to secure their livelihoods, regardless of the impacts on the natural resource base [8,28,29,31,32,47]. Despite decades of land rehabilitation efforts by governmental and non-governmental organizations that have addressed land degradation and the associated loss of production and productivity, as well as improving rural livelihoods, the efficiency and adoption rates of promoted land management practices have shown mixed results [48–52].

Many studies related to sustainable land management (SLM) practices and livelihood diversification have focused either on processes at the farm level [49,52,53] or the extent of adoption as influenced by socioeconomic and behavioral factors [51,53,54]. With a growing number of alternative non- and off-farm livelihood activities in rural economies, like those available in Ethiopia, little is known about the relationship between livelihood diversification efforts and the extent of adoption of SLM practices. Moreover, some efforts at livelihood diversification seem to have had a deagrarianization effect in some parts of Ethiopia [55,56]. We, therefore, think that looking at the relationship between livelihood diversification and uptake of SLM practices is imperative, because engaging in diverse livelihood activities could be associated more with earning additional income than with sustainable agriculture.

Hence, the main purpose of this study was to elucidate both the motivations that drive rural households to diversify livelihood activities and their probable links with the implementation of SLM practices in the Upper Blue Nile basin of Ethiopia. The specific objectives were to: (1) explore the extent and determinants of household livelihood diversification and (2) investigate the relationship between livelihood diversification and adoption intensity of SLM practices. On the basis of household survey data collected in November and December 2017 from three rural watersheds located in different agro-ecological zones, we analyzed the extent of livelihood diversification by using the Herfindahl–Simpson index and applied a multivariate probit model to estimate the probability of choosing certain livelihood activities. Moreover, an ordered probit model was estimated to determine the effects of livelihood diversification on intensity of adoption of SLM practices.

The rest of the paper proceeds as follows. Section 2 presents details of the methods used and describes the data and summary statistics of the explanatory variables. Our main results and discussion about livelihood diversification and its effect on SLM adoption are presented in Section 3. Concluding remarks and policy implications are discussed in Section 4.

2. Materials and Methods

2.1. Study Area

This study was carried out in three watersheds of the Amhara and Benishangul-Gumuz national regional states of Ethiopia, in the Upper Blue Nile basin (Figure 1). People in the three study communities in the watersheds, Aba Gerima (in the Bahir Dar Zuria district), Guder (in the Fagita Lekoma district), and Dibatie (in the Dibatie district), primarily make their livelihoods from a mixed crop–livestock production system. The study area ranged from 1479 to 2900 m above sea level and from the highland to lowland climatic zones representing three different agro-ecological regions (Dega, Woyena Dega, and Kolla) of the Upper Blue Nile basin [53]. The major crops grown are barley (*Hordeum vulgare* L.), teff (*Eragrostis tef* Zucc.), wheat (*Triticum aestivum* L.), and potato (*Solanum tuberosum* L.) [53]. Cattle, sheep, goats, donkeys, and horses are the dominant livestock raised.

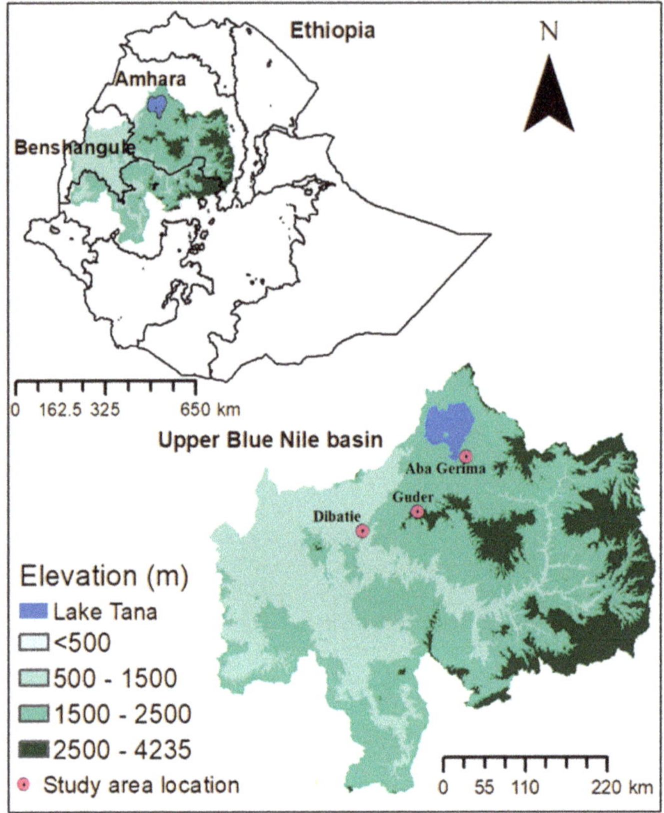

Figure 1. Location map of the study area.

Aba Gerima watershed is categorized as humid sub-tropical with an annual rainfall of 895–2037 mm; Guder is moist tropical with an annual rainfall of 1951–3424 mm, and Dibatie is tropical hot humid with an annual rainfall of 850–1200 mm [53,57,58]. The Swiss Development Cooperation Water and Land Resource Centre in Aba Gerima and the Sustainable Land Management Programme in Guder are notable SLM-related projects in the study areas [53]. In these watersheds, the majority of farmers have subsistence-based livelihoods with additional sources of non- and off-farm income. In the past decade, most farmers in Guder and Aba Gerima have shifted from growing food crops to growing green wattle (*Acacia decurrens*) and khat (*Catha edulis*), respectively [53].

2.2. Data Collection

Primary data were obtained from the sampled respondents by using a structured questionnaire to generate quantitative data on household characteristics, socioeconomic parameters, market access, community institutions, and educational levels of farmers through in-person interviews conducted in November and December 2017. The household was the unit of analysis. Sampling procedures are discussed in the next section.

2.3. Sampling Procedure, Data, and Data Analysis

To select sample households for the study, we used a two-stage sampling procedure. First, on the basis of their specific experiences with SLM practices and varied biophysical and socio-economic characteristics, three watersheds representing different agro-ecological settings were selected: Guder (highland), Aba Gerima (midland), and Dibatie (lowland). Second, household respondents from each watershed were selected randomly by using a probability proportional to size sampling procedure. To gain a better understanding of SLM practices and socioeconomic conditions in these areas, we held a participatory rural appraisal in which we were able to understand the collective dynamics of socio-economic conditions, livelihood shifts and available opportunities in the watersheds. Before the main survey, 15 questionnaires were administered (five in each watershed) in October 2017 to examine the appropriateness of the predesigned set of questions in the selected watersheds. Finally, the household survey was administered to 270 households. The household survey collected detailed information on key socioeconomic and other parameters such as household demographic characteristics, education, asset holdings, livelihoods, income, shock/stress experience, implemented SLM practices, and membership of formal or informal organizations (Table 1). Data management and analyses were performed by using Stata ver. 14.1 (Stata Corp LP, College Station, TX, USA).

Table 1. Description of the independent variables used in the analysis.

Explanatory Variable	Definition/Description	Scale	Hypothesized Relationship with Livelihood Strategies	Hypothesized Relationship with Adoption of SLM Practices	References
Gender	Gender of household head	Binomial, 1 if male	−	+	[19,53]
Age	Age of household head	Metric, in years	+	+/−	[19,52]
Grade	Education level of household head	Metric, in years of schooling	+	+	[19,52]
Household size	Number of individuals in household	Metric, in person	+	+/−	[5,19,52]
Dependency ratio	Ratio of household members aged 0–14 and 65+ years to those aged 15–64 years	Metric, in person	+	+	[9,19]
Distance to market	Distance from home to nearest district market	Metric, in km	+/−	−	[19,52]
Land size	Land size operated by household	Metric, in ha	+/−	+	[19,27,53]
Tenure	Land ownership or tenure type	Binomial, 1 if owned by farmer	+/−	+	[12,19,29,53]
Land for food security	Perception of and's contribution to household's food security	Binomial, 1 if yes	−	+	[29,59]
Access to credit	Household received credit	Binomial, 1 if yes	+	+/−	[19,23,24,52,60]
Access to extension service	Household received agricultural and non-agricultural extension service	Binomial, 1 if yes	+/−	+/−	[12,19]
Membership in CBOs	Household is member of local level community-based organizations	Binomial, 1 if member	+	+	[12,19,52]

Table 1. *Cont.*

Explanatory Variable	Definition/Description	Scale	Hypothesized Relationship with Livelihood Strategies	Hypothesized Relationship with Adoption of SLM Practices	References
Household income	Total income obtained by household	Metric, in ETB	+/−	+/−	[19,52]
Asset value	Total monetary value of assets owned by household	Metric, in ETB	+	+	[12,19]
Aggregate stress/shock	Extent of severity of shocks experienced by household during the last 6 years	Metric, in number	−	+/−	[12,19,56]
Livestock size	Livestock size owned by household	Metric, in tropical livestock unit	+/−	+/−	[19,27]
Intensification	Intensification achieved by household during the year	Binomial, 1 if high	+/−	+/−	[19,24,61]
Agro-ecology	Study location (agro-ecological zones representing high-, mid-, and lowlands)	Binomial, 1 if Aba Gerima; 1 if Guder; 1 if Dibatie	+/−	+/−	[19,52]

2.4. Empirical Model

Respondents were asked to indicate the household's major livelihood activities according to six categories; crop production, livestock production, charcoaling, khat cultivation and daily labor. A related question on the amount of income derived from each livelihood activity over the past 1 year gave us additional means to understand livelihood structures at the household level. In accordance with past studies [6,20,62], we classified livelihood activities as on-farm livelihood activities comprising crop and livestock production, off-farm activities (which included earning wages for work on other farms), non-farm activities where income was earned from non-agricultural sources, and self-employment. We also included petty (minor) trading, beverage making, charcoal making, housing rental, and formal and informal transfers. We then calculated the income share of each livelihood activity carried out by the household in a given year, as follows:

$$S_i = \frac{q_i}{\sum_{i=1}^{n} q_i} i = 1, 2, \ldots, n \tag{1}$$

where n represents the number of livelihood activities, q_i is household income from activity i, and S_i is the share of livelihood activity i in a given household in 1 year.

To identify the most remunerative livelihood strategies, we evaluated the variabilities in returns from various livelihood strategies by using stochastic dominance analysis [63]. In addition, the cumulative per capita annual income densities for major livelihood strategies were plotted to approximate the income distribution of households engaged in each livelihood strategy. As noted by Buckley [63], a typical livelihood strategy first-order stochastically dominates another livelihood strategy when it has a lower cumulative density, thereby proving that households are drawing higher incomes from that strategy. Hence, taking each livelihood activity's income distribution, we were able to test for stochastic dominance. One Way Analysis of Variance (ANOVA) was used to test the differences in livelihood income shares. Moreover, the chi-squared test was used for understanding the difference in households livelihood opportunity choices.

To understand the extent of livelihood diversification, we adopted the Herfindahl–Simpson diversity index, which is commonly applied in ecological and marketing research [64,65]. Although there are many different diversity index methods, this measure of index enable us to use the degree of diversification as a measure of the size of each livelihood activity in relation to its containing groups (in our case, a household's total income), while assessing the activity's diversification and dominance at the same time [6,19,66]. In accordance with the methods of Djido and Shiferaw [9,67], the indices were calculated by using the following formula:

$$HHI_i = 1 - \sum_{i=1}^{n} S_i^2 \tag{2}$$

where HHI_i is the Herfindahl–Simpson diversity index, S_i^2 is the squared income share from each livelihood activity, i is the activity and n is the number of livelihood activities. As stated by Smith and Wilson [68], to address limitations related to evenness and dominance characteristics, we used the total number of livelihood activities to normalize the Herfindahl–Simpson diversity index:

$$NHHI_i = 1 - \frac{HHI_i - \left(\frac{1}{n}\right)}{1 - \left(\frac{1}{n}\right)} \tag{3}$$

where HHI_i refers to the normalized Herfindahl–Simpson diversification index, which ranges from zero (specialization in one activity) to one (full or complete diversification). As stated by Djido and Shiferaw [67] and Brezina et al. [69], higher normalized index values indicate a greater amount of diversification.

For the three simultaneous livelihood choices (i.e., on-farm, off-farm, and non-farm livelihood activities) we estimated a multivariate probit model. The model adopted in this particular study has been used extensively in studies of technology adoption, information and knowledge transfer, labor-related decisions for on- and off-farm employment, and participation in agro-environmental programs [70,71].

We first modeled a random utility for the decision to pursue any livelihood activity. In the utility function, we assumed that households decide to implement a certain livelihood strategy on the basis of maximizing utility, i.e., $U^*_j - U_0 > 0$. Hence, in the utility function, the net benefit of a livelihood activity (B^*_{ij}) could be fit as follows;

$$B^*_{ij} = x'_i \beta_j + \mu_i \ (j = on - farm, \ off - farm \& non - farm \ activities, \ i = 1, 2, \ldots n) \tag{4}$$

where B^*_{ij} is the household's net benefit, which is indicated as a function of a vector of exogenous household variables x_i; β refers to parameter estimates; n is the number of households; and u_i is the error term. Therefore, a typical household would choose a given livelihood strategy in the pursuit of gaining higher household income. Hence, a general multivariate probit model could be specified as follows:

$$
\begin{aligned}
B_{ij} &= x'_{ij} \beta_j + \mu_{ij}, \\
B_{ij} &= \begin{cases} 1 \ if \ y^*_{ij} > 0 \\ 0 \ Otherwise \end{cases} \quad (j = on - farm, off - farm \& non - farm)
\end{aligned} \tag{5}
$$

where B_{ij} ($j = 1, \ldots, m$) is a vector of livelihood activities (in our case $m = 3$) performed by the ith household, x'_{ij} is vector of observed variables that affect the decision to choose any type of livelihood activity, β_j is vector of unknown parameters, and μ_i is the error term.

As indicated by Greene [72], the multivariate probit model follows a series of independent probit models for each alternative livelihood activity j. Note that rural households are likely to undertake multiple livelihood activities simultaneously; thus, it is likely that the decisions among choices are correlated. As a result, the unobserved error terms for the estimated probit models would not be independent. If we were to ignore this characteristic of the outcome variables, the result would be a biased estimate of the probabilities and parameters. In the multivariate probit approach to estimating the unknown parameters in Equation (5), the error terms (across $j = 1, \ldots, m$ alternatives) of the latent equation are assumed to have multivariate normal distributions, and this results in a model with a mean vector equal to zero and a covariance matrix R with diagonal elements equal to one.

With the assumption of multivariate normality, the unknown parameters in Equation (5) can be estimated by using maximum likelihood procedures. The probabilities were computed by using the Geweke–Hajivassiliou–Keane simulation procedure [73].

To model the association between livelihood diversification and implemented SLM practices, reported practices were taken as outcome variables. The respondents were asked about a total of 18 of SLM practices, including soil fertility management, soil and water conservation, and gully rehabilitation. We summarized the results and took the 10 most frequently practiced SLMs as outcome variables for the model estimation. We estimated an ordered probit regression model [49,52,74], which was suitable for count data like ours.

In analyses of the adoption of some technologies, it is essential to note that they will be implemented to different extents by different farmers. In this study, the outcome variable adoption of SLM practices could take values ranging from 0 to 10. The households are heterogeneous because of differences in socio-economic, community, education, and other factors, so the likelihood of any household adopting the first SLM practice might vary from that of other households. In accordance with the method of Wollni et al. [74], the model is specified as:

$$y^* = \beta'X + \mu,$$ (6)

where y^* is the latent variable (number of SLM practice) and takes the values 1 through 10, β' is a vector of unknown parameters to be estimated, X is a vector of explanatory variables, and μ is the error term, which is assumed to be normally distributed with zero mean and a variance of one. The number of observed technologies (y) used is related to the underlying latent variable y^* through the threshold μ_n ($n = 1, \dots, 10$) and the probability that any given number of technologies (y) is used is calculated as follows:

$$prob(y = n) = \varphi(\mu_n - \beta'X) - (\mu_{n-1}\beta'X).\forall_n = 1,\dots, 10$$ (7)

The ordered probit estimation will give the threshold μ and vector parameter β. The threshold μ shows the range of the normal distribution associated with the specific values of the response variables. The remaining parameters (β) represent the effect of changes in explanatory variables on the underlying scale.

When estimating the effect of livelihood diversification and adoption of SLM technologies, a potential endogeneity problem may arise. In our case, it may occur when an explanatory variable of choosing a certain livelihood strategy is jointly determined by the decision to adopt in the SLM adoption specification [75]. To address this problem, we tested for it in accordance with the method of Rivers and Vuong [76] by using a two-stage linear regression, and we confirmed that there was no endogeneity problem between livelihood diversification and adoption of a specific SLM technology.

3. Results and Discussion

3.1. Summary of Socioeconomic Variables

Table 2 summarizes the socio-economic characteristics of the survey respondents (see Table 1 for a description of the variables). For the entire sample, the average age of the household head was 49 years, the dependency ratio was 71%, the average household size was 5.4, and more than 80% of the study households were headed by males. Family size was significantly larger at Dibatie than Aba Gerima and Guder ($p < 0.001$). Most households in the watersheds are characterized as smallholdings, with an average of 1.03 ha of land (which sustains an average household size of about 5.4 people) and an average livestock holding of 3.97 TLUs. The average land holding was somewhat smaller than the national average farm (1.22 ha) [74]. Farmers in Aba Gerima held significantly more land than famers in Guder and Dibatie, while Dibatie farmers owned more livestock compared to Aba Gerima and Guder ($p < 0.05$) More than half (52%) of the sampled households reported having experienced anthropogenic (e.g., price inflation, poor access to social services) and naturally driven (e.g., drought, pest infestation, soil erosion, animal disease) stresses. Experience of these types of stressors differed significantly between watersheds with households in Aba Gerima reporting significantly more stress than Guder and Dibatie ($p < 0.001$).

Table 2. Descriptive statistics of socio-economic characteristics, by watershed (n = 270).

Explanatory Variables	Whole Sample	Aba Gerima	Guder	Dibatie	Test
Gender	0.811(0.41)	79 (11)	60 (30)	79 (11)	b ***
Age	49 (12.9)	47 (11.5)	51 (12.3)	50 (14.3)	a *
Grade	1.3 (2.9)	0.53 (1.56)	1.6 (3.20)	1.76 (3.47)	a **
Household size	5.38 (2.34)	4.65 (2.61)	5.55 (2.08)	5.92 (2.11)	a ***
Dependency ratio	71.19 (12.79)	72.67 (11.87)	70.42 (13.33)	70.46 (13.11)	
Distance to market	9.5 (6.7)	14.33 (5)	6.84 (4.40)	7.37 (7.35))	a ***
Land size	1.03 (0.76)	1.25 (0.70)	0.91(0.58)	0.93 (0.92)	a ***
Tenure	0.77 (0.42)	76 (14)	78 (12)	55 (35)	b ***
Land for food security	0.59 (0.49)	69 (21)	67 (23)	68 (22)	b ***
Access to credit	0.56 (0.50)	55 (35)	53 (37)	42 (28)	
Access to extension service	0.70 (0.46)	76 (14)	48 (42)	64 (26)	b ***
Membership in CBOs	0.51 (0.50)	54 (36)	63 (27)	20 (70)	b ***
Household Income	10,758 (13,021)	9109 (8502)	12,425 (18,649)	10,742 (9321)	a **
Asset value	1919 (4191)	2771 (4224)	1598 (3638)	1389.58 (4572)	a ***
Aggregate stress/shock	0.52 (0.20)	0.60 (0.14)	0.45 (0.23)	0.51 (0.18)	a ***
Livestock size	3.97 (0.49)	3.84 (2.20)	3.52 (2.38)	4.52 (2.67)	a **
Intensification	0.42 (0.50)	0.69 (0.47)	0.21 (0.41)	0.37 (0.41)	b ***

Note: * Significant at 10%, ** significant at 5%, *** significant at 1%; standard deviations in parentheses; [a] non-parametric two-sample test: Wilcoxon's rank-sum test, [b] Chi-squared test.

3.2. Livelihoods in the Study Areas

The majority (81%) of surveyed households were engaged in crop and livestock production (Figure 2). The result is closely in line with a report showing that an average of 79% of Ethiopian rural smallholding farmers earn income from agriculture [77]. In addition to engaging in their mixed crop–livestock farming system, a considerable number of households engage in off-farm and non-farm livelihood activities. Moreover, there was a statistically significant difference of mean income across watersheds as determined by one-way ANOVA ($p < 0.001$) for charcoal production (F(2, 270) = 13.03, $p = 0.000$), khat plantation income (F(2, 270) = 39.96, $p = 0.000$) and crop production (F(2, 270) = 7.50, $p = 0.0007$)

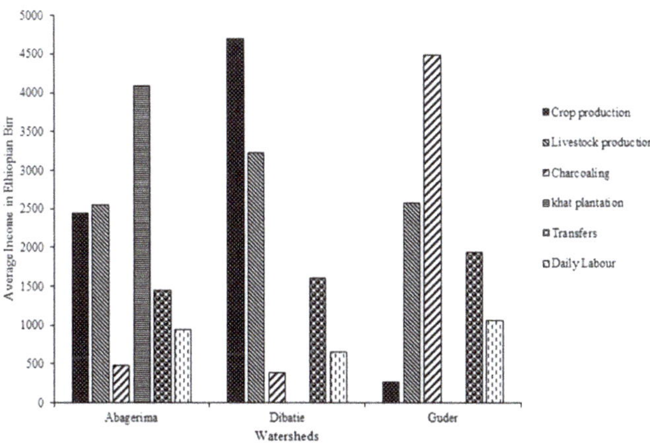

Figure 2. Average annual income from major livelihood activities in the study watersheds.

High-value cash crops such as khat in Aba Gerima and *Acacia decurrens* for charcoaling in Guder are the dominant household livelihood activities in those watersheds. Because of increasing demand, khat cultivation has become a notable remunerative income source for households in the Aba Gerima watershed. Similar studies have shown that, despite its controversial social significance, khat makes a substantial contribution to the country's national income [78,79]. *Acacia decurrens* was the dominant cash crop in the Guder watershed, where the charcoal is usually destined for markets [80]. In Dibatie, crop and livestock farming made a higher contribution to overall income.

Households were asked to mention the opportunities that were available to them to improve their livelihoods. Five opportunities were noted, but three of these are agriculture related, suggesting that there is little diversification beyond farming in rural Ethiopia. Except establishing retailing business, there was significant difference ($p < 0.001$) on the household's choices of existing livelihood opportunities across watersheds. The majority of the households reported that, if initial investment capital or credit were made available, they would prefer to invest the money in livestock fattening (92% of households in Guder, 72% in Aba Gerima, and 37% in Dibatie) (Figure 3). A possible explanation for this could be, alike other parts of the rural Ethiopia, farmers' limited entrepreneurial competence would not be able to allow them to pursue other lucrative opportunities than livestock rearing, or it may be related to its multipurpose role to livelihoods of farmers through provision of food and income from products, employment, insurance against drought, fuel for cooking, manure for crops, and draught power for farming. On the other hand, it could also be related to their perception that livestock rearing outweighs others because of its capacity to optimally use available resources (e.g., crop residue, grass) that could not otherwise be utilized by them.

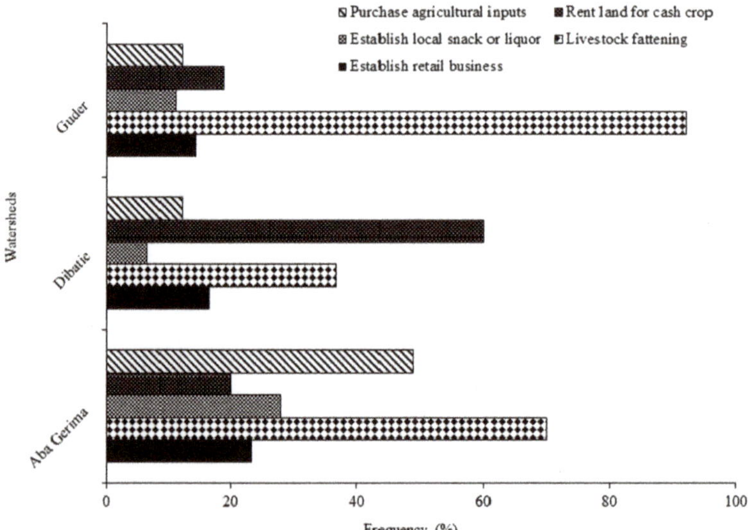

Figure 3. Household livelihood opportunity preferences in the study watersheds.

3.3. Remunerative Livelihood Activities in the Study Areas

On the basis of the stochastic dominance criterion, khat production was the first-order stochastically dominant activity in Aba Gerima (Figure 4). In Dibatie, crop production was dominant and livestock production was the second most dominant. In Guder, growing *Acacia decurrens* for charcoaling was the dominant remunerative livelihood strategy. According to Achamyeleh [81], in comparison with other on-farm sources, acacia plantations made a very high contribution to the overall income of households in Guder. Similarly, Nigussie et al. [80] indicated that 84.6% of households reported income as their major motivation to plant *Acacia decurrens*. In addition, our findings showed that charcoaling and khat production were the most inferior (i.e., least lucrative) of the six livelihood strategies in Aba Gerima and Dibatie, respectively.

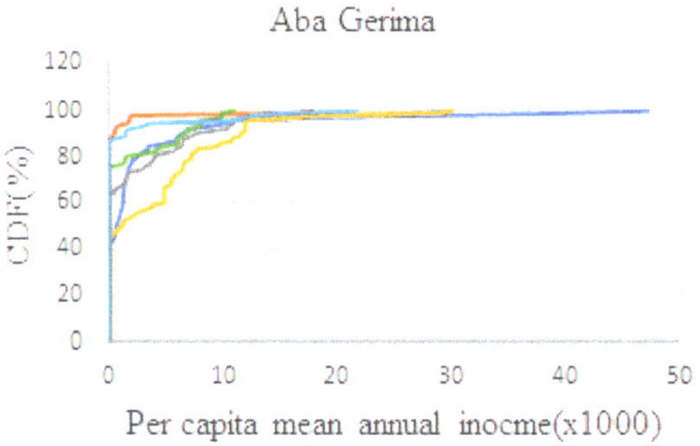

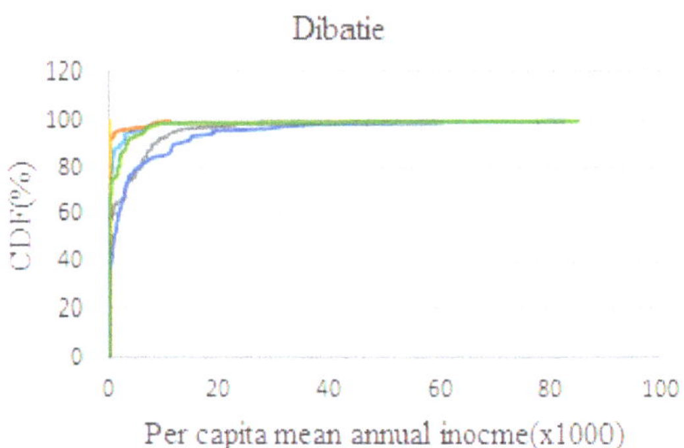

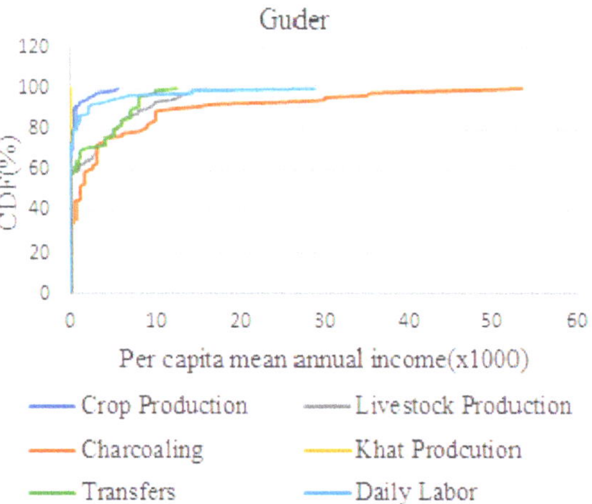

Figure 4. Cumulative density curve (CDV) for major livelihood activities in the study watersheds.

3.4. SLM Practices in the Watersheds

Among the surveyed SLM practices, the most commonly reported SLM technologies included crop rotation, chemical fertilizer use, gabion check dams, wooden check dam, gully filling, fencing, residue management, traditional terracing, soil bunds, diversion channels, and waterways. Crop rotation, fertilizer application, use of soil bunds, traditional terracing, and residue management were reported to be the most extensively applied SLM practices. Aba Gerima watershed had the highest percentage of households implementing SLM activities (Figure 5). Similar results were reported by Nigussie et al. [53], who reported that agroforestry, drainages, and application of manure and fertilizers were commonly implemented in Aba Gerima. In this study, fencing, gabion check dams, and gully filling were the least used SLM practices. Note, however, that the abovementioned technologies are community-wide practices, and households were asked for their participation in these operations.

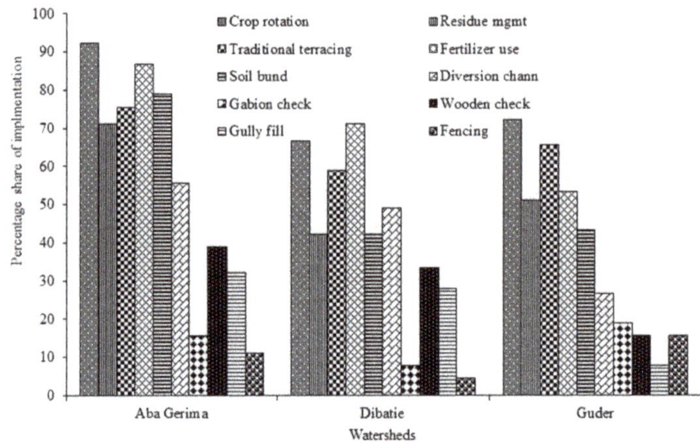

Figure 5. Major sustainable land management (SLM) practices in the study watersheds.

3.5. Drivers of Livelihood Diversification

We used a multivariate probit model to analyze the likelihoods of households engaged in a certain livelihood activity associated with a set of factors related to household, socioeconomic, location, and asset features. On-farm activity was negatively correlated with both off- and non-farm activities, indicating that households that diversified into with in on-farm activity may not have had a chance to engage in the other activities. Conversely, off-farm activity was positively correlated with non-farm activity, implying that households that had engaged in wage employment activities may also have engaged in non-farm activities ($\varrho32$ in Table 3). The likelihood ratio test of the independence of the error terms of the various livelihood activity equations was strongly rejected (χ^2 (3) = 46.9552, $p < 0.001$) (Table 4). We therefore adopted the alternative hypothesis of mutual interdependence among livelihood strategies.

Table 3. Pair-wise correlation coefficients across livelihood strategies.

Livelihood Strategy	Coefficient	Standard Error	*p*-Value
$\varrho21$	−0.588	0.097	0.000
$\varrho31$	−0.704	0.085	0.000
$\varrho32$	0.233	0.148	0.034

Note: $\varrho21 = \varrho$(off-farm, on-farm), $\varrho31 = \varrho$(non-farm, on-farm), $\varrho32 = \varrho$(non-farm, off-farm).

Gender of the household head significantly influenced the probability of adopting certain livelihood strategies (Table 4), and males were more likely to participate in on-farm livelihood activities ($p < 0.05$). Our result is in line with that of Ragasa et al. [82], who also reported that male-headed households have more productive labor and asset ownership than their female counterparts when it comes to on-farm activities. Female-headed households can be characterized by a lack of access to asset ownership and adequate labor to pursue on-farm activities. The probability of participation in non-farm livelihood activities decreased with increasing household size ($p < 0.05$), perhaps because larger households had more dependents. Likewise, Babatunde et al. [18] reported that the larger the household, the less likely it was to support activities other than agriculture. As expected, the dependency ratio significantly influenced all livelihood diversification choices. A negative effect was observed with on-farm livelihood activities ($p < 0.01$); this may indicate that households with more dependents were less likely to choose on-farm activities because of a shortage of active working labor. Eswaran and Kotwal [83] reported that households with more active working labor yielded higher levels of productivity for smallholder farmers. Conversely, households with more dependents tended to choose off-farm ($p < 0.01$) and non-farm ($p < 0.05$) livelihood activities. A possible explanation for this could be that a larger household could shift available labor to alternative off- and non-farm livelihood activities.

Table 4. Multivariate probit regression results.

	Outcome Variables		
Variables	On-farm	Off-farm	Non-farm
Gender	0.557 ** (0.281)	0.191 (0.284)	−0.252 (0.245)
Age	0.006 (0.008)	0.008 (0.008)	−0.005 (0.007)
Grade	−0.045 (0.033)	0.056 (0.035)	0.030 (0.031)
Household size	0.033 (0.050)	0.021 (0.050)	−0.083 * (0.044)
Dependency ratio	−0.040 *** (0.009)	0.027 *** (0.008)	0.020 ** (0.008)
Distance to market	−0.136 (0.122)	−0.275 ** (0.114)	0.205 ** (0.102)
Land size	−0.124 (0.140)	−0.232 (0.149)	−0.102 (0.136)
Tenure	0.118 (0.259)	−0.049 (0.236)	0.090 (0.214)
Land for food security	−0.0271 (0.241)	0.200 (0.220)	−0.225 (0.200)
Access to credit	0.480 ** (0.195)	−0.133 (0.188)	−0.036 (0.178)
Access to extension services	−0.262 (0.214)	0.442 ** (0.216)	−0.056 (0.204)
Membership in CBOs	0.251 (0.214)	−0.188 (0.204)	0.700 *** (0.206)
Household income	0.185 * (0.103)	−0.106 (0.097)	−0.221 ** (0.090)
Asset value	0.002 (0.105)	−0.080 (0.103)	−0.002 (0.100)
Aggregate stress/shock	−0.164 (0.552)	0.956 ** (0.502)	−0.655 (0.468)
Livestock size	−0.029 (0.045)	0.005 (0.046)	0.096 ** (0.045)
Intensification	0.536 ** (0.253)	−0.231 (0.207)	−0.156 (0.202)
Aba Gerima	−0.393 (0.361)	0.614 ** (0.301)	0.591 ** (0.294)
Dibatie	−0.010 (0.340)	−0.216 (0.302)	1.253 *** (0.291)
Constant	1.762 (1.350)	−2.187 * (1.303)	−0.352 (1.192)
Wald χ^2 (*df*)	134.44 (400)		
Prob. $> \chi^2$	<0.001		
n	270		

Note: Standard errors are given in parentheses. Likelihood ratio test of overall error terms correlation: $\varrho21$ (off-farm, on-farm) = $\varrho31$ (non-farm, on-farm) = $\varrho32$ (non-farm, off-farm) = 0: χ^2 (3) = 47.8868, Prob > χ^2 < 0.001. Significance levels are indicated as follows: * 10%, ** 5%, *** 1%.

Market distance significantly affected the choices of off-farm ($p < 0.05$) and non-farm ($p < 0.05$) livelihood activities, but in opposite directions (Table 4). The regression result had a negative value for the choice of off-farm activity, which may suggest that, as the labor market distance increased, households were less likely to choose this option. Our finding that market distance affected off-farm income is well substantiated by Ellis and Bahiigwa [84], but differs from the results of Kung and Lee [85]. In contrast, market distance had a positive influence on the choice of non-farm activities. In our discussion with the residents of Aba Gerima and Guder, we learned that when some products, such as khat and charcoal, were directly sold at distant marketplaces they might have achieved higher prices than could have been obtained from local buyers. In addition, demand for these products might have been higher in distant markets than in closer ones. Using panel data, Jiao et al. [86] revealed that access to infrastructure would help households to diversify to more remunerative strategies.

Contrary to our initial expectation, access to credit showed a significant effect only for on-farm activities ($p < 0.05$) (Table 4). These results could be related to targeting and efficiency issues related to the credit service. It is also possible that the sampled households may have used this credit for on-farm inputs, such as fertilizers, improved seeds, or consumption smoothing. A more profound study done in Ghana revealed a similar result, which indicated that households who received credit found to be effective in terms of improving agricultural productivity and it also helped them to diversify their livelihoods [24]. This idea is in line with the work of Mulwa et al. [87], who found that access to credit allowed households to adopt soil and water conservation activities that help them to invest more in agricultural inputs. Credit users also may not use the financing for the intended purpose of diversifying their income sources [47]. A similar result was found in rural Niger, where access to financial institutions in the community seemed to negatively affect the probability of participating in businesses activities [59]. In contrast, Mentamo and Geda [88] reported that credit access increased the extent of livelihood diversification in Kadida Gamela district, Ethiopia.

Access to extension services had a significant positive effect on the choice of off-farm livelihood activities ($p < 0.05$) (Table 4). A similar result was also reported by Chikobola and Sibusenga [89]. Extension services customarily help households increase production and productivity within the farming system itself [20,22,47]. However, the present findings could be attributed to the fact that these services may include small-scale employment opportunities (e.g., off-farm activities). For example, in northern Ethiopia, agricultural extension programs that made use of public employment schemes such as "food for work" helped farmers to shift to off-farm livelihood sources [27].

As anticipated, membership in rural cooperatives had a significant positive effect on diversification to non-farm livelihood activities ($p < 0.01$). This finding underscores the importance attached to the entry barrier households face in terms of initial capital, because these cooperatives might help them to access credit [26]. Previous studies have also reported that membership in formal and informal community organizations can help smallholder farmers to address financial constraints, gain social cohesion and skills, and increase their market networking in selling and buying products [20,25,44,84,90,91].

Household income had a significant positive effect on the choice of on-farm activities ($p < 0.1$), but it had a significant negative influence on non-farm livelihood activities ($p < 0.01$) (Table 4). This finding may suggest that, despite the increased income, these households still do not have a strong incentive to diversify their livelihood sources. As discussed earlier, access to credit also seemed to make households focus on on-farm livelihood activities. A similar finding was reported by Ellis and Bahiigwa [84], who found that a greater income may aid in the timely purchase of farm inputs, such as fertilizers, improved seeds, or the ability to hire wage labor, leading to enhanced cultivation practices and higher productivity.

The aggregate stress/shock index had a significant positive affect on a household's decision to participate in off-farm livelihood activities ($p < 0.05$). For example, exposure to seasonal rainfall may force households to look for short-term solutions, such as engaging in daily labor. In a study carried out in Zambia, Gautam and Andersen [92] argued that households tended to choose livelihoods related to short-term gains when they experience shocks like the lack of seasonal rainfall. Moreover, our result supports the theory of distress-driven livelihood diversification, as opposed to progressive-driven diversification, in the region [19,21,84,93,94]. Woldehanna [27], Kasie et al. [94], Dercon [93], and Block and Webb [39] reported a similar result in Ethiopia, where the amount and variability of rainfall had a significant effect on the decision by households to engage in any type of off-farm work. Likewise, studies conducted in Uganda [95] and in Indonesia [96] confirmed that rural household labor adjustment decisions serve as a coping mechanism in response to agricultural shocks.

Livestock holding had a significant positive effect on the choice of non-farm livelihood activities ($p < 0.05$). Similarly, Ellis [19] and Kassie et al. [20] found that increasing livestock holding is an essential financial safeguard to starting a new livelihood and helps farmers to diversify both within and outside agriculture. In addition, higher agricultural intensification had a significant positive effect on the choice of on-farm livelihood activities ($p < 0.05$). This could mean that, as more households adopt agricultural technologies and inputs, more will stay on the farm. This result is in line with that of Sanders and McKay [97] and Verkaart et al. [98], who reported that households with greater agricultural intensification and productivity were less likely to be driven to diversify to other sources of income. In terms of study locations, significant differences were observed in the probability of adopting non-farm and off-farm livelihoods among the different agro-ecological areas (Table 4). Despite the fact that most households in the watersheds practice crop–livestock mixed farming, a substantial difference was observed in activities such as khat and charcoal production (Figure 3). Similarly, Tesfaye et al. [99], Ellis [19] and Peng et al. [100] indicated that geographical locations, and policies related to ecologies determines the choices of livelihood strategies.

3.6. Effects of Extent of Livelihood Diversification on SLM Practices

The livelihood diversification index had a mean of 0.10 and a standard deviation of 0.18, indicating that households tended to be relatively concentrated in their main sources of income. The maximum estimation was 0.64 and the minimum was 0. A relatively high diversification was found in Dibatie (0.13), whereas in Guder it was lower (0.07). The values of livelihood diversification were roughly similar in studies carried out in different parts of Ethiopia [20,39,47,99,101,102]. Conversely, a higher extent of livelihood diversification was revealed in studies conducted in Tigray and Gamebella [103,104].

We also estimated an ordered probit model to investigate the marginal effect of each covariate on the probability of adopting SLM practices by the respective households. The joint test of all slope coefficients proved that our null hypothesis was rejected (Table 5). Overall, the results showed that livelihood diversification had a positive effect on a small number of SLM practices, but there was a negative relationship when the number SLMs increased (Table 5).

As shown in Table 5, an increase in a household's livelihood diversification status increases the probability of selecting zero, one, two, or three SLM practices by 3% ($p < 0.1$), 5.3% ($p < 0.05$), 7.3% ($p < 0.05$), and 9.3% ($p < 0.05$), respectively. Conversely, an increase in the livelihood diversification status decreases the probability of adopting five, six, seven, or eight SLM practices by 7.3% ($p < 0.05$), 5.6% ($p < 0.05$), 9% ($p < 0.05$), and 2.6% ($p < 0.1$), respectively. Generally, livelihood diversification shows mixed results in that it favors a lower level of SLM adoption intensity and disfavors higher level of adoption of SLM. These results clearly show the complementary nature of having diversified livelihood strategies and carrying out SLM activities in these watersheds for a relatively low number of SLM adoptions. A slightly similar result was revealed in Gozamin district, Ethiopia [20] where livelihood diversification was positively associated with farmland management strategies. Compared with the case when the household head has any level of education, the probability of adopting five, six, seven, and eight SLM practices was higher by 0.6 ($p < 0.05$), 0.5 ($p < 0.05$), 0.7 ($p < 0.05$), and 0.2 ($p < 0.1$) percentage points, respectively, among household heads of higher education levels (Table 5). Similar studies in the study watersheds and elsewhere in Ethiopia have revealed that more educated heads of households have a higher probability of adopting SLM practices [37,53].

Table 5. Estimates of the ordered probit model and marginal effects.

SLM Intensity	Marginal Effects for Each Outcome											Coeff
	Pr(Y = 0/X)	Pr(Y = 1/X)	Pr(Y = 2/X)	Pr(Y = 3/X)	Pr(Y = 4/X)	Pr(Y = 5/X)	Pr(Y = 6/X)	Pr(Y = 7/X)	Pr(Y = 8/X)	Pr(Y = 9/X)	Pr(Y = 10/X)	
Livelihood diversity	0.030 * (0.020)	0.053 ** (0.028)	0.073 ** (0.037)	0.093 ** (0.045)	0.024 (0.018)	−0.073 ** (0.036)	−0.056 ** (0.029)	−0.090 ** (0.044)	−0.026 * (0.016)	−0.007 (0.006)	−0.010 (0.008)	−0.701 ** (0.332)
Gender	−0.017 * (0.010)	−0.032 * (0.017)	−0.044 * (0.024)	−0.056 ** (0.030)	−0.014 (0.0116)	0.044 * (0.024)	0.034 ** (0.018)	0.054 ** (0.028)	0.015 * (0.009)	0.004 (0.004)	0.006 (0.005)	0.424 ** (0.211)
Age	−0.000 (0.000)	−0.000 (0.000)	−0.000 (0.000)	−0.000 (0.001)	−0.000 (0.000)	0.000 (0.001)	0.000 (0.000)	0.000 (0.001)	0.000 (0.000)	0.000 (0.000)	0.000 (0.000)	0.003 (0.005)
Grade	−0.002 ** (0.001)	−0.004 ** (0.002)	−0.006 ** (0.003)	−0.008 ** (0.003)	−0.002 (0.001)	0.006 ** (0.003)	0.005 ** (0.002)	0.007 ** (0.003)	0.002 * (0.001)	0.001 (0.000)	0.001 (0.001)	0.057 ** (0.025)
Household size	−0.003 * (0.002)	−0.005 ** (0.003)	−0.007 ** (0.003)	−0.009 ** (0.004)	−0.002 (0.002)	0.007 ** (0.003)	0.005 ** (0.003)	0.009 ** (0.004)	0.002 (0.002)	0.001 (0.001)	0.001 (0.001)	0.067 ** (0.031
Dependency ratio	0.000 (0.000)	0.000 (0.000)	0.001 (0.001)	0.001 (0.001)	0.000 (0.000)	−0.001 (0.001)	−0.000 (0.000)	−0.001 (0.001)	−0.000 (0.000)	−0.000 (0.000)	−0.000 (0.000)	−0.005 (0.005)
Distance to market	0.004 (0.003)	0.008 (0.006)	0.011 (0.008)	0.013 (0.011)	0.003 (0.003)	−0.011 (0.008)	−0.008 (0.006)	−0.013 (0.010)	−0.004 (0.003)	−0.001 (0.001)	−0.001 (0.001)	−0.101 (0.077)
Land size	−0.002 (0.004)	−0.003 (0.007)	−0.005 (0.010)	−0.006 (0.013)	−0.002 (0.003)	0.005 (0.010)	0.004 (0.008)	0.006 (0.012)	0.002 (0.003)	0.000 (0.001)	0.001 (0.002)	0.047 (0.095
Tenure	0.010 (0.008)	0.018 (0.013)	0.026 (0.018)	0.033 (0.023)	0.008 (0.008)	−0.026 (0.018)	−0.020 (0.014)	−0.031 (0.022)	−0.009 (0.007)	−0.002 (0.002)	−0.004 (0.003)	−0.246 (0.167)
Land for food security	−0.001 (0.007)	−0.001 (0.012)	−0.002 (0.017)	−0.002 (0.022)	−0.001 (0.006)	0.002 (0.017)	0.001 (0.013)	0.002 (0.021)	0.001 (0.006)	0.000 (0.002)	0.000 (0.002)	0.018 (0.164)
Access to credit	−0.016 ** (0.009)	−0.030 ** (0.012)	−0.041 ** (0.016)	−0.053 *** (0.020)	−0.014 (0.010)	0.041 ** (0.016)	0.032 ** (0.013)	0.051 ** (0.020)	0.015 ** (0.008)	0.004 (0.003)	0.006 (0.004)	0.398 *** (0.14)
Access to extension services	−0.006 (0.007)	−0.011 (0.011)	−0.016 (0.016)	−0.020 (0.020)	−0.005 (0.006)	0.016 (0.016)	0.012 (0.012)	0.019 (0.019)	0.005 (0.006)	0.001 (0.002)	0.002 (0.003)	0.151 (0.149)
Membership in CBOs	−0.004 (0.006)	−0.008 (0.011)	−0.011 (0.015)	−0.014 (0.019)	−0.004 (0.005)	0.011 (0.015)	0.009 (0.011)	0.014 (0.018)	0.004 (0.005)	0.001 (0.002)	0.002 (0.002)	0.108 (0.14)
Household income	0.004 (0.003)	0.007 (0.005)	0.009 (0.008)	0.012 (0.010)	0.003 (0.003)	−0.009 (0.008)	−0.007 (0.006)	−0.011 (0.009)	−0.003 (0.003)	−0.001 (0.001)	−0.001 (0.001)	−0.088 (0.072)
Asset value	0.002 (0.003)	0.004 (0.005)	0.006 (0.007)	0.008 (0.009)	0.002 (0.003)	−0.006 (0.007)	−0.005 (0.005)	−0.008 (0.008)	−0.002 (0.002)	−0.001 (0.001)	−0.001 (0.001)	−0.06 (0.065)
Aggregate stress/shock	−0.046 ** (0.018)	−0.083 ** (0.037)	−0.116 ** (0.050)	−0.147 ** (0.063)	−0.038 (0.028)	0.116 ** (0.050)	0.089 ** (0.040)	0.142 ** (0.056)	0.041 ** (0.021)	0.011 (0.008)	0.016 (0.011)	1.113 *** (0.422)
Livestock size	−0.002 (0.001)	−0.004 * (0.002)	−0.006 * (0.003)	−0.008 * (0.004)	−0.002 (0.002)	0.006 * (0.003)	0.005 ** (0.003)	0.007 ** (0.004)	0.002 (0.001)	0.001 (0.000)	0.001 (0.001)	0.057 * (0.03)
Aba Gerima	−0.023 ** (0.011)	−0.043 ** (0.021)	−0.059 ** (0.029)	−0.075 ** (0.035)	−0.019 (0.014)	0.059 ** (0.028)	0.046 ** (0.022)	0.073 ** (0.033)	0.021 * (0.011)	0.006 (0.005)	0.008 (0.006)	0.568 ** (0.247)
Dibatie	−0.025 ** (0.011)	−0.046 ** (0.020)	−0.064 ** (0.026)	−0.081 ** (0.031)	−0.021 (0.014)	0.064 ** (0.027)	0.049 ** (0.021)	0.0780 ** (0.030)	0.022 ** (0.010)	0.006 (0.005)	0.009 (0.006)	0.611 *** (0.217)

$n = 270$; Wald $\chi^2(20) = 113.92$; Prob > $\chi^2 < 0.001$; Pseudo $R^2 = 0.0888$; Robust standard errors are given in parentheses. *** $p < 0.01$, ** $p < 0.05$, * $p < 0.1$. Marginal effects (dy/dx) are calculated at the mean for continuous variables and for a discrete change from 0 to 1 for dummy variables.

Access to credit was found to have a positive effect on adopting a higher number of SLM practices. Compared with the case when respondents had no access to credit, the probability of adopting five, six, seven, or eight SLM practices was higher by 4.1 ($p < 0.05$), 3.2 ($p < 0.05$), 5.1 ($p < 0.05$), and 1.5 ($p < 0.05$) percentage points, respectively, for respondents who had access to credit. This could possibly be associated with owning more livestock, as it is a liquid asset and provide financial safe guarding [105]. Likewise, in India, Aryal et al. [106] revealed that more ownership of livestock showed a higher intensity of adoption of climate-smart agricultural practices. These findings are in line with those of Nigussie et al. [53], who reported that credit access could encourage the adoption of manure application.

With regard to the shock index, the more severe its impact, the lower adoption intensity for greater intensity level of SLM adoption. For example, the probability of adopting five, six, seven, and eight practices was higher by 11.6 ($p < 0.05$), 8.9 ($p < 0.05$), 14.2 ($p < 0.05$), and 4.1 ($p < 0.05$) percentage points, respectively, for those who experienced a lower shock. Likewise, in their comparative study of Thailand and Vietnam, Nguyen et al. [107] revealed that farmers' experience with weather shocks affected their decision priorities relative to what type of land management practices they should engage in. Nigussie et al. [53] revealed that a higher erosion risk promoted the adoption of SLM practices in a similar area in the Upper Blue Nile basin. This clearly indicates that, when there are few or no shocks, rural households will have more time and labor to invest in SLM practices instead of looking for coping mechanisms to deal with a crisis. Overall, these results showed the varied magnitude and effect of the explanatory variables on the different SLM adoption intensities.

4. Conclusions and Implications

In the Northwestern part of Ethiopia, despite efforts to improve livelihoods of smallholder farmers through promoting their uptake of SLM practices, there is still marked inadequacies of the level of adoption of these strategies and their effect on the sustainability of livelihoods. Rural households always try to diversify their sources of income within and outside of agriculture, and this could be associated with the decision and extent of adopting of certain SLM measures on their farms. There has been a wealth of research evidence available in terms of looking at livelihood diversification and adoption of SLM practices, treating them as separate research topics. However, earlier research works have paid little attention to the links of these two research domains. As a result, this particular study was conducted to be able to extend existing knowledge on this issue.

We used data obtained from a cross-sectional household survey to examine the extent and drivers of livelihoods diversification in the Upper Blue Nile basin of Ethiopia, as well as the effects of diversification on the adoption intensity of SLM practices. In the study sites, for majority of the smallholder farmers, non-farm and off-farm livelihood activities were accounted as supplementary sources to the on-farm livelihoods. The study revealed that crop production, livestock production, charcoaling, khat plantation, transfers, and less capital-intensive activities such as casual daily labor are the most prominent livelihood activities. With more dependents, households tend to choose off-farm and non-farm livelihood strategies. In addition, households' access to credit was found to favor the probability of selecting on-farm livelihoods. Owning more livestock and being a member of community-based organizations increase the probability of participating in non-farm livelihood strategies. Therefore, for more diversified livelihood strategies, it is important to focus on policies and programs that enhances household's livestock production and degree of participation in community-based organizations; and for those who opted to stay in agriculture, facilitating financial supports could contribute to sustaining their agricultural livelihood. Better-off households preferred to stay in on-farm, while poor farmers opted for non-farm livelihoods. The results clearly showed the complementary nature of having diversified livelihood strategies and carrying out SLM activities in the study watersheds for a relatively low number of SLM adoptions. While there is heterogeneity of factors which influence the adoption intensity of any of the 10 SLM practices, findings of this study underscore the importance of higher level of education, better access to credit, shock/stress experience, and more livestock wealth on higher adoption intensity of SLM practices. Likewise, agro-ecological heterogeneity also found to have an influence for higher adoption rate in Aba Gerima (mid land) and Dibatie (low land) watersheds.

Overall, the following important conclusions can be drawn. First, despite the availability of few different livelihood activities in the study areas, diversification as such does not contribute to higher overall income. More important is a household's ability to engage into lucrative sectors with better returns into its livelihood portfolio (e.g., khat production in Aba Gerima and *Acacia decurrens* charcoal production in Guder). Second, decisions to diversify to another livelihood activity are dependent on the gender of the household head, distance to markets, access to credit and extension services, membership in community-based organizations, dependency ratio, level of income, agro-ecological setting, household size, exposure to shocks, livestock holding, and agricultural intensification. Third, a higher extent of livelihood diversification favors a lower level of adoption of SLM practices, whereas the reverse is true for a higher level of SLM adoption. A prospective look at livelihoods in the context of these factors would likely help households to be able to exploit new economic opportunities more effectively in the future. Experiencing stress/shocks tended to push households toward earning short-term economic gains; this can be seen as a coping strategy, but it is not a sustainable way of promoting livelihoods in these watersheds.

Results have implications for development planners in the Upper Blue Nile basin and elsewhere in the country where rain fed agriculture is predominantly make up the livelihoods of the majority of its population there by diversifying livelihood sources is essentially needed to sustain and promote livelihoods. More emphasis should be given to remunerative livelihoods like *Acacia decurrens* and khat plantations while not disregarding their environmental and social feasibility. Sustainable livelihood initiatives that focus on increasing access to financial support mechanisms, improved livestock production, quality extension services, and shock/stress resilience mechanisms, while also accounting for agro-ecological differences, are much needed. In this regard, livelihood transformations to cash crops like *Acacia decurrens* in Guder and khat plantation in Aba Gerima should profoundly be studied as to how they are ecologically, socially and economically feasible. Strengthening policies that encourage positive interplay between diversifying livelihood strategies and SLM practices could help in the attempt to achieve a sustainable agriculture system. This cross-sectional study contributes to the literature of the nexus between rural livelihoods and management of natural resources using three selected watersheds of Upper Blue Nile basin. Hence, further lines of research on a broader location, dynamic links between livelihood diversification and adoption intensity of SLM practices, and economic benefits of SLM practices will be helpful.

Author Contributions: Conceptualization, M.T.A.; Data curation, M.T.A.; Formal analysis, M.T.A.; Funding acquisition, A.T., E.A. and N.H.; Investigation, M.T.A.; Methodology, M.T.A. and Z.N.; Project administration, A.T., E.A., N.H. and Z.N.; Resources, A.T. and N.H.; Supervision, A.T., E.A., N.H., Z.N., Z.A., A.E., D.M. and D.B.; Validation, A.T.; Writing—original draft, M.T.A.; Writing—review & editing, A.T., E.A., N.H., Z.N. and Z.A.

Funding: This research was funded by the Science and Technology Research Partnership for Sustainable Development (SATREPS)—Development of a Next-Generation Sustainable Land Management (SLM) Framework to Combat Desertification project, Grant Number JPMJSA1601, Japan Science and Technology Agency (JST)/Japan International Cooperation Agency (JICA).

Acknowledgments: The authors are grateful to all respondents for their willingness to provide data, and to Anteneh Wubet and Nigus Tadesse for their field assistance.

Conflicts of Interest: The authors declare no potential conflict of interest.

References

1. FAO. *The State of Food Insecurity in the World 2015. Meeting the 2015 International Hunger Targets: Taking Stock of Uneven Progress*; FAO: Rome, Italy, 2015; pp. 1–62.
2. UN. *The 2015 Revision of the UN's World Population Projections*; Wiley Black well: New York, NY, USA, 2015; pp. 557–561.

3.   Rosegrant, M.W.; Cai, X.; Cline, S.A.; Nakagawa, N. *The Role of Rainfed Agriculture in the Future of Global Food Production*; IFPRI: Washington, DC, USA, 2002.

4.   Alobo Loison, S. Rural livelihood diversification in Sub-Saharan Africa: A literature review. *J. Dev. Stud.* **2015**, *51*, 1125–1138. [CrossRef]

5.   Ellis, F. *Rural Livelihoods and Diversity in Developing Countries*; Oxford University Press: Oxford, UK, 2000.

6.   Ellis, F. Household strategies and rural livelihood diversification. *J. Dev. Stud.* **1998**, *35*, 1–38. [CrossRef]

7.   Baffoe, G.; Matsuda, H. Why do rural communities do what they do in the context of livelihood activities? Exploring the livelihood priority and viability nexus. *Community Dev.* **2017**, *48*, 715–734. [CrossRef]

8.   Assan, J.K.; Beyene, F.R. Livelihood impacts of environmental conservation programmes in the Amhara region of Ethiopia. *J. Sustain. Dev.* **2013**, *6*, 87–105. [CrossRef]

9.   Kassie, G.W. The Nexus between livelihood diversification and farmland management strategies in rural Ethiopia. *Cogent Econ. Financ.* **2017**, *5*, 1–16. [CrossRef]

10.   Cordingley, J.E.; Snyder, K.A.; Rosendahl, J.; Kizito, F.; Bossio, D. Thinking outside the plot: Addressing low adoption of sustainable land management in sub-Saharan Africa. *Curr. Opin. Environ. Sustain.* **2015**, *15*, 35–40. [CrossRef]

11.   Daregot, B.; Ayalneh, B.; Belay, K.; Degnet, A. Poverty and Natural Resources Degradation: Analysis of their Interactions in Lake Tana Basin, Ethiopia. *J. Int. Dev.* **2015**, *27*, 516–527. [CrossRef]

12.   Kassie, M.; Jaleta, M.; Shiferaw, B.; Mmbando, F.; Mekuria, M. Adoption of interrelated sustainable agricultural practices in smallholder systems: Evidence from rural Tanzania. *Technol. Forecast. Soc. Chang.* **2013**, *80*, 525–540. [CrossRef]

13.   Kelly, P.; Huo, X. Land Retirement and Nonfarm Labor Market Participation: An Analysis of China's Sloping Land Conversion Program. *World Dev.* **2013**, *48*, 156–169. [CrossRef]

14.   Lee, D.R. Agricultural sustainability and technology adoption: Issues and policies for developing countries. *Am. J. Agric. Econ.* **2005**, *87*, 1325–1334. [CrossRef]

15.   Liu, Z.; Lan, J. The sloping land conversion program in China: Effect on the livelihood diversification of rural households. *World Dev.* **2015**, *70*, 147–161. [CrossRef]

16.   Harcourt, C.; Sayer, J.A. *Conservation Atlas of Tropical Forests: The Americas*; World Conservation Union and World Conservation Monitoring Centre: Cambridge, UK, 1996.

17.   Salafsky, N.; Wollenberg, E. Linking livelihoods and conservation: A conceptual framework and scale for assessing the integration of human needs and biodiversity. *World Dev.* **2000**, *28*, 1421–1438. [CrossRef]

18.   Babatunde, R.; Olagunju, F.; Fakayode, S.; Adejobi, A. Determinants of participation in off-farm employment among small-holder farming households in Kwara State, Nigeria. *Prod. Agric. Technol.* **2010**, *6*, 1–14.

19.   Ellis, F. The determinants of rural livelihood diversification in developing countries. *J. Agric. Econ.* **2000**, *51*, 289–302. [CrossRef]

20.   Kassie, G.W.; Kim, S.; Fellizar, F.P., Jr. Determinant factors of livelihood diversification: Evidence from Ethiopia. *Cogent Soc. Sci.* **2017**, *3*, 1–16. [CrossRef]

21.   Martin, S.M.; Lorenzen, K. Livelihood diversification in rural Laos. *World Dev.* **2016**, *83*, 231–243. [CrossRef]

22.   Rahut, D.B.; Ali, A.; Kassie, M.; Marenya, P.P.; Basnet, C. Rural livelihood diversification strategies in Nepal. *Poverty Public Policy* **2014**, *6*, 259–281. [CrossRef]

23.   Baffoe, G.; Matsuda, H. Understanding the Determinants of Rural Credit Accessibility: The Case of Ehiaminchini, Fanteakwa District. *Ghana J. Sustain. Dev.* **2015**, *8*, 183–195. [CrossRef]

24.   Baffoe, G.; Matsuda, H.; Nagao, M.; Akiyama, T. The dynamics of rural credit and its impacts on agricultural productivity: An empirical study in rural Ghana. *OIDA Int. J. Sustain. Dev.* **2014**, *7*, 19–34.

25.   Riithi, A.N.; Irungu, P.; Munei, K. *Determinants of Choice of Alternative Livelihood Diversification Strategies in Solio Resettlement Scheme, KENYA*; University of Nairobi: Nairobi, Kenya, 2015.

26.   Dercon, S.; Krishnan, P. Income portfolios in rural Ethiopia and Tanzania: Choices and constraints. *J. Dev. Stud.* **1996**, *32*, 850–875. [CrossRef]

27.   Woldehanna, T. Rural farm/nonfarm income linkages in northern Ethiopia. In *Promoting Farm/Nonfarm Linkages for Rural Development: Case Studies from Africa and Latin America*; FAO: Rome, Italy, 2002; pp. 121–144.

28.   Asfaw, A.; Simane, B.; Hassen, A.; Bantider, A. Determinants of non-farm livelihood diversification: Evidence from rainfed-dependent smallholder farmers in northcentral Ethiopia (Woleka sub-basin). *Dev. Stud. Res.* **2017**, *4*, 22–36. [CrossRef]

29. Bezu, S.; Holden, S. Are rural youth in Ethiopia abandoning agriculture? *World Dev.* **2014**, *64*, 259–272. [CrossRef]

30. Carswell, G. Livelihood diversification: Increasing in importance or increasingly recognized? Evidence from southern Ethiopia. *J. Int. Dev.* **2002**, *14*, 789–804. [CrossRef]

31. Davis, B.; Di Giuseppe, S.; Zezza, A. Are African households (not) leaving agriculture? Patterns of households' income sources in rural Sub-Saharan Africa. *Food Policy* **2017**, *67*, 153–174. [CrossRef]

32. Kassie, G.W. Agroforestry and farm income diversification: Synergy or trade-off? The case of Ethiopia. *Environ. Syst. Res.* **2018**, *6*, 1–14. [CrossRef]

33. Belay, M.; Bewket, W. Farmers' livelihood assets and adoption of sustainable land management practices in north-western highlands of Ethiopia. *Int. J. Environ. Stud.* **2013**, *70*, 284–301. [CrossRef]

34. Deininger, K.; Jin, S. Tenure security and land-related investment: Evidence from Ethiopia. *Eur. Econ. Rev.* **2006**, *50*, 1245–1277. [CrossRef]

35. Holden, S.; Shiferaw, B. Land degradation, drought and food security in a less-favoured area in the Ethiopian highlands: A bio-economic model with market imperfections. *Agric. Econ.* **2004**, *30*, 31–49. [CrossRef]

36. Jagger, P.; Pender, J. *Impacts of Programs and Organizations on the Adoption of Sustainable Land Management Technologies in Uganda*; Environmentand Production Technology Division Discussion Paper No. 101; IFPRI: Washington, DC, USA, 2003; pp. 277–307.

37. Pender, J.; Gebremedhin, B. Determinants of agricultural and land management practices and impacts on crop production and household income in the highlands of Tigray, Ethiopia. *J. Afr. Econ.* **2007**, *17*, 395–450. [CrossRef]

38. Berhanu, W.; Colman, D.; Fayissa, B. Diversification and livelihood sustainability in a semi-arid environment: A case study from southern Ethiopia. *J. Dev. Stud.* **2007**, *43*, 871–889. [CrossRef]

39. Block, S.; Webb, P. The dynamics of livelihood diversification in post-famine Ethiopia. *Food Policy* **2001**, *26*, 333–350. [CrossRef]

40. Canali, M.; Slaviero, F. Food Insecurity and Risk Management of Smallholder Farming Systems in Ethiopia. In Proceedings of the Ninth European IFSA Symposium, Vienna, Austria, 4–7 July 2010; pp. 4–7.

41. Vaitla, B.; Tesfay, G.; Rounseville, M.; Maxwell, D. *Resilience and Livelihoods Change in Tigray, Ethiopia*; Feinstein International Center, Tufts University: Somerville, MA, USA, 2012.

42. García-Fajardo, B.; Orozco-Hernández, M.E.; McDonagh, J.; Álvarez-Arteaga, G.; Mireles-Lezama, P. Land management strategies and their implications for Mazahua farmers' livelihoods in the Highlands of Central Mexico. *Misc. Geogr.* **2016**, *20*, 5–12. [CrossRef]

43. World Bank. *Sustainable Land Management Sourcebook*; World Bank: Washington, DC, USA, 2008.

44. Escobal, J. The determinants of nonfarm income diversification in rural Peru. *World Dev.* **2001**, *29*, 497–508. [CrossRef]

45. Kowalski, J.; Lipcan, A.; McIntosh, K.; Smida, R.; Sørensen, S.J.; Seff, I.; Jolliffe, D. Nonfarm enterprises in rural Ethiopia: Improving livelihoods by generating income and smoothing consumption? *Ethiop. J. Econ.* **2016**, *25*, 171–204.

46. Yona, Y.; Mathewos, T. Assessing challenges of non-farm livelihood diversification in Boricha Woreda, Sidama zone. *J. Dev. Agric. Econ.* **2017**, *9*, 87–96.

47. Carswell, G. *Livelihood Diversification in Southern Ethiopia*; IDS Working Paper 117; Institute of Development Studies: Brighton, UK, 2000.

48. Adgo, E.; Teshome, A.; Mati, B. Impacts of long-term soil and water conservation on agricultural productivity: The case of Anjenie watershed, Ethiopia. *Agric. Water Manag.* **2013**, *117*, 55–61. [CrossRef]

49. D'souza, G.; Cyphers, D.; Phipps, T. Factors affecting the adoption of sustainable agricultural practices. *Agric. Resour. Econ. Rev.* **1993**, *22*, 159–165. [CrossRef]

50. Haregeweyn, N.; Tsunekawa, A.; Nyssen, J.; Poesen, J.; Tsubo, M.; Tsegaye Meshesha, D.; Schütt, B.; Adgo, E.; Tegegne, F. Soil erosion and conservation in Ethiopia: A review. *Prog. Phys. Geogr.* **2015**, *39*, 750–774. [CrossRef]

51. Schmidt, E.; Zemadim, B. Expanding sustainable land management in Ethiopia: Scenarios for improved agricultural water management in the Blue Nile. *Agric. Water Manag.* **2015**, *158*, 166–178. [CrossRef]

52. Teklewold, H.; Kassie, M.; Shiferaw, B. Adoption of multiple sustainable agricultural practices in rural Ethiopia. *J. Agric. Econ.* **2013**, *64*, 597–623. [CrossRef]

53. Nigussie, Z.; Tsunekawa, A.; Haregeweyn, N.; Adgo, E.; Nohmi, M.; Tsubo, M.; Aklog, D.; Meshesha, D.T.; Abele, S. Factors influencing small-scale farmers' adoption of sustainable land management technologies in north-western Ethiopia. *Land Use Policy* **2017**, *67*, 57–64. [CrossRef]

54. Adimassu, Z.; Kessler, A.; Hengsdijk, H. Exploring determinants of farmers' investments in land management in the Central Rift Valley of Ethiopia. *Appl. Geogr.* **2012**, *35*, 191–198. [CrossRef]

55. Bezu, S.; Barrett, C.B.; Holden, S.T. Does the nonfarm economy offer pathways for upward mobility? Evidence from a panel data study in Ethiopia. *World Dev.* **2012**, *40*, 1634–1646. [CrossRef]

56. World Bank. *Determinants of the Adoption of Sustainable Land Management Practices and Their Impacts in the Ethiopian Highlands Public*; World Bank: Washington, DC, USA, 2007; Volume 1, pp. 1–23.

57. Ebabu, K.; Tsunekawa, A.; Haregeweyn, N.; Adgo, E.; Meshesha, D.T.; Aklog, D.; Masunaga, T.; Tsubo, M.; Sultan, D.; Fenta, A.A. Analyzing the variability of sediment yield: A case study from paired watersheds in the Upper Blue Nile basin, Ethiopia. *Geomorphology* **2018**, *303*, 446–455. [CrossRef]

58. Sultan, D.; Tsunekawa, A.; Haregeweyn, N.; Adgo, E.; Tsubo, M.; Meshesha, D.T.; Masunaga, T.; Aklog, D.; Fenta, A.A.; Ebabu, K. Efficiency of soil and water conservation practices in different agro-ecological environments in the Upper Blue Nile Basin of Ethiopia. *J. Arid Land* **2018**, *10*, 249–263. [CrossRef]

59. Dedehouanou, S.F.A.; Araar, A.; Ousseini, A.; Harouna, A.L.; Jabir, M. Spillovers from off-farm self-employment opportunities in rural Niger. *World Dev.* **2018**, *105*, 428–442. [CrossRef]

60. Shiferaw, K.; Gebremedhin, B.; Legesse, D. *Factors Determining Household Allocation of Credit to Livestock Production in Ethiopia*; LIVES Working Paper 21; International Livestock Research Institute (ILRI): Nairobi, Kenya, 2016; pp. 1–33.

61. Mutyasira, V.; Hoag, D.; Pendell, D.; Manning, D. Is sustainable intensification possible? Evidence from Ethiopia. *Sustainability* **2018**, *10*, 4174. [CrossRef]

62. Berjan, S.; El Bilali, H.; Sorajic, B.; Driouech, N.; Despotovic, A.; Simic, J. Off-farm and non-farm activities development in rural south-eastern Bosnia. *Int. J. Environ. Rural Dev.* **2013**, *4*, 130–135.

63. Buckley, J.J. Stochastic dominance: An approach to decision making under risk. *Risk Anal.* **1986**, *6*, 35–41. [CrossRef]

64. Johny, J.; Wichmann, B.; Swallow, B.M. Characterizing social networks and their effects on income diversification in rural Kerala, India. *World Dev.* **2017**, *94*, 375–392. [CrossRef]

65. Nagendra, H. Opposite trends in response for the Shannon and Simpson indices of landscape diversity. *Appl. Geogr.* **2002**, *22*, 175–186. [CrossRef]

66. Khatun, D.; Roy, B.C. Rural livelihood diversification in West Bengal: Nature and extent. *Agric. Econ. Res. Rev.* **2016**, *29*, 183–190. [CrossRef]

67. Djido, A.I.; Shiferaw, B.A. Patterns of labor productivity and income diversification–Empirical evidence from Uganda and Nigeria. *World Dev.* **2018**, *105*, 416–427. [CrossRef]

68. Smith, B.; Wilson, J.B. A consumer's guide to evenness indices. *Oikos* **1996**, *76*, 70–82. [CrossRef]

69. Brezina, I.; Pekár, J.; Čičková, Z.; Reiff, M. Herfindahl–Hirschman index level of concentration values modification and analysis of their change. *Cent. Eur. J. Oper. Res.* **2016**, *24*, 49–72. [CrossRef]

70. Kau, P.; Hill, L. Application of Multivariate Probit to a Threshold Model of Grain Dryer Purchasing Decisions. *Am. J. Agric. Econ.* **1973**, *55*, 19–27.

71. Velandia, M.; Rejesus, R.M.; Knight, T.O.; Sherrick, B.J. Factors affecting farmers' utilization of agricultural risk management tools: The case of crop insurance, forward contracting, and spreading sales. *J. Agric. Appl. Econ.* **2009**, *41*, 107–123. [CrossRef]

72. Greene, W.H. *Econometric Analysis*, 5th ed.; Prentice-Hall: Englewood Cliffs, NJ, USA, 2003.

73. Cappellari, L.; Jenkins, S.P. Multivariate probit regression using simulated maximum likelihood. *Stata J.* **2003**, *3*, 278–294. [CrossRef]

74. Wollni, M.; Lee, D.R.; Thies, J.E. Conservation agriculture, organic marketing, and collective action in the Honduran hillsides. *Agric. Econ.* **2010**, *41*, 373–384. [CrossRef]

75. Abdulai, A.; Huffman, W. The adoption and impact of soil and water conservation technology: An endogenous switching regression application. *Land Econ.* **2014**, *90*, 26–43. [CrossRef]

76. Rivers, D.; Vuong, Q.H. Limited information estimators and exogeneity tests for simultaneous probit models. *J. Econ.* **1988**, *39*, 347–366. [CrossRef]

77. FAO. *Small Family Farms Country Factsheet*; FAO: Rome, Italy, 2012; pp. 1–2.

78. Alemu, K.S. Contribution of khat kellas and the impacts of its closure to Ethiopian economy (the case of Hararghe khat kella). *Glob. J. Manag. Bus. Res.* **2015**, *XV*, 21–29.

79. Dachew, B.A.; Bifftu, B.B.; Tiruneh, B.T. Khat use and its determinants among university students in northwest Ethiopia: A multivariable analysis. *Int. J. Med. Sci. Public Health* **2015**, *4*, 319–323. [CrossRef]

80. Nigussie, Z.; Tsunekawa, A.; Haregeweyn, N.; Adgo, E.; Nohmi, M.; Tsubo, M.; Aklog, D.; Meshesha, D.T.; Abele, S. Factors affecting small-scale farmers' land allocation and tree density decisions in an acacia decurrens-based taungya system in Fagita Lekoma District, North-Western Ethiopia. *Small-scale For.* **2017**, *16*, 219–233. [CrossRef]

81. Achamyeleh, K. *Integration of Acacia Decurrens (J.C. Wendl.) Willd. into the Farming System, It's Effects on Soil Fertility and Comparative Economic Advantages in North Western Ethiopia*; Bahir Dar University: Bahir Dar, Ethiopia, 2015.

82. Ragasa, C.; Berhane, G.; Tadesse, F.; Taffesse, A.S. Gender differences in access to extension services and agricultural productivity. *J. Agric. Educ. Ext.* **2013**, *19*, 437–468. [CrossRef]

83. Eswaran, M.; Kotwal, A. Access to capital and agrarian production organisation. *Econ. J.* **1986**, *96*, 482–498. [CrossRef]

84. Ellis, F.; Bahiigwa, G. Livelihoods and rural poverty reduction in Uganda. *World Dev.* **2003**, *31*, 997–1013. [CrossRef]

85. Kung, J.K.; Lee, Y.-F. So what if there is income inequality? The distributive consequence of nonfarm employment in rural China. *Econ. Dev. Cult. Chang.* **2001**, *50*, 19–46. [CrossRef]

86. Jiao, X.; Pouliot, M.; Walelign, S.Z. Livelihood strategies and dynamics in rural Cambodia. *World Dev.* **2017**, *97*, 266–278. [CrossRef]

87. Mulwa, C.; Marenya, P.; Kassie, M. Response to climate risks among smallholder farmers in Malawi: A multivariate probit assessment of the role of information, household demographics, and farm characteristics. *Clim. Risk Manag.* **2017**, *16*, 208–221. [CrossRef]

88. Mentamo, M.; Geda, N.R. Livelihood diversification under severe food insecurity scenario among smallholder farmers in Kadida Gamela District, Southern Ethiopia. *Kontakt* **2016**, *18*, e258–e264. [CrossRef]

89. Chikobola, M.M.; Sibusenga, M. Employment and Income Sources: Key Determinants of Off-Farm Activity Participation Among Rural Households in Northern Zambia. *J. Agric. Econ.* **2016**, *1*, 91–98.

90. Alobo Loison, S.; Bignebat, C. Patterns and determinants of household income diversification in rural Senegal and Kenya. *J. Poverty Alleviation Int. Dev.* **2017**, *8*, 93–126.

91. Sallawu, H.; Tanko, L.; Nmadu, J.; Ndanitsa, A. Determinants of income diversification among farm households in niger State, Nigeria. *Russ. J. Agric. Socio-Econ. Sci.* **2016**, *50*, 55–65. [CrossRef]

92. Gautam, Y.; Andersen, P. Rural livelihood diversification and household well-being: Insights from Humla, Nepal. *J. Rural Stud.* **2016**, *44*, 239–249. [CrossRef]

93. Dercon, S. Income risk, coping strategies, and safety nets. *World Bank Res. Obs.* **2002**, *17*, 141–166. [CrossRef]

94. Kasie, T.A.; Adgo, E.; Botella, A.G.; García, I.G. Measuring resilience properties of household livelihoods and food security outcomes in the risky environments of Ethiopia. *Iberoam. J. Dev. Stud.* **2018**, *7*, 52–80. [CrossRef]

95. Kijima, Y.; Matsumoto, T.; Yamano, T. Nonfarm employment, agricultural shocks, and poverty dynamics: Evidence from rural Uganda. *Agric. Econ.* **2006**, *35*, 459–467. [CrossRef]

96. Newhouse, D. The persistence of income shocks: Evidence from rural Indonesia. *Rev. Dev. Econ.* **2005**, *9*, 415–433. [CrossRef]

97. Sanders, C.; McKay, K. Where have all the young men gone? Social fragmentation during rapid neoliberal development in Nepal's Himalayas. *Hum. Organ.* **2014**, *73*, 25–37. [CrossRef]

98. Verkaart, S.; Orr, A.; Harris, D.; Claessens, L. *Intensify or Diversify? Agriculture as a Pathway from Poverty in Eastern Kenya*; Series Paper Number 40; ICRISAT: Nairobi, Kenya, 2017.

99. Tesfaye, Y.; Roos, A.; Campbell, B.M.; Bohlin, F. Livelihood strategies and the role of forest income in participatory-managed forests of Dodola area in the bale highlands, southern Ethiopia. *For. Policy Econ.* **2011**, *13*, 258–265. [CrossRef]

100. Peng, W.; Zheng, H.; Robinson, B.; Li, C.; Wang, F. Household livelihood strategy choices, impact factors, and environmental consequences in Miyun reservoir watershed, China. *Sustainability* **2017**, *9*, 175. [CrossRef]

101. Amare, D. Determinants of livelihood diversification strategies in Borena pastoralist communities of Oromia regional state, Ethiopia. *Agric. Food Secur.* **2018**, *7*, 41.

102. Robaa, B.; Tolossa, D. Rural livelihood diversification and its effects on household food security: A case study at Damota Gale Woreda, Wolayta, Southern Ethiopia. *East. Afr. Soc. Sci. Res. Rev.* **2016**, *32*, 93–118. [CrossRef]
103. Addisu, Y. Livelihood strategies and diversification in western tip pastoral areas of Ethiopia. *Pastor. Res. Policy Pract.* **2017**, 1–9. [CrossRef]
104. Gebrehiwot, G. Determinants of livelihood diversification strategies in Eastern Tigray Region of Ethiopia. *Agric. Food Secur.* **2018**, *7*, 1–9.
105. DFID of UK. *Sustainable Livelihoods Guidance Sheets*; DFID: London, UK, 1999; p. 445.
106. Aryal, J.P.; Jat, M.; Sapkota, T.B.; Khatri-Chhetri, A.; Kassie, M.; Rahut, D.B.; Maharjan, S. Adoption of multiple climate-smart agricultural practices in the Gangetic plains of Bihar, India. *Int. J. Clim. Chang. Strateg. Manag.* **2018**, *10*, 407–427. [CrossRef]
107. Nguyen, T.T.; Nguyen, L.D.; Lippe, R.S.; Grote, U. Determinants of farmers' land use decision-making: Comparative evidence from Thailand and Vietnam. *World Dev.* **2017**, *89*, 199–213. [CrossRef]

Article

Soil Structure Stability under Different Land Uses in Association with Polyacrylamide Effects

Amrakh I. Mamedov [1,*], Atsushi Tsunekawa [1], Nigussie Haregeweyn [2], Mitsuru Tsubo [1], Haruyuki Fujimaki [1], Takayuki Kawai [1], Birhanu Kebede [3], Temesgen Mulualem [4], Getu Abebe [3], Anteneh Wubet [4] and Guy J. Levy [5]

[1] Arid Land Research Center, Faculty of Agriculture, Tottori University, Tottori 668-0001, Japan; tsunekawa@tottori-u.ac.jp (A.T.); tsubo@tottori-u.ac.jp (M.T.); fujimaki@tottori-u.ac.jp (H.F.); kawai_alrc@tottori-u.ac.jp (T.K.)

[2] International Platform for Dryland Research and Education, Tottori University, Tottori 668-0001, Japan; nigussie_haregeweyn@yahoo.com

[3] The United Graduate School of Agricultural Sciences, Tottori University, Tottori 680-8553, Japan; birhanukg@gmail.com (B.K.); gabebe233@gmail.com (G.A.)

[4] College of Agriculture & Environmental Sciences, Bahir Dar University, Bahir Dar P.O. Box 1289, Ethiopia; tommm2001@yahoo.com (T.M.); a.wubet@yahoo.com (A.W.)

[5] Institute of Soil, Water and Environmental Sciences, ARO, Rishon LeZion 7505101, Israel; vwguy@volcani.agri.gov.il

[*] Correspondence: amrakh03@yahoo.com

Academic Editor: Alejandro Rescia

Received: 1 January 2021; Accepted: 26 January 2021; Published: 29 January 2021

Abstract: Soil structural stability is a vital aspect of soil quality and functions, and of maintaining sustainable land management. The objective of this study was to compare the contribution of four long-term land-use systems (crop, bush, grass, and forest) coupled with anionic polyacrylamide (PAM = 0, 25, and 200 mg L^{-1}) application on the structural stability of soils in three watersheds of Ethiopia varying in elevation. Effect of treatments on soil structural stability indices were assessed using the high energy moisture characteristic (HEMC, 0–50 hPa) method, which provides (i) water retention model parameters α and n, and (ii) soil structure index (SI). Soil (watershed), land use and PAM treatments had significant effects on the shape of the water retention curves (α, n) and SI, with diverse changes in the macropore sizes (60–250; >250 µm). Soil organic carbon (SOC) content and SI were strongly related to soil pH, CaCO$_3$ soil type-clay mineralogy, exchangeable Ca^{2+}, and Na$^+$ (negatively). The order of soil SI (0.013–0.064 hPa^{-1}) and SOC (1.4–8.1%) by land use was similar (forest > grass > bush > cropland). PAM effect on increasing soil SI (1.2–2.0 times), was inversely related to SOC content, being also pronounced in soils from watersheds of low (Vertisol) and medium (Luvisol) elevation, and the cropland soil from high (Acrisol) elevation. Treating cropland soils with a high PAM rate yielded greater SI (0.028–0.042 hPa^{-1}) than untreated bush- and grassland soils (0.021–0.033 hPa^{-1}). For sustainable management and faster improvement in soil physical quality, soil properties, and land-use history should be considered together with PAM application.

Keywords: land use; soil organic carbon; structure stability; soil type; polyacrylamide; dryland

---

1. Introduction

1.1. Soil Structure Stability and Its Importance

Soil structure and aggregation are central physical properties of soil that control a wide array of soil properties and functions including water retention and infiltration [1], susceptibility to erosion and the movement of associated contaminants [2], aeration, gaseous exchanges, and greenhouse gas emission [3],

C sequestration, soil organic carbon (SOC) protection [4], soil organic matter mineralization [5], and biogeochemical cycling of essential elements such as macro- and micronutrients [6]. Hence, monitoring of soil structure and stability is vital in determining the sustainability of land use and management practices in both agricultural and natural ecosystems [7].

In Ethiopian highlands, like to similar regions, soils under long-term grassland and forest may contain higher levels of SOC content, which is characterized by persistent particulate and mineral-associated fractions, enhanced microbial activity, and a positive C balance [8]. All those, in turn, may improve soil hydraulic properties, and yield a better soil structure with a heterogenic aggregate and pore size distribution (PSD), compared with soils under long-term cropping, where soil structure is substantially modified by soil management [9,10]. Changing land use from grassland or forest to cropland significantly impacts soil quality, decreases SOC, and negatively affects soil functions such as soil aggregation and gas exchange, water and nutrients availability, and plant growth [7,11,12]. Restoring good soil structure and the proper functioning of degraded cropland soils, by using traditional conservation measures such as no-till practices, are highly dependent on climate and soil type, and texture, and may take over 20–40 years; yet significant positive trends could be noted already after 5–10 years of conversion of cropped land to grassland [12–16].

1.2. Ethiopian Highland Soils: Land Use and Soil Degradation

The Ethiopian highlands, an important region of natural biodiversity and agriculture production, are characterized by high erosion rates (interrill, rill, and gully, >40 t ha^{-1}) and of CO_2 efflux, and the deterioration of soil hydrological characteristics and cycles. The issue of land degradation is further complicated by global climate change associated with monsoon–type behavior, delivering significant June–September rainfall and flooding, that turn the soil resources vulnerable to physical and chemical degradation [17,18]. Sustainable watershed management and regreening projects (e.g., conservation systems, building terraces, the establishment of agroforestry), promoted by the Ethiopian government about 20 years ago, have led to satisfactorily integrated sustainable land management and livelihoods activities. Yet, erosion in many areas remains challenging. Moreover, the predicted changes in climate and land use indicate great variability and high levels of erosion rates in the upstream areas and increasing sediment loads in the Blue Nile [19].

Numerous studies in natural and farmed field plots in Ethiopian watersheds, on the impacts of short- and long-term land use and management (e.g., agroforest, grass, and croplands), and conservation measures (e.g., soil bunds by grass, use of enclosure, restrained grazing) on the improvement of soil quality, have recently been performed. The results for SOC, cation exchange capacity (CEC), exchangeable cations (Ca^{2+}, Mg^{2+}, Na^+), pH, and available nutrients (P and K), soil microbial activity, aggregate stability, crop productivity, and erosion, showed that the efficiency of change in land use and implementation of conservation practices were related to the prevailing agro-ecology and the nature and duration of the used conservation measures [20,21]. Results of these local studies were in line with results from international ones [11,14–16,22]. Moreover, the decline in soil quality by land-use change and under high rain intensity, were related to CEC and clay depletion, translocation and leaching or losses of particulate and dissolved organic matter by runoff and erosion, and clogging of subsoil macropores by the suspended clay-size material. Heavy rains may also cause soil-saturation-related runoff and subsequent shallow lateral flow that increases the loss of fine particles from diverse land uses [8,23,24].

1.3. Polyacrylamide (PAM) as a Soil Stabilizing Agent

The future of sustainable land management should be based on appropriate practices under different land-use types, which allow the existence of agricultural production in balance with crop and soil systems to support adequate drainage capacity and soil quality [12]. It should be noted that, despite its importance, the contribution of land use, conservation measures, and amendments (e.g., polyacrylamide, lime, manure, biochar) on soil hydraulic properties (water retention, saturated-unsaturated hydraulic conductivity, effective porosity, infiltration rate) and structure stability of Ethiopian soils, has received merely limited attention [25–27].

Application of polyacrylamide (PAM) can considerably (i) enhance soil structure, pore continuity, and aeration via binding of soil particles and eliminating loss of clay particles [28], (ii) sustain SOC accumulation through increasing microbial activity, protecting of SOC in macro- and micro-aggregates, and slowing down of residue decomposition and mineralization rate [29,30], (iii) increase SOC accumulation in subsurface soil layers by providing an appropriate medium for the growth of plants (root penetration, uptake of water and nutrients) [31,32], and (iv) mitigate on-site and off-site impacts of erosion by decreasing runoff generation and erosion rate. PAM addition can, therefore, control the quality of water in various land use areas and could be used for restoring marginal, grass, and forest land [33,34].

Recent studies also revealed that application of PAM could improve soil structure stability and protect SOC, and increase the diversity of soil bacterial communities by increasing soil moisture content and regulating the C to N ratio, regardless of soil with and without plants [35,36]. Therefore, it is expected that PAM application to the Ethiopian highland soils could improve soil structure stability that in turn may facilitate SOC protection [32] and possibly increase SOC storage. It is further anticipated that the use of PAM could replace, at least partly, traditional conservation measures, which often include high-cost conservation practices whose impact may emerge only after 3–4 decades [13,16]. The objective of this study was, therefore, to compare the contribution of four long-term land-use systems (crop, bush, grass, and forest) resulting in significantly varying soil properties and SOC content, with and without anionic PAM application, on structure stability of soils in three watersheds of Ethiopian highlands varying in elevation level.

2. Materials and Methods

2.1. Study Area and Soil Sampling

The Upper Blue Nile basin of Ethiopia is characterized by a monsoon climate with a dry (November to April) and a wet (May to October) season, with >80% of the annual precipitation occurring from June to September. Three watersheds in this basin, Guder, Abagerima, and Dibatie, differing in soils, elevation (2500–2900, 1900–2100, and 1500–1700 m), mean annual rainfall (2450, 1340, and 1020 mm), and temperature (19, 23, and 24 °C), respectively, represent three different agro-ecological zones, that should be considered for sustainable land management [37]. At these three watersheds, livestock types are similar, and croplands occupy a larger percentage of the area than grazing, bushlands, and forest [37]. In Guder, soil type was Acrisol (Alfisol/Ultisol by US taxonomy) with an acidic reaction, and mixed clay mineralogy with a high content of kaolinite (and Fe, Al oxides), and a small fraction of smectite. In Abagerima, the soil type was Luvisol (Cambisol/Alfisol by US taxonomy) with an acidic reaction and mixed kaolinite-smectite clay mineralogy. In Dibatie, the soil type was Vertisol with a slightly acidic reaction and predominantly smectite clay mineralogy. The parent material of the studied area is related to volcanic rocks (e.g., mainly basalt, then rhyolite, tuff, ignimbrites). Both Acrisol and Luvisol develop agric horizons resulting from clay dispersion, transport, and accumulation [23,38].

Soil samples (0–15 cm) were collected from experimental field plots (~0.1–0.2 ha) located in each of the three watersheds. The plots were placed in four long-term land-use systems (cropland, degraded bushland, grassland for grazing, and forest, hereafter referred to as crop, grass, bush, and forest land) that exhibited varying soil properties, including significantly differing SOC contents (Table 1). In each of the 12 sites (3 watersheds × 4 land uses), three randomized samples (used as replicates), each of ~0.5 kg were taken from the same experimental plot, using a soil probe sampler. The samples were brought to the laboratory, air dried, and ground to pass through a 2-mm sieve and then analyzed. The soils were characterized for (i) electrical conductivity (EC) and pH in a 1:2.5 soil:water extraction; (ii) particle size distribution using the hydrometer method, (iii) cation exchange capacity by sodium acetate, (iv) exchangeable cations by ammonium acetate, (v) calcium carbonate content using the volumetric calcimeter method, and (vi) organic matter content by wet combustion [39]. The determined properties of the soils are presented in Table 1. Also, aggregate separation to group sizes of 0.5–1.0 mm by dry sieving was done for HEMC measurement, and <0.25 mm by wet sieving was performed for SOC determination in micro-aggregates to be compared with SOC of bulk soil (<2 mm) [39].

Table 1. Selected physical-chemical properties (mean ± standard deviation) of the soils sampled from 0–15 cm depth in the study sites. In each column, means labeled with the same letter are not significantly different at $p < 0.05$ based on the Tukey–Kramer HSD test.

Soil Watershed Elevation	Land Use	pH	EC	Particle Size Class, %			SOC
		1:2.5	dS/m	Sand	Silt	Clay	%
Acrisol Guder High	Crop	5.18 ± 0.09 f	0.18 ± 0.01 ij	42	40	18 ± 0.9 f	1.83 ± 0.09 f
	Bush	5.73 ± 0.11 d	0.21 ± 0.02 hi	56	32	12 ± 1.0 g	5.67 ± 0.27 b
	Grass	5.68 ± 0.10 d	0.25 ± 0.01 g	38	42	20 ± 0.7 ef	5.42 ± 0.30 b
	Forest	6.16 ± 0.12 c	0.23 ± 0.01 gh	72	18	10 ± 0.8 g	8.03 ± 0.38 a
Luvisol Abagerima Medium	Crop	5.21 ± 0.10 f	0.20 ± 0.01 hi	52	26	22 ± 1.1 de	1.39 ± 0.07 g
	Bush	5.43 ± 0.11 e	0.53 ± 0.04 e	32	40	28 ± 1.2 c	2.32 ± 0.13 e
	Grass	5.38 ± 0.10 e	0.64 ± 0.03 d	34	42	24 ± 1.4 d	2.61 ± 0.12 d
	Forest	6.43 ± 0.13 b	0.94 ± 0.06 a	60	16	24 ± 1.3 d	8.07 ± 0.40 a
Vertisol Dibatie Low	Crop	6.05 ± 0.12 c	0.16 ± 0.01 j	16	18	66 ± 3.3 a	1.69 ± 0.09 fg
	Bush	6.59 ± 0.12 a	0.77 ± 0.02 c	16	28	56 ± 3.4 b	2.28 ± 0.13 e
	Grass	6.15 ± 0.14 c	0.29 ± 0.05 f	12	20	68 ± 2.8 a	2.50 ± 0.11 de
	Forest	6.67 ± 0.11 a	0.87 ± 0.05 b	18	24	58 ± 2.9 b	5.11 ± 0.26 c

Soil Watershed Elevation	Land use	Exchangeable cations, cmol$_c$/kg				CEC	CaCO$_3$
		Ca	Mg	Na	K	cmol$_c$/kg	%
Acrisol Guder High	Crop	9.4 ± 0.47 g	3.6 ± 0.14 fg	4.9	1.5	26.4	0.28 ± 0.01 g
	Bush	16.1 ± 0.51 b	3.3 ± 0.15 hi	5.1	1.6	29.6	0.48 ± 0.01 g
	Grass	10.2 ± 0.81 f	3.7 ± 0.10 ef	4.7	1.6	37.5	0.32 ± 0.02 g
	Forest	14.0 ± 0.70 d	3.2 ± 0.13 i	4.5	1.3	39.6	1.28 ± 0.05 f
Luvisol Abagerima Medium	Crop	10.1 ± 0.50 f	3.8 ± 0.15 d	3.3	1.4	25.1	4.01 ± 0.16 c
	Bush	11.3 ± 0.59 e	3.8 ± 0.18 de	3.4	1.5	34.0	2.72 ± 0.22 e
	Grass	11.8 ± 0.56 e	4.3 ± 0.16 b	4.1	2.2	35.4	5.60 ± 0.11 a
	Forest	17.8 ± 0.89 a	3.5 ± 0.14 fg	3.9	3.0	33.6	5.80 ± 0.23 a
Vertisol Dibatie Low	Crop	9.9 ± 0.52 fg	3.5 ± 0.14 gh	6.3	1.9	26.5	3.52 ± 0.12 d
	Bush	13.5 ± 0.67 d	4.5 ± 0.16 a	2.9	2.5	32.4	4.56 ± 0.15 b
	Grass	13.4 ± 0.64 d	4.0 ± 0.18 c	2.4	2.0	29.5	3.44 ± 0.18 d
	Forest	14.7 ± 0.73 c	4.2 ± 0.17 b	3.4	1.8	36.0	4.42 ± 0.20 b

pH and EC were determined in the solution part of 1:2.5 soil:water extract using deionized water. EC: electrical conductivity, CEC: cation exchange capacity, SOC: soil organic matter.

2.2. Preparation of PAM-Treated Soil Aggregates

Treating aggregates with PAM solution (0, 25, and 200 mg L^{-1}) was conducted in accordance with the procedure detailed in Mamedov et al. (2010) [40]. The tested PAM concentration were based on previous relevant studies, where various PAM rates were used to understand the mechanism of soil structure stabilization, and to evaluate a strategy of PAM application along with other conservation measures [32–34]. An anionic PAM of high molecular weight (~18 × 10^6 Da) and 30% hydrolysis with a trading name of Superfloc A–110 (Kemira, Lakeland, Florida, USA) was used. Solutions of 25 and 200 mg L^{-1} PAM were prepared with tap water (electrical conductivity, [EC] = 0.4 dS m^{-1}, sodium adsorption ratio, [SAR] of = 1.2 [mmolc/L]$^{0.5}$, and pH = 6.5) under constant stirring and slow addition of PAM granules over 4 h. Plastic boxes (30 × 60 × 3 cm) were filled with very coarse sand to form a 5-mm thick layer that was then covered with a high porosity (>100 µm pore size) filter paper allowing PAM molecules to diffuse to the aggregates from the coarse sand layer. Aggregates (0.5–1.0 mm) from a given soil sample were gently spread on the filter paper to form a monolayer of aggregates, and then saturated from below with 25 or 200 mg L^{-1} PAM solution for 1 h (at a rate of 4 mm h^{-1}) using a peristaltic pump, and the boxes were then covered and kept in their respective solution for 24 h to reach equilibrium. Thereafter, the boxes were drained and the aggregates were placed in an oven to dry at 60 °C for 24 h, and then the aggregates were sieved to eliminate broken aggregates.

2.3. Determination of Soil Structural Stability Indices

Soil pore size distribution (PSD) and structure stability indices were determined for 108 samples (3 watersheds × 4 land uses × 3 PAM rates × 3 replicates), using the high–energy moisture characteristics (HEMC) method. In the HEMC method, aggregates' wetting rate is accurately controlled, and energy of hydration, differential swelling, and compression of entrapped air are the main forces responsible for breaking down of aggregates. This method enables the detection of small changes in soil structure and had been successfully applied in the determination of structure stability indices of soils from humid and arid regions with a wide range of stability [41,42].

Briefly, 15 g of aggregates (0.5–1.0 mm) were placed in a 60 mm I.D. funnel with a fritted disc (pore size: 20–40 μm) to form a bed ~5 mm thick with a bulk density of ~1.05 g cm^{-3}. Saturation of the fritted disc was ensured prior to placing aggregates in the funnel. The aggregates were wetted (fast = 100 mm h^{-1}) from the bottom with the deionized water (EC = 0.04 dS m^{-1}) using a peristaltic pump. Then a soil water retention curve (0 to −50 hPa), corresponding to drainable pores of 60 to 2000 μm, using small steps (1–2 hPa), was performed by the hanging column and pipette. Consequently, structure stability indices were then inferred from differences among the water retention and specific water capacity (dθ/dψ) curves (i.e., differences in PSD) of the treatments (Figure 1) by using the modified van Genuchten model [43,44]:

$$\theta = \theta_r + (\theta_s - \theta_r)\left[1 + (\alpha\,\psi)^n\right]^{(1/n-1)} + A\psi^2 + B\psi + C \tag{1}$$

$$d\theta/d\psi = (\theta_s - \theta_r)\left[1 + (\alpha\,\psi)^n\right]^{(1/n-1)}(1/n - 1)(\alpha\,\psi)^n n/\left[\psi(1 + (\alpha\,\psi)^n)\right] + 2A\psi + B \tag{2}$$

$$SI = VDP/MS \tag{3}$$

In Equations (1)–(3), ψ is the matric potential (hPa), θs and θr are pseudo saturated and residual gravimetric water contents, respectively. The parameters α (hPa^{-1}), and n (dimensionless) represent the location of the inflection point and the steepness of the water retention curve, and the reciprocal of α is often equated with the air entry suction; A, B, and C are quadratic terms used to improve the fitting of the model to the water retention curve [43]. Equation (2) is used to determine the volume of drainable pores (VDP, kg kg^{-1}), defined as the integral of the area under the specific water capacity curve (dθ/dψ) and above its baseline; MS (hPa) is the modal suction corresponding to the matric potential at the peak of the specific water capacity curve (Figure 1d) and relates to the most frequent pore size. The SI in Equation (3) is used to evaluate the susceptibility of the tested samples to breakdown; the higher the value of SI the less susceptible the aggregates are to slaking. The coefficient of variation between replicates of water content (θ, kg kg^{-1}) was <5%.

2.4. Statistical Analysis

Analysis of variance (ANOVA) tests were conducted using the SAS Proc GLM procedure [45] to assess the effects of soil type (watershed elevation), land use and PAM treatments and their interactions on soil properties, near saturation water retention curve parameters (α, n), and HEMC stability indices (θs, θr, SI). Treatments mean comparisons were done using the Tukey-Kramer HSD test at $p < 0.05$ (Supplementary material: Table S1). Pearson pairwise correlation, regression analysis ($p < 0.05$), and stepwise regression analysis were used to examine the effects of soil properties (e.g., SOC, EC, pH, CEC, cations) on structure stability indices (Table S2).

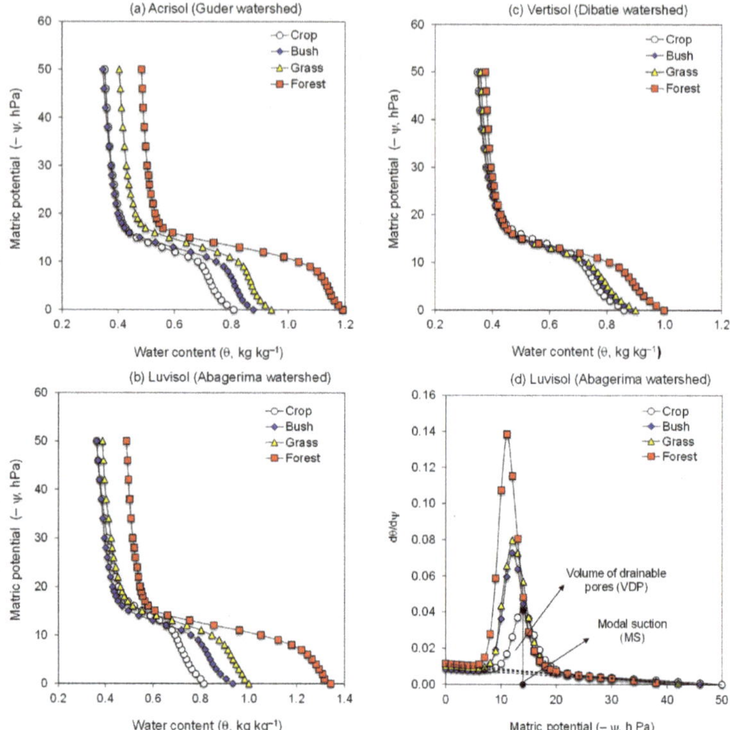

Figure 1. Water retention curves of the soils form (**a**) Guder (Acrisol), (**b**) Abagerima (Luvisol), and (**c**) Dibatie (Vertisol) watersheds used under crop, grass, bush, and forest land, and (**d**) the specific water capacity curves of soils from Abagerima (Luvisol). The dashed baseline in the specific water capacity curve represents the soil shrinkage line.

3. Results

Results of the ANOVA tests revealed that all analyzed soil properties (Table 1), structural stability indices, and parameters (α, n) were significantly ($p < 0.001$) affected by soil type (watershed elevation), land use, and their interaction (Table S1). The VDP and SI, n and MS were highly correlated [41], and thus we opted to focus the discussion mostly on the SI and the model parameter α in the presented analyses.

3.1. Land Use and Soil Properties

Studying soil properties in different land uses is crucial for understanding the current prevailing conditions in soils from different agro-ecological zones from conservation measures strategy perspective [37,45]. Soil properties showed large variation in pH (5.2–6.7), SOC (1.4–8.1%), cation exchange capacity (CEC, 25–40 $cmol_c kg^{-1}$), exchangeable Ca^{2+} (9.4–16.1 $cmol_c kg^{-1}$), soil texture (sandy loam to clay), electrical conductivity (EC, 0.2–0.9 dS m^{-1}), and $CaCO_3$ (0.3–5.8%) (Table 1). Some clear associations of certain soil properties (e.g., SOC, pH, EC), with elevation and land use were noted. Content of SOC was highest and pH was lowest in Guder watershed (highest elevation and rainfall), followed by Abagerima (medium elevation and rainfall) and Dibatie (lowest elevation and rainfall). Differences in soil clay content (texture) were also linked to watersheds: Guder (loam) < Abagerima (clay loam) < Dibatie (clay). The Clay/CEC ratio (i.e., indication of the clay mineralogy [40], also decreased with the increase in elevation: Guder (0.3–0.7) < Abagerima (0.7–0.9) < Dibatie (1.6–2.5)

(Table 1). In each watershed, soil pH and SOC were related to land use, and decreased in the following order: Forest > Grass ≥ Bush > Crop. Largely, soil CEC, exchangeable Ca^{2+} and K^+, and EC and $CaCO_3$ were lowest under cropland and highest under forest, grass, or bushland. The changes in exchangeable Mg^{2+} (3.2–4.5 $cmol_c kg^{-1}$) and Na^+ (2.4–6.3 $cmol_c kg^{-1}$) with variations in land use were not consistent, yet the latter mostly was higher under cropland or bushland (Table 1).

In general, only in 10 out of 85 cases, meaningful correlations (i.e., $r \geq 0.7$) were noted among the various soil properties studied (Table 2a). However, in each watershed, strong correlations (50% of cases) between SI or SOC and the mean of soil properties were found (Table 2b). For instance, soil pH or EC were weakly related to SOC and SI ($r < 0.3$–05) when all samples were involved (Table 2a). Nevertheless, when separating correlation analysis by watersheds, pH was very strongly associated ($r > 0.9$) with SOC and SI in the acidic Acrisol and Luvisol, and moderately with slightly acidic Vertisol; soil EC was closely related to SOC and SI in Luvisol and Vertisol ($r > 0.7$–0.9), but not in Acrisol ($r < 0.7$) (Table 2b). Correlation analysis for all soils studied indicated that exchangeable Ca^{2+} was closely related to SOC ($r = 0.7$), while it was not related to $CaCO_3$ content ($r = 0.3$) (Table 2a). However, analysis for each watershed alone showed better relation between Ca^{2+} and $CaCO_3$ ($r = 0.5$–0.6) (Table S2).

Table 2. Pearson pair-wise correlation coefficients for properties and structural index (SI) of the soils used. The correlations with $r \geq 0.7$ are in bold. Units of properties are in Table 1.

Variables	SI	pH	EC	Sand	Silt	Clay	Ca	Mg	Na	K	CEC	CaCO₃	CCR	SOC
(a) between the indicators of soil properties for all watersheds														
SI	1													
pH	0.38 *	1												
EC	0.54 *	0.57 *	1											
Sand	0.51 *	−0.3 *	−0.18	1										
Silt	−0.3 *	−0.6 *	−0.11	0.05	1									
Clay	−0.3 *	0.56 *	0.23	**−0.9 ****	−0.41 *	1								
Ca	**0.70 ***	**0.70 ****	0.57 **	0.29	−0.47 *	−0.02	1							
Mg	−0.28	0.23	0.55 **	−0.6 *	0.24	0.51 *	0.03	1						
Na	−0.04	−0.20	−0.5 *	0.27	0.08	−0.25	−0.20	−0.6 *	1					
K	0.45 *	0.53 *	**0.73 ****	−0.22	−0.26	0.34 *	0.52 **	0.43 **	−0.24	1				
CEC	0.56 *	0.40 *	0.42 **	0.21	0.12	−0.19	0.38 *	0.15	0.06	0.12	1			
CaCO₃	0.24	0.38 *	**0.73 ****	−0.30	−0.36 *	0.45 *	0.30 *	0.59 *	−0.4 **	**0.72 ****	0.03	1		
CCR	−0.4 *	0.44 *	−0.06	**0.84 ****	−0.44 *	**0.97 ****	−0.13	0.38 *	−0.16	0.27	−0.4 *	0.39 *	1	
SOC	**0.84 ****	0.46 *	0.27	0.63 **	−0.33 *	−0.41 *	**0.72 ****	−0.5 *	0.14	0.15	0.63 *	−0.11	−0.49 *	1

	Variable	SI	pH	EC	Sand	Silt	Clay	Ca	Mg	Na	K	CEC	CaCO₃	CCR	ESP
(b) between soil properties and SOC or SI for each watershed: Guder (G), Abagerima (A), and Dibatie (D)															
G	SOC	0.91 **	**0.99 ****	**0.69 ****	0.66 *	**−0.76 ****	−0.65 *	0.68 *	−0.61 *	−0.4 *	−0.4 *	**0.81 ****	0.80 **	**−0.96 ****	−0.85 **
G	SI	1	**0.91 ****	0.53 *	**0.79 ****	**−0.85 ****	−0.65 *	0.43 *	−0.65 *	−0.6 *	**−0.7 ****	**0.80 ****	0.93 **	**−0.90 ****	−0.70 **
A	SOC	0.97 **	**0.99 ****	**0.88 ****	0.64 *	**−0.70 ****	−0.11	**0.99 ****	−0.55 *	0.49 *	**0.91 ****	0.37 *	0.62 *	-0.56 **	−0.92 **
A	SI	1	**0.94 ****	**0.96 ****	0.43 *	−0.52 *	0.10	**0.96 ****	−0.47 *	0.55 *	**0.92 ****	0.52 *	0.60 *	−0.67 *	−0.86 **
D	SOC	0.95 **	**0.70 ****	**0.72 ****	0.53 *	0.32 *	−0.39 *	**0.74 ****	0.41 *	−0.3 *	−0.4 *	**0.82 ****	0.52 *	**−0.72 ****	−0.40 *
D	SI	1	0.63 *	**0.72 ****	0.33	0.35 *	−0.48 *	**0.78 ****	0.45 *	−0.5 *	−0.4 *	**0.73 ****	0.47 *	**−0.71 ****	−0.56 *

EC: electrical conductivity, CEC: cation exchange capacity, SOC: soil organic carbon, CCR: clay to CEC ratio, ESP: exchangeable sodium percentage; *, **: significant at the 0.01 (0.05) and 0.001 level accordingly.

3.2. Water Retention and Structure Stability of Untreated and PAM-Treated Soils

3.2.1. Water Retention of Untreated and PAM-Treated Soils

An evaluation of the shape of the water retention curves indicates the existence of differences associated with the distribution of the studied pores (>60 μm), with pore diameter being calculated from the matric potential ($d = -300/\psi$, where d is the equivalent pore diameter, μm). Land use and PAM treatments had considerable effects on the shape of the near saturation water retention curves, and consequently on its model parameters (θs, θr, α, n), and the soil structural stability indices (VDP, MS, SI) (Figures 1–3, Table 3).

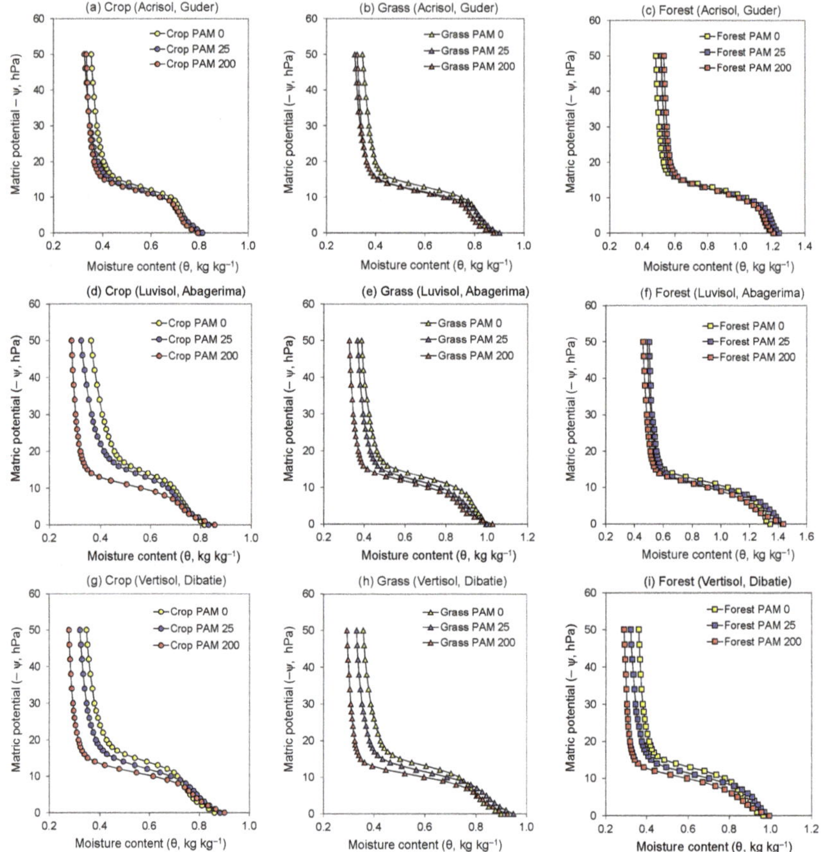

Figure 2. Water retention curves of untreated (PAM 0) and polyacrylamide (PAM) treated (25 and 200 mg L^{-1}) soils used under crop, grass, and forest land for (**a–c**) Guder (Acrisol), (**d–f**) Abagerima (Luvisol), and (**g–i**) Dibatie (Vertisol) watersheds.

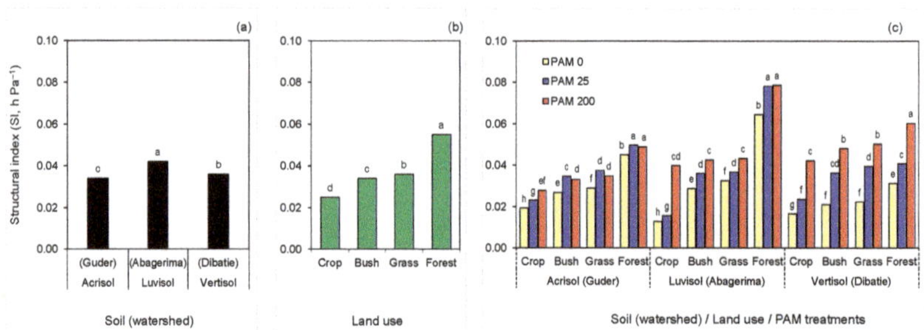

Figure 3. Structural index (SI) of the soils as affected by (**a**) soil type (watershed), (**b**) land use, and (**c**) polyacrylamide (PAM) treatments (0—untreated, 25 and 200 mg L^{-1}). Columns labeled with the same letter are not significantly different at $p < 0.05$ level: (**a**), within the soil (watershed); (**b**), within land use; and (**c**), within each soil (watershed).

Table 3. Effects of the treatments on the HEMC structure stability indices (VDP, MS, and SI) and near saturation water retention model parameters (θs, θr, α, n). Within each factor, the columns labeled with the same letter are not significantly different at $p < 0.05$ level.

Factors	Treatments	θ_s	θ_r	α	n	VDP	MS	SI
		kg kg^{-1}	kg kg^{-1}	hPa^{-1}		kg kg^{-1}	hPa	hPa^{-1}
Soil	Acrisol (Guder)	1.04 a	0.39 a	0.083 b	8.52 a	0.405 b	11.92 a	0.034 c
	Luvisol (Abagerima)	0.96 b	0.36 b	0.088 a	8.23 b	0.453 a	11.22 b	0.042 a
	Vertisol (Dibatie)	0.92 c	0.32 c	0.089 a	8.56 a	0.385 c	11.15 b	0.036 b
Land use	Crop	0.83 d	0.32 c	0.082 b	9.73 a	0.285 d	12.10 a	0.025 d
	Bush	0.91 c	0.32 c	0.085 c	8.04 c	0.389 c	11.59 c	0.034 c
	Grass	0.96 b	0.35 b	0.087 b	8.22 b	0.402 b	11.28 b	0.036 b
	Forest	1.19 a	0.43 a	0.092 a	7.76 d	0.581 a	10.75 d	0.055 a
PAM mg L^{-1}	0	0.96 b	0.33 c	0.079 c	9.67 a	0.359 c	12.62 a	0.029 c
	25	0.98 a	0.36 b	0.088 b	8.43 b	0.412 b	11.21 b	0.038 b
	200	0.99 a	0.38 a	0.094 a	7.22 c	0.472 b	10.46 c	0.046 a

VDP: volume of drainable pores, MS: modal suction, SI: structural index, θs and θr: saturated and residual water contents, α and n: represent the location and steepness of the water retention curve.

In all watersheds, the water retention curves of untreated soils for the different land uses exhibited similar trends, but of different magnitudes (Figure 1, Table 3). In the relative wet ($-\psi > 12$ hPa; pore size > 250 μm) and dry ($-\psi < 24$ hPa; pore size 60–125 μm) parts of the HEMC curves, soil moisture content (θ) at a given matric potential (ψ) was positioned in the following order of land use: forest > grass > bush > crop (excluding bushland soil at the dry end which was similar to the crop one). In the medium matric potential range ($-\psi = 12$–24 hPa; pore size 125–250 μm) such differences for the different land uses were inconsistent (Figure 1, Table 3).

Treating the soils with PAM modified the shape of water retention curves in the entire range of the matric potential studied ($\psi = 0$–50 hPa) compared with the untreated ones, with the effect of PAM being dependent on PAM rate, land use, and soil type (watershed) (Table 3). In all three watersheds, water retention curves of PAM-treated soils were mostly to the left side of (or below) those of the untreated soils in crop and grassland soils (results for bushland were similar to grassland—data not presented). Moreover, a trend was noted whereby, the difference between the water retention of untreated and PAM-treated soil was higher in the cropland soils compared with the grass or forest ones. In the forest soil, the water retention curves of the untreated and PAM-treated samples were comparable in Guder and Abagerima watersheds, whereas in the Dibatie watershed (with very sparse forest) water retention of samples was somewhat similar to those observed in the grassland soils (Figure 2).

The contribution of the PAM rate in Acrisol (Guder), yielded only small differences between the retention curves of the two PAM rates at the entire range of the matric potential (Figure 2a–c). By contrast, in Luvisol (Abagerima), a distinct difference between the retention curves for the two rates of PAM was noted, especially at $\psi > 10$–12 hPa in the crop and grassland, where water retention curves of high rate PAM-treated soils were on the left side of the low rate PAM-treated soils, revealing to more draining water at a given matric potential (Figure 2d–f). In Vertisol (Dibatie), the impact of PAM rate on the water retention curves was similar to that noted in the Luvisol (Abagerima), but it extended also to the forest soil (Figure 2g–i).

3.2.2. Structure Stability of Untreated and PAM-Treated Soils

Structure stability indices and model parameters by soil type (watershed), land use, or PAM differed: the n (and MS) decreased, while θs, θr, α, and SI (and VDP), increased with the following order of land use: crop < bush < grass < forest (Table 3). The SI values of the individual treatments varied over a wide range, between 0.013 to 0.079 hPa^{-1}, which could be related to the difference in soil type and properties among the watersheds and land use. However, within each watershed, and regardless of land use, PAM addition significantly increased soil SI. Moreover, similar to SOC order, untreated and PAM-treated soils had the same order of SI by land use: crop < bush < grass < forest (Figure 3).

In each watershed, the PAM effect on SI was highest in cropland and lowest in the forest (Figure 3). Relative to the untreated soils, PAM application increased soil SI by 1.2–1.3, 1.2–1.8, and 1.6–2.3 times in the soils of Guder (Acrisol), Abagerima (Luvisol), and Dibatie (Vertisol) watersheds respectively. At a high rate, PAM increased SI in the following order: Vertisol ≥ Luvisol > Acrisol for each type of land use (expect forest in Abagerima), while at a low rate of PAM and for untreated samples, the order of SI for same land use between the watersheds (or soil types) was inconsistent, although untreated soils under bush, grass, and forest land from Guder and Abagerima were more stable than from Dibatie watershed (Figure 3). It should be noted that treating cropland soils with a high PAM rate yielded greater or comparable SI (0.028–0.042 hPa^{-1}) than untreated grass and bushland soils (0.021–0.033 hPa^{-1}) in all watersheds (Figure 3). However, increase in PAM rate was (i) effective only in cropland soil with the lowest SOC (<2%) in Acrisol, (ii) not significant only in forest soil with highest SOC (~8.0%) in Luvisol, and (iii) significant for all land use samples in Vertisol (SOC = 1.7–5.1%). In the Guder watershed, the PAM rate has not changed the SI of bush, grass, and forest land samples (Figure 3).

A strong linear relation (R^2 = 0.72) between soil SI and SOC for the untreated soils, and for the low rate PAM-treated soil (R^2 = 0.64), was noted; for the high rate PAM-treated soils, no meaningful relation (R^2 = 0.26) between SI and SOC was obtained (Figure 4a). The relation between SOC of micro-aggregates (<0.25 mm) and that of bulk soil exhibited a strong linear relation that could be associated with the effect of micro-aggregate stabilization by SOC on soil SI (Figure 4a,b).

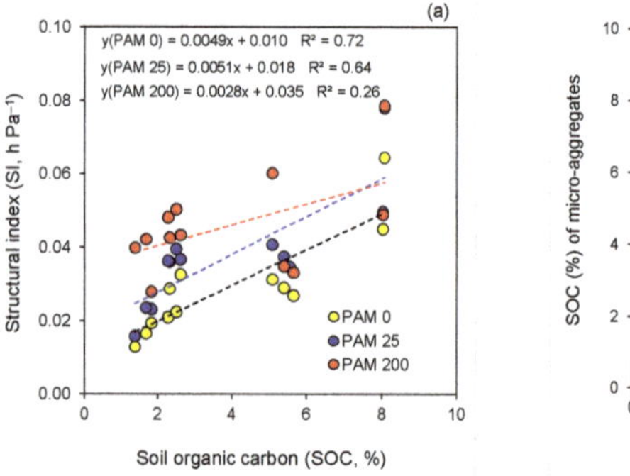

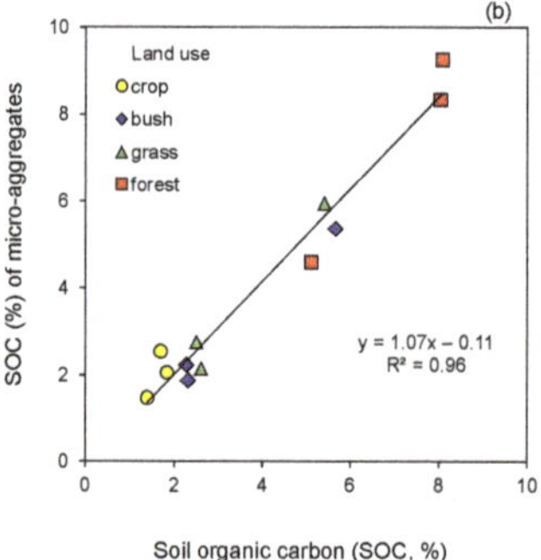

Figure 4. Relations (p < 0.001) between (**a**) soil structural index (SI) and soil organic carbon (SOC) for untreated (PAM 0) and polyacrylamide (PAM)-treated soils (25, 200 mg L^{-1}), and (**b**) SOC of micro-aggregates (<0.25 mm) and bulk soil (<2 mm).

A stepwise regression analysis revealed that in untreated soils, 92% of the variation in SI was associated mainly with SOC (71%), and CaCO$_3$ (+12%). For the low rate PAM-treated soils, variation in SI was also associated with SOC (62%) and EC (+17%). However, for the high PAM application rate, 96% of the variation was related to EC, silt, SOC, pH, and clay content (Table 4). There was an exponential relation (3 watersheds × 4 land use × 3 PAM rate = 36 treatments) between soil SI and water retention model parameter α; the strength of this relation was inversely related to SOC level (Figure 5).

Table 4. Stepwise regression analysis of the effect of soil properties on the structural index (SI), for the untreated (PAM 0) and PAM-treated (25, 200 mg L^{-1}) samples.

Treatment	Parameters	$p < F$	R^2	Treatment	Parameters	$p < F$	R^2
All	SOC	0.001	0.43	PAM 0	SOC	0.001	0.71
	CaCO$_3$	0.001	0.62		CaCO$_3$	0.001	0.83
	EC (or CEC)	0.078	0.64		pH	0.001	0.86
					EC	0.008	0.89
					CCR	0.002	0.92
PAM 25 mg L^{-1}	SOC	0.001	0.62	PAM 200 mg L^{-1}	EC	0.001	0.56
	EC	0.001	0.79		Silt	0.001	0.88
	CEC	0.028	0.80		SOC	0.001	0.90
	CaCO$_3$	0.120	0.82		pH	0.001	0.93
					Clay	0.001	0.95
					Ca + Mg	0.074	0.97

EC: electrical conductivity, CEC: cation exchange capacity; SOC: soil organic carbon, CCR: clay to CEC ratio.

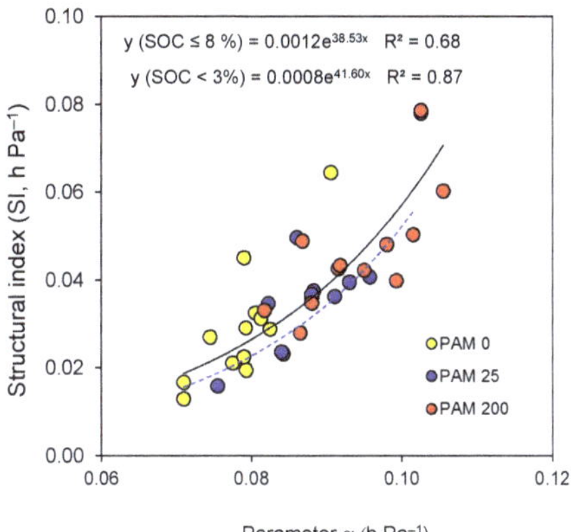

Figure 5. Relations ($p < 0.001$) between soil water retention model parameter α and soil structural index (SI) for all samples (SOC ≤ 8.1%, solid line) and samples with SOC ≤ 3% (dash line). Soil samples include untreated (PAM 0) and polyacrylamide (PAM)-treated ones (25, 200 mg L^{-1}).

4. Discussion

4.1. Land Use and Soil Type (Elevation) Effects

The impact of land use on the soil properties (Table 1) was in agreement with other studies on similar soils or from those in Ethiopia, revealing that topsoil attributes, such as CEC and exchangeable cations (Ca^{2+}, Mg^{2+}, or Na$^+$), SOC, pH, and aggregate stability were considerably higher, in soils under forest, grass or pasture lands than under cropland [8,20,21]. Results of the pair-wise correlation analysis for all the soils studied (Table 2a) highlight the considerable contribution of land use type coupled with soil type/elevation on the examined soil properties. These observations were in agreement with recent studies [37,46]. Moreover, the variation in the soil properties and structure stability indices under various land use (Tables 1–3) could be related to the sensitivity of characteristics of the studied soil types to structure-modifying processes [12,14,16], which is also associated with elevation or watersheds. Soil structure stabilization in the studied watersheds of the Ethiopian highland under long-term conservation practices leads to a reduction of runoff and soil loss, minimizing loss of crop residues and organic components, and to SOC accumulation [37,46,47]. However, the role of soil $CaCO_3$ on soil biogeochemistry (e.g., pH, EC, Ca^{2+}, CEC) in general and in particular its contribution to improving soil structure stability as observed in our study (Table 4), is still greatly overlooked in acidic soils, including in soils from Ethiopian highland [48].

Aggregate-structure stability (expressed in terms of SI or α) was directly associated with SOC irrespective of elevation (Table 2). The distinct relation between SOC and aggregate stability has been recognized long ago and has been ever since verified in many studies [21,44,47]. Aggregate stability was directly related to exchangeable Ca^{2+} in Luvisol and Vertisol (Abagerima and Dibatie) and $CaCO_3$ content in Acrisol and Luvisol (Guder and Abagerima) (Table 2b). Evidentially, to maintain stable structure in soils of Ethiopian highlands, agricultural practices should include, in addition to means for increasing soils SOC, timely monitoring of soil Ca^{2+} and, if needed, adding some external source of Ca^{2+} (e.g., lime), especially at high altitudes, to balance Ca^{2+} deficiency and storage [27,34,48]. Similar to soils from semi-arid and arid regions, the stability of the soils in the Ethiopian highlands was susceptible to exchangeable Na^+ or sodicity (i.e., ESP). Even though the negative role of sodicity on soil structure stability was evident in our study (Table 2), this topic, as well as the inverse relation between sodicity and SOC, have not been receiving their due attention in studies related to these acidic soils [25,26], although elevated exchangeable Na^+ was also recently noted by other studies performed under natural condition or conservation practices in the highland region [8,46].

4.2. Soil Organic Carbon and Polyacrylamide Effect on Soil Structure Stability Indices

The shape of the water retention curves and hence the range of SI in untreated soils is associated with aggregation and can be explained by the conceptual aggregate hierarchy model [49], postulating that aggregates are sequentially formed through the action of organic binding agents leading to the formation of micro-aggregates (20–250 µm) and then macro-aggregates (>250 µm). Micro-aggregates could be formed within macro-aggregates by roots and microbial activity; aggregates and pores provide physical protection of SOC and mineralization [4,5,50]. Aggregate size influences the emissions of CO_2, and correlations exist between aggregate size and macroporosity, number of pores, and pore size [3,28]. It should be noted that in the untreated soils, SOC content (and thus SI) is controlled by soil type or clay mineralogy (adsorption of SOC on clay minerals) and it is linked to CEC, pH, and $CaCO_3$, i.e., soil attributes affected by the consequences of land use [4,7,9,51].

The water retention curves for forest, grass or bushland soils differed from that of cropland in the range of macro-pores (>250 µm; $-\psi > 12$ hPa) and micropores (60–125 µm; $12 < -\psi < 24$ hPa) (Figure 2). The deviation of the grass and forest soil water retention curves from that of the cropland one at the micropore size range (60–125 µm) was associated with micro-aggregate stabilization through avoiding soil disturbance by tillage, and higher SOC content [4,6]. The observed strong correlations between SI and SOC ($r > 0.9$) for each soil type, linear relations between SOC and SI ($R^2 = 0.72$), and between SOC of bulk soil and micro-aggregates ($R^2 = 0.92$) (Table 2, Figure 4) emphasize that good soil structure could support soil SOC accumulation at micro-and macro-aggregate level [7]. However, stable micro-aggregates play a critical role in the long-term stabilization of soil organic matter, whereas less stable macro-aggregates provide only a small physical protection [50]. In turn, good and stable soil structure facilitate the supply of food source for soil microbes and subsequently SOC development, and hence, positive feedback exists between microbial activity, SOC accumulation, and soil structure formation [36,42,51,52].

Treating soils with anionic PAM was effective in stabilizing existing aggregates and improving bonding between and aggregation of soil particles. The magnitude of the PAM effects, similar to the effects of SOC, were soil type-clay mineralogy and land use (e.g., SOC, EC, pH, Ca^{2+}) dependent (Table 3, Figure 4a) and could be related to the integrated PAM–Soil (Ca^{2+}, Mg^{2+})–SOC interaction [51,53]. Positively charged cations (adsorbed on negatively charged mineral surfaces) facilitate the adsorption of negatively charged long-chain organic molecules through cation bridging. Aggregation depends on the strength of bonding governed by the chain length of the organic molecules and the type of cations on the exchange sites. Aggregates can be held together more strongly by PAM rather than by electrolytes because a single PAM chain is linked to multiple soil particles [54,55]. Thus, the higher SI from PAM treatments could be attributable to the adsorption of the long-chain PAM molecules to soil particles, which act as cementing material, and minimize aggregate slaking, and enhances soil aggregate and structure stability more than organic or inorganic amendments [40,56].

The deviation of the water retention curves of the PAM-treated cropland soils from the untreated one was less evident than the aforementioned deviations observed in the case of the changes in land use (Figures 1 and 2); yet, at times changes took place in the regions of the micro- and meso-size pores (60–250 μm; $-\psi < 12$ hPa). This phenomenon could possibly be linked to the fact that, in contrast to the non-tilled land uses (forest, grass, and bushland), application of PAM does not lead to the buildup of new soil aggregates with time, but to preserving and strengthening existing aggregates [29,32,41]. Moreover, in the aggregate size range studied (0.5–1.0 mm), adsorption of PAM of high molecular weight used on soils takes place mostly on exterior surfaces of soil material, because the narrow pores in small–size aggregates may not allow penetration of the large PAM molecules into the aggregates [57]. This pattern of PAM adsorption to the aggregates explains the significant impact of the PAM rate on the SI (Table 3); the larger the rate, the greater the SI [28,34,58].

The favorable impact of PAM, irrespective of its rate, on SI of the slightly acidic soils under cropland (Figure 3), was significantly greater than on the other land uses. This observation was also noted in soils from a semi-arid region [58,59]. Moreover, it is worthwhile noting that the contribution of PAM to improving SI (Acrisol < Luvisol < Vertisol) by watershed was opposite to the trend of SOC accumulation in these soils (Acrisol > Luvisol > Vertisol); yet all soil samples with SOC < 3–5% used under crop, grass, and bushland were significantly affected by PAM application (Figure 3). This fact could be seen as analogous to findings from former studies, where the positive effect of PAM on SI was related to (i) soil texture, with PAM being more effective in loam with initially much weaker soil structure than in clay soils [28,29], and (ii) soil clay mineralogy, with PAM being more effective in unstable smectitic soils than in the stable kaolinitic ones [40,59–61]. Our observations regarding PAM contribution to improving aggregate stability being more pronounced in soils with lower SOC than with higher SOC content, further highlight that PAM is more effective in soils of a priori lesser stability. Moreover, at a low PAM rate, SOC and PAM could jointly contribute (Table 4, Figure 4a) to soil quality and macropore stabilization for short or long-term PAM application strategy [32,40].

The exponential relation between soil SI and model parameter α (Figure 5), reveals that an increase in α by SOC or PAM treatment may significantly increase soil SI due to the increase in the mean aggregate and pore size, and resistance of aggregates to slaking [10,34]. Excluding samples with high SOC content (>5%) from the relation increased the coefficient of determination R^2 from 0.68 to 0.87 (Figure 5), yet regression curves with the close exponents were "parallel". Thus, exponential relations between SI and α (which include untreated and PAM-treated samples), could be considered as important and might suggest that along with SI, parameter α could also be used for evaluating changes in aggregate-structure stability following changes in soil intrinsic or extrinsic conditions associated with land use and management, including PAM application [59,60]. Finally, it transpires from our findings that PAM, in combination with other conservation methods, can integrally regulate soil aggregate-structure stability. Thus, PAM can be applied for restoration of crop, grass, bushland soil, or even forest soils, particularly in "hot spot" areas associated with enhanced translocation and leaching of the clay particles and cations in degraded lands from tropical, humid, and arid regions [8,23,32–34,59,62].

5. Conclusions

It emerged from our study that change of land use from cropland soils (annually tilled) to land uses that do not require tillage (bush, grass, and forest) has substantial positive effects on aggregate-structure stability indices in the studied acidic soils from the Ethiopian highlands. The observed positive effect of $CaCO_3$, and the negative effects of sodicity on SOC accumulation and soil structure development in the acidic soils studied, which have previously been mostly overlooked, need greater attention. The positive effects of the non-tilled land uses were ascribed to SOC accumulation and a lesser extent to $CaCO_3$ content; the magnitude of the effects depended on soil type. A viable alternative in the form of using small amounts of PAM (e.g., a soil amendment), was found effective in stabilizing aggregates and improving soil structure in all the studied soils. The contribution of PAM to improving aggregate stability was more pronounced in soils of lower SOC content, thus emphasizing PAM efficacy as a soil stabilizing agent in soils of a priori weak structure stability. The mechanisms by which SOC and PAM improve aggregate stability seems to differ as predominant changes were observed in the pore size range of macropores (>250 μm) for changes in land use and hence SOC effect, and in the pore size range of 60–250 μm in the case of PAM application.

The efficacy of PAM application suggests that improving soil aggregate-structure stability is obtained almost immediately without the need to go through the time-consuming process of changing land use or using long-term no-till practices, to achieve the Sustainable Development Goals of the United Nations Land Degradation Neutrality challenge [63]. The results of PAM addition to the Ethiopian highland soils met our expectations that PAM addition (i) improves soil structure stability, and (ii) could replace, at least partly, traditional conservation measures, which include high-cost cultivation practices whose impact may emerge only after several decades. As the soil types and type of clay minerals and soil texture cannot be changed by management practices, employing practices such as PAM addition and SOC driven aggregation is crucial for short- and long-term soil structure stabilization under various land uses.

Supplementary Materials: The following are available online at http://www.mdpi.com/2071-1050/13/3/1407/s1, Table S1: Analysis of variance (ANOVA) of the effect of the treatments on soil properties, and near saturation water retention model parameters (θs, θr, α, n), and the HEMC structure stability indices (VDP, MS, and SI). W: watershed, LU: land use, PAM: polyacrylamide, VDP: volume of drainable pores, MS: modal suction, SI: structural index, α and n, are the location of the inflection point and the steepness of the water retention curve. θr and θs, are the residual and saturated water content, EC: electrical conductivity (1:2.5), CEC: cation exchange capacity, SOC: soil organic carbon; CCR: the ratio of clay content to CEC (indication of clay mineralogy); *: $p < 0.001$, Table S2: Pearson pairwise correlation coefficients for properties of the soils used. Units of the respective properties are as in Table S1. The coefficients higher than 0.5 are significant at $p < 0.01$–0.001. EC: electrical conductivity (1:2.5); CEC: cation exchange capacity, SOC: soil organic carbon.

Author Contributions: Conceptualization, A.I.M., A.T., N.H., and G.J.L.; methodology, A.I.M., G.J.L., T.K., B.K., T.M., G.A., and A.W.; software, A.I.M. and H.F.; validation, A.I.M., A.T., and G.J.L.; formal analysis, A.I.M., M.T., and N.H.; investigation, A.I.M., H.F., B.K., T.M., A.W., and M.T.; resources, A.T., H.F., and T.K.; data curation, M.T. and H.F.; writing-original draft preparation, A.I.M. and G.J.L.; writing—review and editing, A.I.M., G.J.L., and A.T.; visualization, A.I.M.; supervision, A.T., A.I.M., and N.H.; project administration, A.T.; funding acquisition, A.T. and H.F. All authors have read and agreed to the published version of the manuscript.

Funding: This research was funded by the Science and Technology Research Partnership for Sustainable Development (SATREPS)—Development of a Next-Generation Sustainable Land Management (SLM) Framework to Combat Desertification project, Grant Number JPMJSA1601, Japan Science and Technology Agency (JST)/Japan International Cooperation Agency (JICA).

Institutional Review Board Statement: Not applicable.

Informed Consent Statement: Not applicable.

Acknowledgments: The authors thank Arid Land Research Center, Tottori University, The Science and Technology Research Partnership for Sustainable Development (SATREPS), and Japan Science and Technology Agency (JST)/Japan International Cooperation Agency (JICA). We also would like to thank the Kemira Company (Lakeland, Florida, USA) for providing anionic polyacrylamide, SUPERFLOC A-110.

Conflicts of Interest: The authors declare no conflict of interest.

References

1. Chandrasekhar, P.; Kreiselmeier, J.; Schwen, A.; Weninger, T.; Julich, S.; Feger, K.H.; Schwärzel, K. Why we should include soil structural dynamics of agricultural soils in hydrological models. *Water* **2018**, *10*, 1862. [CrossRef]

2. Mamedov, A.I.; Bar–Yosef, B.; Levkovich, I.; Fine, P.; Silber, A.; Levy, G.J. Physicochemical mechanisms underlying soil and organic amendment effects on runoff P losses. *Land Degrad. Dev.* **2020**, *31*, 2395–2404. [CrossRef]

3. Mangalassery, S.; Sjögersten, S.; Sparkes, D.L.; Sturrock, C.J.; Mooney, S.J. The effect of soil aggregate size on pore structure and its consequence on emission of greenhouse gases. *Soil Tillage Res.* **2013**, *132*, 39–46. [CrossRef]

4. Kravchenko, A.N.; Negassa, W.C.; Guber, A.K.; Rivers, M.L. Protection of soil carbon within macro-aggregates depends on intra-aggregate pore characteristics. *Sci. Rep.* **2015**, *5*, 16261. [CrossRef]

5. Juarez, S.; Nunan, N.; Duday, A.C.; Pouteau, V.; Schmidt, S.; Hapca, S.; Chenu, C. Effects of different soil structures on the decomposition of native and added organic carbon. *Eur. J. Soil Biol.* **2013**, *58*, 81–90. [CrossRef]

6. Totsche, K.U.; Amelung, W.; Gerzabek Martin, H.; Guggenberger, G.; Klumpp, E.; Knief, C.; Lehndorff, E.; Mikutta, R.; Peth, S.; Kögel-Knabner, I.; et al. Microaggregates in soils. *J. Plant. Nutr. Soil Sci.* **2017**, *181*, 104–136. [CrossRef]

7. Rabot, E.; Wiesmeier, M.; Schlüter, S.; Vogel, H.J. Soil structure as an indicator of soil functions: A review. *Geoderma* **2018**, *314*, 122–137. [CrossRef]

8. Kassa, H.; Dondeyne, S.; Poesen, J.; Frankl, A.; Nyssen, J. Impact of deforestation on soil fertility, soil carbon and nitrogen stocks: The case of Gacheb catchment in the White Nile basin, Ethiopia. *Agric. Ecosyst. Environ.* **2017**, *247*, 273–282. [CrossRef]

9. Kodešová, R.; Jirku, V.; Kodeš, V.; Mühlhanselová, M.; Nikodem, A.; Žigová, A. Soil structure and soil hydraulic properties of Haplic Luvisol used as arable land and grassland. *Soil Tillage Res.* **2011**, *111*, 154–161. [CrossRef]

10. de Oliveira, J.A.T.; Cássaro, F.A.M.; Pires, L.F. Estimating soil porosity and pore size distribution changes due to wetting-drying cycles by morphometric image analysis. *Soil Tillage Res.* **2021**, *25*, 104814. [CrossRef]

11. Schweizer, S.A.; Fischer, H.; Haring, V.; Stahr, K. Soil structure breakdown following land use change from forest to maize in Northwest Vietnam. *Soil Tillage Res.* **2017**, *166*, 10–17. [CrossRef]

12. Wiesmeier, M.; Urbanski, L.; Hobley, E.; Lang, B.; von Lutzow, M.; Marin-Spiotta, E.; van Wesemael, B.; Rabot, E.; Ließ, M.; Garcia-Franco, N.; et al. Soil organic carbon storage as a key function of soils-a review of drivers and indicators at various scales. *Geoderma* **2019**, *333*, 149–162. [CrossRef]

13. Jacobs, A.; Rauber, R.; Ludwig, B. Impact of reduced tillage on carbon and nitrogen storage of two Haplic Luvisols after 40 years. *Soil Tillage Res.* **2009**, *102*, 158–164. [CrossRef]

14. Ogle, S.M.; Alsaker, C.; Baldock, J.; Bernoux, M.; Breidt, F.J.; McConkey, B.; Regina, K.; Vazquez-Amabile, G.G. Climate and soil characteristics determine where no-till management can store carbon in soils and mitigate greenhouse gas emissions. *Sci. Rep.* **2019**, *9*, 11665. [CrossRef] [PubMed]

15. Jensen, J.L.; Schjønning, P.; Watts, C.W.; Christensen, B.T.; Obour, P.B.; Munkholm, L.J. Soil degradation and recovery—Changes in organic matter fractions and structural stability. *Geoderma* **2020**, *364*, 114181. [CrossRef] [PubMed]

16. Liang, C.; Vanden Bygaart, A.; Macdonald, J.; Cerkowniak, D.; McConkey, B.; Desjardins, R.; Angers, D. Revisiting no-till's impact on soil organic carbon storage in Canada. *Soil Tillage Res.* **2020**, *198*, 104529. [CrossRef]

17. Haregeweyn, N.; Tsunekawa, A.; Nyssen, J.; Poesen, J.; Tsubo, M.; Meshesha, D.T.; Schutt, B.; Adgo, E.; Tegegne, F. Soil erosion and conservation in Ethiopia: A review. *Prog. Phys. Geogr.* **2015**, *39*, 750–774. [CrossRef]

18. Tesfaye, M.A.; Bravo, F.; Ruiz–Peinado, R.; Pando, V.; Bravo–Oviedo, A. Impact of changes in land use, species and elevation on soil organic carbon and total nitrogen in Ethiopian Central Highlands. *Geoderma* **2016**, *261*, 70–79. [CrossRef]

19. Lemann, T.; Roth, V.; Zeleke, G.; Subhatu, A.; Kassawmar, T.; Hurni, H. Spatial and temporal variability in hydrological responses of the Upper Blue Nile basin, Ethiopia. *Water* **2019**, *11*, 21. [CrossRef]

20. Tesfahunegn, G.B. Soil quality indicators response to land use and soil management systems in Northern Ethiopia's catchment. *Land Degrad. Dev.* **2016**, *27*, 438–448. [CrossRef]

21. Delelegn, Y.T.; Purahong, W.; Blazevic, A.; Yitaferu, B.; Wubet, T.; Goransson, H.; Godbold, D.L. Changes in land use alter soil quality and aggregate stability in the highlands of northern Ethiopia. *Sci. Rep.* **2017**, *7*, 13602. [CrossRef] [PubMed]

22. Dignac, M.F.; Derrien, D.; Barré, P.; Barot, S.; Cecillon, L.; Chenu, C.; Chevallier, T.; Freschet, G.T.; Garnier, P.; Roumet, C.; et al. Increasing soil carbon storage: Mechanisms, effects of agricultural practices and proxies. A review. *Agron. Sustain. Dev.* **2017**, *37*, 14. [CrossRef]

23. Tebebu, T.Y.; Bayabil, H.K.; Stoof, C.R.; Giri, S.K.; Gessess, A.A.; Tilahun, S.A.; Steenhuis, T.S. Characterization of degraded soils in the humid Ethiopian highlands. *Land Degrad. Dev.* **2017**, *28*, 1891–1901. [CrossRef]

24. Chaplot, V.; Darboux, F.; Alexis, M.; Cottenot, L.; Gaillard, H.; Quenea, K.; Mutema, M. Soil tillage impact on the relative contribution of dissolved, particulate and gaseous (CO2) carbon losses during rainstorms. *Soil Tillage Res.* **2019**, *187*, 31–40. [CrossRef]

25. Shabtai, I.A.; Shenker, M.; Edeto, W.L.; Warburg, A.; Ben–Hur, M. Effects of land use on structure and hydraulic properties of Vertisols containing a sodic horizon in northern Ethiopia. *Soil Tillage Res.* **2014**, *136*, 19–27. [CrossRef]

26. Bayabil, H.K.; Stoof, C.R.; Lehmann, J.C.; Yitaferu, B.; Steenhuis, T.S. Assessing the potential of biochar and charcoal to improve soil hydraulic properties in the humid Ethiopian Highlands: The Anjeni watershed. *Geoderma* **2015**, *243–244*, 115–123. [CrossRef]

27. Bekele, A.; Kibret, K.; Bedadi, B.; Yli–Halla, M.; Balemi, T. Effects of lime, vermicompost, and chemical P fertilizer on selected properties of acid soils of Ebantu District, Western Highlands of Ethiopia. *Appl. Environ. Soil Sci.* **2018**, *2018*, 8178305. [CrossRef]

28. Mamedov, A.I.; Huang, C.; Aliev, F.A.; Levy, G.J. Aggregate stability and water retention near saturation characteristics as affected by soil texture, aggregate size and polyacrylamide application. *Land Degrad. Dev.* **2017**, *28*, 543–552. [CrossRef]

29. Caesar-Tonthat, T.; Busscher, W.; Novak, J.; Gaskin, J.; Kim, Y. Effects of polyacrylamide and organic matter on microbes associated to soil aggregation of Norfolk loamy sand. *Appl. Soil Ecol.* **2008**, *40*, 240–249. [CrossRef]

30. Awad, Y.M.; Lee, S.S.; Kim, K.H.; Ok, Y.S.; Kuzyakov, Y. Carbon and nitrogen mineralization and enzyme activities in soil aggregate-size classes: Effects of biochar, oyster shells, and polymers. *Chemosphere* **2018**, *198*, 40–48. [CrossRef]

31. Lee, S.S.; Shah, H.S.; Awad, Y.M.; Kumar, S.; Ok, Y.S. Synergy effects of biochar and polyacrylamide on plants growth and soil erosion control. *Environ. Earth Sci.* **2015**, *74*, 2463–2473. [CrossRef]

32. Ma, B.; Ma, B.L.; McLaughlin, N.B.; Mi, J.; Yang, Y.; Liu, J. Exploring soil amendment strategies with polyacrylamide to improve soil health and oat productivity in a dryland farming ecosystem: One-time versus repeated annual application. *Land Degrad. Dev.* **2020**, *31*, 1176–1192. [CrossRef]

33. Sojka, R.E.; Orts, W.J.; Entry, J.A. Soil physics and hydrology: Conditioners. In *Encyclopedia of Soils in the Environment*; Elsevier: Oxford, UK, 2005; pp. 301–306.

34. Mamedov, A.I.; Fujimaki, H.; Tsunekawa, A.; Tsubo, M.; Levy, G. Structure stability of acidic Luvisols: Effects of tillage type and exogenous additives. *Soil Tillage Res.* **2021**, *206*, 104832. [CrossRef]

35. Tian, X.; Fan, H.; Wang, J.; Ippolito, J.; Li, Y.; Feng, S.; An, M.; Zhang, F.; Wang, K. Effect of polymer materials on soil structure and organic carbon under drip irrigation. *Geoderma* **2019**, *340*, 94–103. [CrossRef]

36. Tian, X.; Wang, K.; Liu, Y.; Fan, H.; Wang, J.; An, M. Effects of polymer materials on soil physicochemical properties and bacterial community structure under drip irrigation. *Appl. Soil Ecol.* **2020**, *150*, 103456. [CrossRef]

37. Abebe, G.; Tsunekawa, A.; Haregeweyn, N.; Takeshi, T.; Wondie, M.; Adgo, E.; Masunaga, T.; Tsubo, M.; Ebabu, K.; Berihun, M.L.; et al. Effects of land use and topographic position on soil organic carbon and total nitrogen stocks in different agro–ecosystems of the Upper Blue Nile Basin. *Sustainability* **2020**, *12*, 2425. [CrossRef]

38. IUSS Working Group WRB. World reference base for soil resources 2014, update 2015. In *International Soil Classification System for Naming Soils and Creating Legends for Soil Maps. World Soil Resource Reports No. 106*; FAO: Rome, Italy, 2015; p. 239.

39. Klute, A. *Methods of Soil Analysis*, 2nd ed.; Agronomy Monograph 9; ASA and SSSA: Madison, WI, USA, 1986.

40. Mamedov, A.I.; Wagner, L.E.; Huang, C.; Norton, L.D.; Levy, G.J. Polyacrylamide effects on aggregate and structure stability of soils with different clay mineralogy. *Soil Sci. Soc. Am. J.* **2010**, *74*, 1720–1732. [CrossRef]

41. Mamedov, A.I.; Levy, G.J. High energy moisture characteristics: Linking between some soil physical processes and structure stability. In *Quantifying and Modeling Soil Structure Dynamics: Advances in Agricultural Systems Modeling. Trans-disciplinary Research, Synthesis, Modeling and Applications*; SSSA: Madison, WI, USA, 2015; Volume 3, pp. 41–74. [CrossRef]
42. Saffari, N.; Hajabbasi, M.A.; Shirani, H.; Mosaddeghi, M.R.; Mamedov, A.I. Biochar type and pyrolysis temperature effects on soil quality indicators and structural stability. *J. Environ. Manag.* **2020**, *221*, 110190. [CrossRef]
43. Pierson, F.B.; Mulla, D.J. An Improved method for measuring aggregate stability of a weakly aggregated loessial soil. *Soil Sci. Soc. Am. J.* **1989**, *53*, 1825–1831. [CrossRef]
44. Levy, G.J.; Mamedov, A.I. High-Energy-Moisture-Characteristic aggregate stability as a predictor for seal formation. *Soil Sci. Soc. Am. J.* **2002**, *66*, 1603–1609. [CrossRef]
45. SAS Institute. *SAS User's Guide*; Version 9.2; SAS Institute: Cary, NC, USA, 2008.
46. Ebabu, K.; Tsunekawa, A.; Haregeweyn, N.; Adgo, E.; Meshesha, D.T.; Aklog, D.; Masunaga, T.; Tsubo, M.; Sultan, D.; Fenta, A.A.; et al. Exploring the variability of soil properties as influenced by land use and management practices: A case study in the Upper Blue Nile basin, Ethiopia. *Soil Tillage Res.* **2020**, *200*, 104614. [CrossRef]
47. Welemariam, M.; Kebede, F.; Bedadi, B.; Birhane, E. Effect of community-based soil and water conservation practices on soil glomalin, aggregate size distribution, aggregate stability and aggregate-associated organic carbon in northern highlands of Ethiopia. *Agric. Food Secur.* **2018**, *7*, 42. [CrossRef]
48. Rowley, M.C.; Grand, S.; Verrecchia, E.P. Calcium–mediated stabilization of soil organic carbon. *Biogeochemistry* **2018**, *137*, 27–49. [CrossRef]
49. Tisdall, J.M.; Oades, J.M. Organic matter and water stable aggregates in soils. *J. Soil Sci.* **1982**, *33*, 141–163. [CrossRef]
50. Six, J.; Bossuyt, H.; Degryze, S.; Denef, K. A history of research on the link between (micro)aggregates, soil biota, and soil organic matter dynamics. *Soil Tillage Res.* **2004**, *79*, 7–31. [CrossRef]
51. Singh, M.; Sarkar, B.; Biswas, B.; Churchman, J.; Bolan, N.S. Adsorption–desorption behavior of dissolved organic carbon by soil clay fractions of varying mineralogy. *Geoderma* **2016**, *280*, 47–56. [CrossRef]
52. Gui, J.; Holden, N.M. The relationship between soil microbial activity and microbial biomass, soil structure and grassland management. *Soil Tillage Res.* **2015**, *146*, 32–38. [CrossRef]
53. Levy, G.J.; Warrington, D.N. Polyacrylamide addition to soils: Impacts on soil structure and stability. In *533 Functional Polymers in Food Science: From Technology to Biology, Volume 2: Food Processing*; John Wiley and Sons: Hoboken, NJ, USA, 2015; pp. 9–32. [CrossRef]
54. Bhardwaj, A.K.; McLaughlin, R.A.; Shainberg, I.; Levy, G.J. Hydraulic characteristics of depositional seals as affected by exchangeable cations, clay mineralogy, and polyacrylamide. *Soil Sci. Soc. Am. J.* **2009**, *73*, 910–918. [CrossRef]
55. Shainberg, I.; Goldstein, D.; Mamedov, A.I.; Levy, G.J. Granular and dissolved polyacrylamide effects on hydraulic conductivity of a fine sand and a silt loam. *Soil Sci. Soc. Am. J.* **2011**, *75*, 1090–1098. [CrossRef]
56. Miller, W.P.; Willis, R.L.; Levy, G.J. Aggregate stabilization in kaolinitic soils by low rates of anionic polyacrylamide. *Soil Use Manag.* **1998**, *14*, 101–105. [CrossRef]
57. Mamedov, A.I.; Beckmann, S.; Huang, C.; Levy, G.J. Aggregate stability as affected by polyacrylamide molecular weight, soil texture, and water quality. *Soil Sci. Soc. Am. J.* **2007**, *71*, 1909–1918. [CrossRef]
58. Kebede, B.; Tsunekawa, A.; Haregeweyn, N.; Mamedov, A.I.; Tsubo, M.; Fenta, A.A.; Meshesha, D.T.; Masunaga, T.; Adgo, E.; Abebe, G.; et al. Effectiveness of polyacrylamide in reducing runoff and soil loss under consecutive rainfall storms. *Sustainability* **2020**, *12*, 1597. [CrossRef]
59. Mamedov, A.I.; Tsunekawa, A.; Tsubo, M.; Fujimaki, H.; Ekberli, I.; Şeker, C.; Öztürk, H.S.; Cerdà, A.; Levy, G.J. Structure stability of cultivated soils from semi-arid region: Comparing the effects of land use and anionic polyacrylamide application. *Agronomy* **2020**, *10*, 2010. [CrossRef]
60. Porebska, D.; Slawinski, C.; Lamorski, K.; Walczak, R.T. Relationship between van Genuchten's parameters of the retention curve equation and physical properties of soil solid phase. *Int. Agrophys.* **2006**, *20*, 153–159.
61. de Melo, D.V.M.; de Almeida, B.G.; de Souza, E.R.; Silva, L.S.; Jacomine, P.K.T. Structural quality of polyacrylamide-treated cohesive soils in the coastal tablelands of Pernambuco. *Rev. Brasil. Ciênc. Solo* **2014**, *38*, 476–485. [CrossRef]

62. Inbar, A.; Ben–Hur, M.; Sternberg, M.; Lado, M. Using polyacrylamide to mitigate post fire soil erosion. *Geoderma* **2014**, *239*, 107–114. [CrossRef]
63. Keesstra, S.; Nunes, J.P.; Novara, A.; Finger, D.C.; Avelar, D.; Kalantari, Z.; Cerda, A. The superior effect of nature based solutions in land management for enhancing ecosystem services. *Sci. Total Environ.* **2018**, *610*, 997–1009. [CrossRef]

Publisher's Note: MDPI stays neutral with regard to jurisdictional claims in published maps and institutional affiliations.

MDPI
St. Alban-Anlage 66
4052 Basel
Switzerland
Tel. +41 61 683 77 34
Fax +41 61 302 89 18
www.mdpi.com

Sustainability Editorial Office
E-mail: sustainability@mdpi.com
www.mdpi.com/journal/sustainability